INTERFACES IN MEDICINE AND MECHANICS—2

Proceedings of Interfaces 90, the Second International Conference on Interfaces in Medicine and Mechanics held at the Instituti Ortopedici Rizzoli, Bologna, Italy, 9–14 September 1990.

Also published by Elsevier Applied Science Publishers:

Interfaces in Medicine and Mechanics, edited by K. R. Williams and T. H. J. Lesser, being the proceedings of the First International Conference on Interfaces in Medicine and Mechanics held in Swansea in April 1988.

Conference Co-ordinators

K. R. Williams, UK
A. Toni, Italy
J. Middleton, UK
G. Pallotti, Italy

Advisory Committee

J. F. Bates	UK
A. W. Blayney	Ireland
C. A. van Blitterswijk	The Netherlands
E. Czerwinski	Poland
C. Doyle	UK
K. Fujikawa	Japan
P. O. Glantz	Sweden
M. Green	UK
T. H. J. Lesser	UK
A. Meunier	France
T. A. Roberts	UK
A. Rohlmann	Germany
B. R. Simon	USA
K. Tanne	Japan
F. A. Young	USA

INTERFACES IN MEDICINE AND MECHANICS—2

Edited by

K. R. WILLIAMS

Department of Basic Dental Science, University of Wales College of Medicine, Cardiff, UK

A. TONI

Orthopaedic Clinic, University of Bologna, Italy

J. MIDDLETON

Department of Civil Engineering, University College of Swansea, UK

G. PALLOTTI

University of Bologna, Italy

ELSEVIER APPLIED SCIENCE
LONDON and NEW YORK

ELSEVIER SCIENCE PUBLISHERS LTD
Crown House, Linton Road, Barking, Essex IG11 8JU, England

Sole Distributor in the USA and Canada
ELSEVIER SCIENCE PUBLISHING CO., INC.
655 Avenue of the Americas, New York, NY 10010, USA

WITH 47 TABLES AND 259 ILLUSTRATIONS

© 1991 ELSEVIER SCIENCE PUBLISHERS LTD

British Library Cataloguing in Publication Data

International Conference on Interfaces in Medicine and
Mechanics (*2nd: 1990: Bologna, Italy*)
Interfaces in medicine and mechanics—2.
1. Medicine. Implants
I. Title   II. Williams, K. R.
617.95

ISBN 1-85166-583-8

Library of Congress CIP data applied for

Preface

The first Interfaces Conference was held at Swansea in April 1988 and represented the then state of the art of the science of implant surgery. The motivation for the initial venture was a supposed need for a closer interaction and dialogue between the clinician and scientist working in this area.

As expressed in the Preface to the first Conference, we felt that the interface was represented graphically, scientifically and psychologically by the drawings of Edgar Rubins (1915), again widely used in the literature to the present Proceedings.

The first Conference, we believe, achieved the aims of the organisers in bringing together scientists and clinicians towards an exchange of ideas by logically pursuing the sequence of events in clinical implant surgery.

The present Conference, in collaboration with our Italian colleagues, has also attempted to achieve the same aims by examining the behaviour of implants constructed of a variety of materials in both hard and soft tissue. Many contributions in the conference employed the technique of finite element analysis, both for design and optimisation purposes, particularly in relation to bone remodelling. Indeed, this particular aspect of the Conference led to much debate and will require a major examination of the many levels of physical, chemical and biomechanical interactive behaviour of the implant and its environment. All this natural behaviour was presented and discussed, but difficulties and failures remain with such procedures and we feel it is only by continuing such meetings that we progress in this difficult area of clinical science.

The editors believe that this unique group of scientists and clinicians should continue with the above-mentioned aims by organising a third conference at a venue to be arranged early in 1993.

K. R. WILLIAMS
A. TONI
J. MIDDLETON
G. PALLOTTI

vii

Contents

2. Mathematical Modelling of Prosthetic Surgery

3. Successes and Failures of Materials in Medicine

4. Tissue Structure, Histories of Applications

5. Numerical Analysis of Biological Implants and Surrounding Tissue

6 and 7. Interactions in Medicine and Mechanics

xiii

8. Medical Physics

POLYACTIVE: A BONE-BONDING POLYMER

C.A. van Blitterswijk, J.J. Grote, S.C. Hesseling, D. Bakker

Laboratory for Otobiology & Biocompatibility

Ear Nose & Throat Dept., Biomaterials Research Group

University of Leiden, 2333 AA Leiden

The Netherlands

ABSTRACT

Polyactive a 55/45 poly(ethylene oxide)/poly(butylene terephthalate) copolyether ester, was investigated as regards its general in vitro and in vivo biocompatibility. Special attention was directed to the interactions of the polymer with bone.

Light microscopy, scanning and transmission electron microscopy, x-ray microanalysis, and morphometry showed a satisfactory general biocompatibility of Polyactive and seemed indicative for the bone-bonding capacity of this material. Push-out experiments confirmed the bonding of Polyactive to bone.

INTRODUCTION

Bone-bonding biomaterials have been available since the introduction of calcium phosphates, glass ceramics and Bioglass to the field of biomaterial science.

Although the exact mechanism of bone-bonding is not known and may vary among the different materials (1-3) (calcium phosphate based versus silicate based) these bonding materials have several characteristics in common. One aspect is their behavior at the interface with bone where both calcium phosphates and bioglass have a calcium and phosphorus rich zone (1, 4,5). Furthermore, bone-bonding materials are characterized by a bone deposition which in case of porous implants frequently starts from the periphery of the pore and then proceeds to its center (3,6). Finally, bone-bonding materials are characterized by an intimate contact with bone but this phenomenon can also be found with materials like aluminum oxide and titanium. The first does not bond to bone and the latter is subject to controversy as far as bonding is concerned. Most other biomaterials however are characterized by a juxtaposed layer of fibrous tissue at the biomaterial surface in a bony implantation bed.

The bone-bonding ceramics that are currently available are already in wide clinical use but their mechanical properties limit their applicability in many surgical applications that would benefit from a more elastomeric bone-bonding material.

In this report the behavior of Polyactive a 55/45 poly(ethylene oxide)/poly(butylene terephthalate) segmented copolyether ester will be reviewed as concerns its general biocompatibility but with special emphasis on bone-bonding capacity.

MATERIALS & METHODS

Material properties

Polyactive was used in three forms. In the form of a dense film (100-125 um thick) cast from solution. As a 95 um thick porous film, prepared according to a saltcasting technique, with an average poresize of 96 microns. Finally it was also used as pressed dense blocks (dimensions:2.5x2.5x2 mm3).

In vitro analysis

The in vitro analysis of Polyactive comprised two experiments. In the first explants of rat middle ear mucosa and serially cultured middle ear epithelium were cultured on dense films, tissue culture polystyrene (TCPS) served as a control. In the second experiment, which will be referred to as artificial aging, Polyactive was exposed to pseudo extracellular fluid (PECF) at 115 C for 48 h. Aliquots of the heat treated PECF were added to serially cultured rat middle ear epithelium. Heat treatment without the addition of any biomaterial was used as a positive control and extracted PVC as a negative control.

In vivo analysis

The in vivo experiment was composed of two experiments. Initially macroporous films were implanted in the middle ear of the rat at three sites, in the tympanic membrane, between the middle ear mucosa and bulla bone, and between the bulla bone and muscle tissue. The implants were evaluated after 1, 2, 4, 13, 26, and 52 weeks in situ. A total of 272 implants were studied. In the other experiment a total of 36 dense blocks of Polyactive were implanted in the tibia of rats and another 36 blocks subcutaneously and evaluated after 3, 6 and 26 weeks. Hydroxyapatite and tetracalcium phosphate served as positive controls, Silastic (silicone rubber) as a negative control.

Analytical techniques

The implants were evaluated with light microscopy, morphometry, scanning electron microscopy, transmission electron microscopy, and x-ray microanalysis. In case of bone ingrowth specimens were routinely decalcified with exception of scanning electron microscopy samples and those samples of which the undecalcified bone/biomaterial interface was studied. In the in vitro experiments cell number was calculated after trypsinisation and quantitation in a Burker chamber.

More detailed information on the materials, the techniques and experimental design has been published elsewhere(7,8,9).

RESULTS

In vitro analysis

In the first in vitro experiment scanning electron microscopy revealed that the epithelium in the explants was predominantly composed of flat polygonal cells and a small number of ciliated cells. Scanning and transmission electron microscopy showed that the cellular outgrowths from these explants were monolayers of thin epithelial cells with varying amounts of microvilli and few ciliated cells. In serial cultivation ciliated cells were no longer found, the gross morphology was rather similar to that of the explant outgrowth. No distinction could be made between the cells cultured on either TCPS or Polyactive. A quantitative analysis of cell growth in the serial cultures showed that both the positive control TCPS and Polyactive caused sigmoid curves, without statistical difference between the curves when analysed with the Wilcoxon's two sample test ($p < 0.05$). The artificial aging experiment showed no significant divergence in morphology between the cells cultured in an extract of Polyactive as against the control without the addition of a biomaterial. The negative control (PVC) showed very few cells, and these were all characterized by a morphology that differed from normal cells. Transmission electron microscopy showed pycnotic nuclei and disrupted cell membranes.

In vivo analysis

Middle ear experiment: All three implantation sites initially mainly showed the presence of exudate in their pores when studied with light microscopy. As implantation time increased fibrous tissue became more prominent. Usually, however, a layer of phagocytes (macrophages and multinucleated cells) was still observed at the interface between the implant and fibrous tissue.

The total amount of surface area occupied by these phagocytes depended on the implantation site. Morphometry revealed that the exudative cells occupied the largest area at the bulla bone/muscle tissue site, whereas the smallest amount was observed in the tympanic implant. The fibrous tissue was composed of randomly oriented collagen fibres, fibroblasts and capillaries. Morphometric analysis of the total implant area showed a gradual decrease of implant surface area, indicating a distinct degradation (fig 1). The presence of implant-derived material in the cytoplasm of the phagocytes suggests that they play a role in the degradation process.

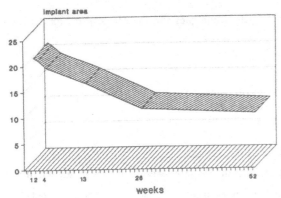

Figure 1. Diagram showing the degradation rate of Polyactive.

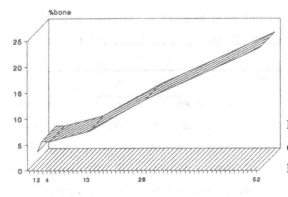

Figure 2. Diagram showing the amount of bone occupying the pores of Polyactive.

Bone deposition in the pores of the implant was seen in both the implants located between the bulla bone and middle ear mucosa and those between the bulla bone and muscle tissue. The amount of bone increased in the course of time (fig 2) and both light and electron microscopy showed a normal morphology, the bone being composed of collagen fibres, osteoblasts, osteocytes, lacunae, canaliculae and a calcified matrix. Transmission electron miocroscopy showed an intimate contact between bone and the implant surface and frequently an electron-dense structure at the bone/Polyactive interface which showed a continuity with the lamina limitans of bone in several cases. Bone seemed to be deposited from the pore wall to the pore center.

Rat tibia experiment:Three weeks after implantation the various techniques showed that large parts of the four biomaterials (hydroxyapatite, tetracalcium phosphate, Polyactive, and Silastic) were covered with fibrous tissue. Usually a very thin layer of phagocytes was still observed at the biomaterial/tissue interface. However, even at this interval part of the implant surface was already covered by bone. In contrast to the other materials, which showed an intimate contact with bone as shown by transmission electron microscopy, Silastic was always characterized by fibrous tissue or at least a non-decalcified zone of collagen fibres at the biomaterial/bone interface. In the course of time the implants were progessively covered with bone and after 26 weeks the bulk of the material had bone near its surface.

Transmission electron microscopy of decalcified material showed an electron-dense layer, with varying morphology, at the bone/biomaterial interface for Polyactive and the two positive controls. Silastic did not possess such a structure. The surface of undecalcified Polyactive was impregnated with tightly packed electron-dense crystals (fig 3). X-ray microanalysis demonstrated the presence of calcium and phosphorus, and a small amount of iron (fig 4), in these crystals.

Push out experiments performed six weeks after implantation exhibited a bonding of hydroxyapatite (12-16 N), tetracalcium phosphate (7-14.5 N), and Polyactive (6-9 N) to bone, Silastic (0.02 N) was not bonded. The first three figures between brackets indicate the force at which implant fracture occurred. Polyactive and the positive controls did not detach at their interface with bone. In contrast to the other materials Silastic detached at its interface with bone at the reported force. At 26 weeks push out experiments could not be performed because of the mechanical weakness of Polyactive at this stage, evidently caused by biodegradation. During all implantation intervals it was interesting to observe that pieces of Polyactive would stay attached at the interface with either bone or fibrous tissue, decalcified or undecalcified even though the major part of the biomaterial was mechanically removed as an artefact during sectioning.

Subcutaneous implantation in rats: In general all biomaterials that were subcutaneously implanted in this study were covered by a thin layer of fibrous tissue with collagen fibers running parallel to the implant surface. On top of this strucure a loosely arranged fibrous tissue was seen. The implant surface was usually characterized by a thin layer of macrophages and multinucleated cells which could be best visualized by transmission electron microscopy and were less prominent than those seen at the surface of the porous middle ear implants. These specimens were undecalcified and when stained with alizarin red (a calcium stain) revealed a positive staining up to 60 um deep directly at the implant periphery or at a slight distance of the implant surface. X-ray microanalysis demonstrated the presence of calcium and phosphorus in these areas. Furthermore, transmission electron microscopy showed similar electron-dense crystals as seen at the bone/Polyactive interface.

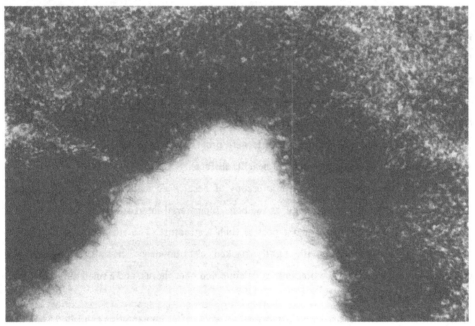

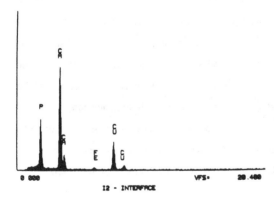

Figure 3. transmission electron micrograph of the Polyactive /bone interface. Note the ingrowth of small crystals in the implant surface.

Figure 4. X-ray microanalysis of the crystals growing into the implant. Calcium phosphorus and iron were detected.

DISCUSSION

The in vitro biocompatibility of Polyactive as investigated in this study seemed to be satisfactory. Explants and serially cultured cells attached to the material and showed a proliferation and morphology that was rather similar to that of the positive control. This type of experiment, where cells are cultured upon a material, primarily provides data on the acute toxicity of a material in contrast to experiments using extracts of biomaterials which are considered to give an indication of long term behavior (10). Extracts of Polyactive added to the culture medium did not cause any noteworthy changes as compared to the positive control suggesting a good long term biocompatibility, while a material like PVC when extracted caused significant cell-deformation.

The implantation of Polyactive in the middle ear, in the tibia and subcutaneously in the rat showed some inflammatory reaction near the surface of the material which depended on the implants location and structure. Most inflammatory response was found at the site where the implant bordered muscle tissue, and the porous implants caused a higher inflammatory reaction than the dense implants. The effects of implantation site and implant texture on the inflammatory response towards a biomaterial have been subject of several studies which at least partially support these findings (8, 11, 12). The inflammatory reaction towards Polyactive corresponds with that found for several other biomedical polymers (9) and also resembled that of the calcium phosphate ceramics hydroxyapatite and tetracalcium phosphate in the subcutaneous and tibia implantation sites. The degradation of Polyactive found in this study showed that Polyactive can be designed to be a resorbable polymer. The fibrous tissue near the implant showed no deviating morphology as compared to that of many other biomaterials with a similar shape.

The interactions between Polyactive and bone were of particular interest during this study. When implanted near the bony middle ear wall a bone deposition in the pores was found similar to that described for hydroxyapatite at this location (5,6). Bone deposition was according to the theory of bonding osteogenesis (3) since there was an intimate contact with bone at the Polyactive/bone interface and bone was deposited from the pore periphery to its center. Furthermore, the interactions between Polyactive and bone in the tibia were similar to those found for hydroxyapatite and tetracalcium phosphate in the same experiment. In both the middle ear and the tibia the Polyactive/bone interface was characterized by an electron-dense zone which has been described before for hydroxyapatite in earlier studies and is known to contain calcium and phosphorus in addition to an organic matrix (5). This structure showed a similarity and continuity with the lamina limitans of bone (13) and seems to play an essential role in the bone bonding process. The morphological indications of the bone-bonding properties of Polyactive were confirmed by the push out experiment. Hydroxyapatite, tetracalcium phosphate and Polyactive all showed implant failure and did not detach at their interface with bone, Silastic did not bond with bone. The forces that were necessary for the push out experiment are no absolute indication for the bonding strength because first of all the implant did fracture and not the bond with bone and second the mechanical properties of the elastomer Polyactive differ too much from those of the two ceramics to allow a comparison. Furthermore, the attachment of fragments of Polyactive at the interface with either fibrous tissue or decalcified bone suggests that more than an epitaxy of hydroxyapatite crystals alone would be the driving force behind the bonding.

According to the authors, bone-bonding properties as found for Polyactive in this study have never been described for a polymer before. The exact mechanism of bone-bonding still has to be elucidated. The presence of the electron-dense layer at the interface suggests a similarity with the bone-bonding mechanism of hydroxyapatite ceramic but the initial absence of calcium and phosphorus in the Polyactive pinpoints to a mechanism that would at least partially deviate from that found for all other bone-bonding biomaterials. An essential element in the bone bonding process might be the hydrogel-like behavior of Polyactive, which swells significantly when in contact with water. A material like Poly-HEMA for instance has been reported to show calcification inside the material even at subcutaneous implantation sites (14), while polyethylene oxide seems to play a vital role in the calcification of certain vascular prostheses (15).

In conclusion the results of this study indicate the bone-bonding properties of Polyactive. Future studies will have to elucidate the mechanism underlying the bone-bonding process and should be directed to the variation between the soft and hard segments

REFERENCES

1.   Hench, L.L., Splinter, R.J., Allen, W.C. and Greenlee, K.T., Bonding mechanisms at the interface of ceramic prosthetic material. Biomed. Mater. Res. Symp., 1972, 2, 117

2.   Jarcho, M., Kay, J.F., Gumaer K.I., Doremus R.H. and Brobeck ,H.P., Tissue, cellular, and subcellular events at a bone ceramic hydroxyapatite interface. J. Bioeng., 1977, 1, 79

3.   Osborn, J.F. and Newesly, H., Dynamic aspects of the implant-bone interface. Dental Implant Materials, Heimke, G.(ed.), Carl Hansen Verlag, Muenchen, 1980, 111

4.   Tracy, B.M. and Doremus, R.H., Direct microscopy studies of the bone-hydroxyapatite interface. J. Biomed. Mater. Res., 1984, 18, 719

5.   van Blitterswijk, C.A., Grote, J.J., Kuijpers, W., Blok-van Hoek, C.J.G. and Daems, W.Th., Bioreactions at the tissue/hydroxyapatite interface. Biomaterials, 1985, 6, 243

6.   van Blitterswijk, C.A., Grote, J.J., Kuipers, W., Daaems, W.Th. and de Groot, K., Macropore tissue ingrowth: A quantitative and qualitative study on hydroxyapatite ceramic, Biomaterials, 1986, 7, 137

7.   Bakker, D., van Blitterswijk, C.A., Daems, W.Th. and Grote, J.J., Biocompatibility of six elastomers in vitro, J. Biomed. Mater. Res., 1988, 22, 423

8.   Bakker, D., van Blitterswijk, C.A., Hesseling, S.C., Grote, J.J. and Daems, W.Th., The

effect of implantation site on phagocyte/polymer interaction and fibrous capsule formation, Biomaterials, 1988, 9, 14

9.   Bakker, D., van Blitterswijk, C.A., Hesseling, S.C., Daems, W.Th. and Grote J.J., Tissue/biomaterial interface characteristics of four elastomers. A transmission electron microscopical study. J. Biomed. Mater. Res., 1990, 24, 277

10.   Homsy, C.A., Biocompatibility in selection of materials for implantation, J. Biomed. Mater. Res., 1970, 4, 341

11.   Behling, C.A., and Spector, M., Quantitative characterization of cells at the interface of long-term implants of selected polymers. J. Biomed. Mater. Res., 1986, 20, 653

12.   Salthouse, T.N., Some aspects of macrophage behavior at the implant interface. J. Biomed. Mater. res., 1984, 18, 395

13.   Scherft, J.P., The lamina limitans of the organic matrix of calcified cratilage and bone. J. Ultrastr. Res., 1972, 38, 318

14.   Winter, G.D. and Simpson, B.J., Heterotopic bone formed in a synthetic sponge in the skin of young pigs. Nature, 1969, 223, 88

15.   Thoma, R.J., Hung, T.Q., Nyialas, E., Haubold, A.D. and Phillips, R.E., Metal ion complexation of poly(ether)urethanes. Advances in Biomedical Polymers, Gebelein, C.G. (ed.), Plenum Press, New York, 1985, 131

BIOMECHANICAL BEHAVIOUR OF ARTICULAR DISC IN ADULT DOGS
: A MEASURING METHOD AND PRELIMINARY RESULTS

KAZUO TANNE, EIJI TANAKA, TATSUYA SHIBAGUCHI, MAMORU SAKUDA
Department of Orthodontics
Osaka University Faculty of Dentistry
1-8 Yamadaoka, Suita, Osaka 565, Japan

SIGEO WADA, MASAO TANAKA, YASUYUKI SEGUCHI
Department of Mechanical Engineering
Osaka University Faculty of Engineering Science
1-1 Machikaneyama, Toyonaka, Osaka 560, Japan

ABSTRACT

Biomechanical behaviour of temporomandibular joint (TMJ) disc was investigated by means of a tensile test. Articular discs derived from eight adult dogs were used as experimental materials. Each disc was further divided into medial, middle and lateral parts, in parallel to the antero-posterior direction. The whole articular disc exhibited a non-linear stress-strain relationship, where the point of inflexion was around 150 gf/mm^2. It was found that elastic modulus of the disc was approximately 3.23 ± 0.85 kgf/mm^2 and 6.34 ± 1.19 kgf/mm^2 in lower and higher stress regions, respectively. In addition, mechanical properties of the articular disc varied in different areas of the disc. It is shown that the disc plays an important role in reducing stress induced in the TMJ space and its susceptibility to internal stress may be different in the lateral, middle and medial regions.

INTRODUCTION

In recent years, disorder of temporomandibular joint (TMJ) has become a great interest and various studies have been conducted in many fields of dentistry. Among various factors on TMJ dysfunctions, structural changes of TMJ components is speculated as one of causes for derangement of the TMJ (1). It is also known that the TMJ is one of stress bearing organs and stress induced in the TMJ is less than in other joints (2), (3). It is clinically significant to understand biomechanical properties of TMJ disc which may play a role in reducing excessive stresses generated on the condyle and the glenoid fossa.

In previous studies (4), (5), it was indicated that histological findings of the articular disc was similar to those of ligaments and tendons, consisting of collagenous fibres. However, the mechanical behaviour of the disc is not fully investigated.

The present study was conducted to investigate the biomechanical behaviour of the articular disc in response to tensile stresses.

MATERIALS AND METHODS

Articular discs derived from eight adult dogs were used for a tensile test (Fig. 1). The discs were subjected to the measurement within 12 hours after the removal to keep freshness. A disc was divided into three specimens along the antero-posterior direction, i.e. medial, middle and lateral regions. Size of specimen was approximately 5.6 mm in length and 2.7 mm^2 in cross-sectional area.

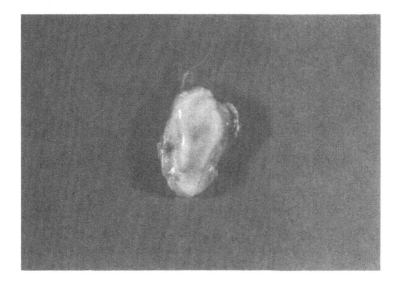

Figure 1. An articular disc derived from an adult dog.

Fig. 2 shows the whole experimental system of tensile test. Tensile forces for specimens, produced by a motor in a constant rate, was measured by use of a digital force gauge (Shimpo Corp., Tokyo, Japan), as shown in Fig. 3. In this experiment, tensile forces ranged in magnitude of stresses from 0 to 350 gf/mm^2. Measured values of stresses and strains for each specimen were simultaneously entered into a personal computer (PC-9801UX, NEC Corp., Tokyo, Japan).

Figure 2. Experimental system of tensile test for articular disc (left) and a specimen during the measurement (right).

Accuracy of the measurement was investigated in terms of reproducibility of the measured values obtained by repeating the tensile test for the same disc five times. Means, standard deviations and coefficients of variation were calculated.

Changes in the length of specimen were measured with stresses applied. In this experiment, two analytical techniques were employed. One was the clamp to clamp measurement, or direct measurement and the other was

Figure 3. A digital force gauge to measure tensile force.

indirect method simulating Video Dimensional Analyser (VDA) system (6). In the clamp to clamp measurement, strains of specimens were measured by a strain gauge placed between the clamps. For indirect method, two points were marked on each specimen with ink (Fig. 2). Photographs were taken from specimens before and during the measurement. By use of a digitizer, the original and deformed lengths were measured on the pictures and the strains were calculated. Stress-strain relationship was evaluated by means of regression analysis and thus modulus of elasticity was obtained.

RESULTS

Table 1 shows results on accuracy of the tensile test. In the range of 0-250 gf/mm^2 , coefficients of variation were less than 6.3% and reproducibility of the measurement was acceptable in the present experiment.

The articular disc experienced a non-linear change in response to stresses, where the point of inflexion was around 150 gf/mm^2 (Fig. 4).

TABLE 1
Strains and their coefficients of variation

Stress (gf/mm^2)	Strain (x 10^{-3})	Coefficient of variation (%)
50	23.2 ± 1.5	6.3
130	33.9 ± 1.1	3.2
190	42.7 ± 2.0	4.7
250	50.3 ± 2.2	4.4

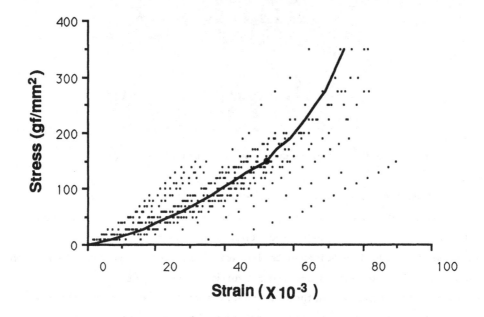

Figure 4. Mean stress-strain curve for articular discs derived from eight dogs. ● indicates the point of inflexion. Scattered dots denote the actual measured values of stress and strain.

Results by regression analyses revealled that inclinations of two-phase stress-strain lines were 3.23 ± 0.85 kgf/mm^2 and 6.34 ± 1.19 kgf/mm^2 in lower and higher stress regions, respectively. Modulus of elasticity was greater in higher stress area than in lower stress region.

Table 2 shows the elastic moduli of TMJ discs in the medial, middle and lateral regions. The modulus of elasticity was greater in the middle region than in the other two areas. The modulus of elasticity was the lowest in the lateral region among those in the three areas, indicating that the lateral part of articular disc is more susceptible to internal stress exerted in the TMJ region, which is in concurrent with actual morphological changes of the disc observed in patients with TMDs.

The elastic modulus of articular disc obtained by the indirect measurement was 4.42 ± 0.78 kgf/mm^2 on average in the lower stress area, which was about 1.37 times that from the clamp to clamp measurement. However, the nature of stress-strain relationship was essentially the same as that by the direct method.

TABLE 2

Elastic moduli of the discs in the medial, middle and lateral regions

Disc region	N	Elastic modulus (kgf/mm^2)	Significance
Medial	12	2.87 ± 0.36	
			* *
Middle	12	3.96 ± 0.90	n.s.
			* *
Lateral	10	2.79 ± 0.62	

* * :significant at 1 % level of confidence
n.s.:not significant

DISCUSSION

Articular disc generally consists of fibrous cartilage, collagenous fibres and elastin. Among fibres, the collagen of the disc is mainly type I, but type II collagen appears with aging (4)(5). Type I collagen is well known as the main collagen in the connective tissues such as skin, fascia and ligament, and thus it is indicated that histological components of the disc are similar to those of the general connective tissues.

With respect to the elastic moduli of connective tissues, various studies have been reported (7), (8), (9), (10), (11), (12). Vogel (8) investigated the biomechanical behaviour of the skin and Butler et al. (7) measured that of the human ligament and fascia. Crofts et al. (12) also evaluated the mechanical property of the bovine pericardium. However, the elastic modulus of the disc has not been investigated at all. In this study, the modulus of elasticity of the disc was about 3.23 ± 0.85 kgf/mm^2 in the stress range of 0-150 gf/mm^2. This value was almost the same as those of the skin and pericardium, and in particular that of intervertebral disc (11). The elastic modulus of the disc was about 1.96 times larger in the stress range of 150-350 gf/mm^2 than that in the lower stress area. This value was similar to that of ligament in regrown animal (13). Thus, the TMJ disc is indicated to have a similar property to other soft connective tissues. On the other hand, elastic moduli of the cancellous and compact bones were much larger than those of the TMJ disc (14), (15), indicating a role of the disc in absorbing excessive stresses acting at the bony surface of the TMJ.

In previous study (7), technical problems in strain measurement for various soft tissues have been discussed. Butler et al. (7) reported that measuring error was caused by differences in the loading rate, freshness of the specimen, the tensile speed and measurement technique. In order to keep the disc fresh, the experiment was performed within 12 hours after the removal of articular discs from dogs, and saline solution was continuously used during the experiment. Loading rate or speed was carefully maintained constant during the whole experiment. Two methods are available for strain measurement of soft tissue. One is the clamp to clamp measurement and the other is VDA system. The former has shortcomings such as the slippage of the specimen at its grips and large deformation of specimen near the grips. In the indirect method from the VDA system, only the strain in the middle area of the specimen, where tensile stress is relatively uniform, can be measured because the slippage between the specimen and its grips does not affect the strain. Therefore, the elastic moduli of the discs measured by the indirect method were about 1.37 times larger than those from the clamp to clamp measurement. It was indicated that 30% difference in the measured values was induced by the VDA system in comparison with the direct measurement (6), which was concurrent with the present finding.

In previous anatomical study of the articular disc (16), deformation and perforation were more frequently observed in the lateral and posterior sides than in the medial and anterior areas. This anatomical finding may have confirmed the results in this study, indicateing that the mechanical strength of the articular disc in the lateral part is less than in other two regions.

Yamada (17) described that tensile strength of tendon in human beings was the most similar to that in dog among many animals. Therefore, the present study would be useful for understanding biomechanical behaviours of the human TMJ disc and further for diagnosis and treatment of patients with TMDs, although the results pertain to dogs, not to human beings.

REFERENCES

1. Ramfjord, S. P., Temporomandibular joint dysfunction, Dysfunctional temporomandibular joint and muscle pain. J. Prosthet. Dent., 1961, 11, 353-74.

2. Brehnan, K., Boyd, R. L., Laskin, J., Gibbs, C. H. and Mahan, P., Direct measurement of loads at the temporomandibular joint in Macaca arctoides. J. Dent. Res., 1981, 60, 1820-4.

3. Hylander, W.L., Mandibular function and temporomandibular joint loading. In Developmental Aspects of Temporomandibular Joint Disorders, eds. D.S. Carlson, J.A. McNamara, Jr. and Ribbens, K.A., Center for Human Growth and Development The University of Michigan, Ann Arbor, 1984, pp. 19-35.

4. Kashima, K., Kino, K., Shiota, S. and Kuboki, Y., Collagen biochemistry of human temporomandibular articular discs- analysis of cross-links and their precursors-. Jpn. J. Oral Biol., 1985, 27(Suppl.), 99.

5. Kashima, K., Biochemical studies on collagen in temporomandibular joint discs. Jpn. J. Oral Maxillofac. Surg., 1988, 34, 424-37.

6. Woo, S. L-Y., Mow, V. C. and Lai, W. M., Biomechanical properties of articular cartilage. In Handbook of Bioengineering, eds.R. Skalak and S. Chien, McGraw-Hill Book Co., New York, 1988, pp. 4.1-44.

7. Butler, D.L., Grood, E.S., Noyes, F.R., Zernicke, R.F. and Brackett, K., Effects of structure and strain measurement technique on the material properties of young human tendons and fascia. J. Biomech., 1984, 17, 8, 579-96.

8. Vogel, H. G., Influence of maturation and aging on mechanical and bio-chemical properties of connective tissue in rats. Mechanisms of Aging and Development, 1980, 14, 283-92.

9. Stromberg, D. D. and Wiederhielm, C. A., Viscoelastic description of a collagenous tissue in simple elongation. J. Appl. Phys., 1969, 26, 857-62.

10. Haut, R.C. and Little, R.W., Rheological properties of canine anterior cruciate ligaments. J. Biomech., 1969, 2, 289-98.

11. Belytschko, T., Kulak, R.F., Schultz, A. B. and Galante, J. O., Finite element stress analysis of an intervertebral disc. J. Biomech., 1974, 7, 277-85.

12. Crofts, C.E. and Trowbridge, E.A., The tensile strength of natural and chemically modified bovine pericardium. J. Biomed. Mater. Res., 1988, 22, 89-98.

13. Walker, P., Amstutz, H.C. and Rubinfeld, M., Canine tendon studies. II. Biomechanical evaluation of normal and regrown canine tendons. J. Biomed. Mater. Res., 1976, 10, 61-76.

14. Currey, J. D., Mechanical properties of bone tissues with greatly differing functions., J. Biomech., 1979, 12, 313-9.

15. Vasu, R., Carter, D. R. and Harris, W. H., Stress distributions in the acetabular region I. Before and after total joint replacement. J. Biomech., 1982, 15, 155-64.

16. Takamura, H. and Maruyama, T., Studies on the structural changes of the temporomandibular joint - macroscopic observations -. J. Jpn. Prothod. Soc., 1984, 28, 49-59.

17. Yamada, H., Strength of Biological Materials. Williams and Wilkins Co., Baltimore, 1970.

THE IMMUNE SYSTEM AT THE METALLIC IMPLANT INTERFACE: METAL IONS INHIBIT IMMUNE FUNCTION BUT ARE NOT CYTOTOXIC

GRAÇA S. CARVALHO[1], ISABEL BRAVO[2] and MÁRIO A. BARBOSA[3]
(1) Department of Biology, Aveiro University, 3800 Aveiro
(2) Mestrado de Imunologia, Institute Abel Salazar,
Porto University, 4000 Porto
(3) Department of Metallurgy, Porto University, 4000 Porto
Portugal

ABSTRACT

Metal ions released from metallic orthopaedic implants can affect biological activities such as immune functions. We studied the effects of cobalt, nickel, iron and molybdenum on the proliferation and viability of stimulated lymphocytes, and on the frequency of sister chromatid exchanges (SCE) of proliferating lymphocytes. Normal human lymphocytes were stimulated through CD2 and CD3 molecules and the effects of the addition to the cell cultures of $CoCl_2$, $NiCl_2$, $FeC_6H_5O_7$, or MoO_3 were observed on lymphocyte proliferation. Various concentrations (between 0.3 pM and 5.0 mM) of the above metallic solutions showed that $CoCl_2$ and $NiCl_2$ were the most effective salts in causing inhibition of lymphocyte proliferation; in contrast MoO_3 did not cause significant effect; and $FeC_6H_5O_7$ showed interesting divergent inhibitory effects on cell proliferation, probably related to the cell donor HLA antigenic specificities. The trypan blue dye exclusion test on lymphocytes cultured in the presence of the above metal solutions showed that the great majority of the cells were viable (more than 86%). Furthermore the frequencies of SCE observed in the chromosomes of proliferating lymphocytes cultured in the presence or in the absence (controls) of the metals were similar, indicating that these metals are neither cytotoxic nor mutagenic.

INTRODUCTION

The cellular interactins occuring at the tissue/biomaterial interface have been recognized as an important factor in determining the *in vivo* biocompatibility. Evaluation of the biocompatibility of a material likely to be used in orthopaedic or odontological implantation has been carried out in *in vivo* and *in vitro* studies. The former show some disadvantages such as the lenght of the experimentation (approximately 2 years), the high number of animals necessary to give a statistical approach of the results, as well as reliability [1]. *In vitro* biocompatibility assays usually employ bone derived cells [2-4] or osteogenic cell lines [5-6]. Migration ability of the cells onto the biomaterial interface and morphology of the colonizing cells by light microscopy, scanning or transmission microscopy are taken as criteria for material biocompatibility [2-3]. Cytotoxicity caused by materials has been assessed by different methods [reviewed in 7]: *(i)* vital staining techniques, such as trypan blue exclusion test, *(ii)* release of radioactive markers (^{51}Cr or ^{3}H) from damaged cells, *(iii)* lack of cell multiplication, and *(iv)* changes in the metabolic activity of the cells.

The biological responses that occur following implantation of a biomaterial in the body are often evaluated by morphologic examination of the tissue at the implant site. Although they are seen most frequently in the vicinity of the implant, the host responses may also occur systemically or at remote sites, such as the regional lymph nodes of the immune system. In traditional terms, the immune system contributes to the maintenance of the physiological integrity of the body by eliminating foreign materials. This process requires the recognition, at the molecular level, of the organism's own molecules, known as "self", and of foreign elements, known as "non-self". The immune cells involved in the process of recognition and elimination of foreign molecules are lymphocytes and monocytes/macrophages. A first event of the action of metal ions on the immune system may be their binding to the lymphocyte surface membrane. In fact, our recent studies have shown that the *in vitro* exposure of human lymphocytes to $CoCl_2$, $NiCl_2$ or $FeC_6H_5O_7$, but not to MoO_3, causes a significant reduction of lymphocytes expressing the surface molecules involved in T lymphocyte activation: CD2 and CD3 [8-10]. As communication between lymphocytes is necessary for lymphocyte activation and involves cell-cell interactions via lymphocyte surface antigens [reviewed in 11], the binding of metal ions to these molecular structures may compromise lymphocyte functions, such as lymphocyte proliferation.

In the present work normal human blood lymphocytes were stimulated through their CD2 and CD3 surface molecules by adding the complex HE-anti-CD3 (anti-CD3 monoclonal antibody coupled to human erythrocytes) to the cultures of mononuclear cells (*i.e.*, blood lymphocytes and monocytes) in the presence or in the absence of metals. Lymphocyte proliferation was assessed by ^{3}H-thymidine incorporation in the cellular DNA. For the determination of the frequency of SCE, lymphocytes in total blood were stimulated with PHA (phytohaemagglutinin), in the presence or in the absence of metals, and the cell cycle stopped at the metaphase stage of mitosis with colcemid. The addition to the medium of 5-bromodeoxyuridine (BrdU) during the last 48 hours allowed the visualization and quantification of SCE.

MATERIALS AND METHODS

Peripheral Blood Mononuclear Cells

Mononuclear cells (lymphocytes and monocytes) were isolated from healthy blood donors from the Blood Service of the Portuguese Oncology Institute (Northern Region) as previously described [8].

HE-anti-CD3 Complex

The anti-CD3 monoclonal antibody (mAb) used was a rat IgG2b subclass YTH 12.5.22 kindly provided by Dr. Herman Waldmann (Cambridge, U.K.). After ammonium sulfate precipitation of the YTH 12.5.22 ascites fluid, the pellet was exhaustively dialised against 0.9% NaCl and the protein concentration adjusted to 1mg/mL. This anti-CD3 mAb preparation was then coupled to papain treated human erythrocytes (HE), using $CrCl_3$ (Sigma, Chemical Company, USA). For the papain treatment the concentration of erythrocytes was adjusted to 20% (V/V) in phosphate buffer saline (PBS) and treated with 0.3 mg/mL of papain (EC 3.4.22.2, Sigma) for 20 min at 37°C in the presence of 0.5% of L-cystein HCl (Sigma). After treatment cells were well washed in 0.9% NaCl and kept at 4°C. For coupling the anti-CD3 to papain treated erythrocyte, 50 μL of well packed HE were added to an equal volume of 1mg/mL of YTH 12.5.22 mAb. While mixing very hard (using a vortex) 100 μL of 0.1% $CrCl_3$ (in NaCl) were added dropwise. After 1 h of incubation at room temperature and under slow rotation the YTH 12.5.22 coupled HE (HE-anti-CD3) were washed three times and resuspended in RPMI 1640 medium (GIBCO Limited, U.K.) to give an erythrocyte concentration of 1.0%.

Culture Conditions

The culture medium was RPMI 1640 supplemented with 500 IU/mL penicillin (Atral, Portugal), 0.5 mg/mL streptomycin (Sigma), 24 μM HEPES buffer (Sigma) and 10% foetal calf serum (GIBCO). To 50 μL of the lymphocyte suspension $(2x10^6/mL)$ was added 50 μL of HE-anti-CD3 at 1.0% (V/V) and 100 μL of the appropriate metal salt concentration.

Cells were cultured in triplicates in round-bottom 96-well microtitre plates (Nunc, Denmark) at $37^{\circ}C$ in a 5% CO_2, humified atmosphere. Cultures were pulsed with 0.2 μCi of ^{3}[H]-thymidine (Radiochemical Centre, Amersham, U.K.) for the last 4 h of the culture. The ^{3}[H]-thymidine incorporation was measured by liquid scintillation counting after harvesting by a semi-automatic multiple harvester (Skatron, Lierbyen, Norway). Percent alteration, from control, of ^{3}[H]-thymidine incorporation was calculated according to the formula:

$$\% \text{ alteration} = \frac{\text{cpm in metal-ion containing cultures}}{\text{cpm in metal-ion free cultures}} \times 100 - 100\%$$

Cell Viability

After 3 days of culture (as described above) the cells were removed and mixed with equal volume of 0.2% W/V trypan blue in PBS and immediately read in a haemocytometer chamber. Trypan blue cannot penetrate the membrane of live cells, therefore only dead cells are stained.

Sister Chromatid Exchange (SCE) Analysis

For the SCE analysis, 0.5 mL of heparinized blood was cultured in 4.5 mL RPMI 1640 supplemented as above and containing the appropriate metal salt concentration. After 24 h of culture at $37^{\circ}C$, 4 drops of 5-bromodeoxyuridine (BrDU) at 1 mg/mL were added to the culture tubes. After an additional 48 h incubation in the dark, 0.1 mL colcemid was added to each 5 mL culture, 3 h prior to harvest. Harvested cells were then treated with hypotonic solution (0.075 M KCl) and fixed with metanol/acetic acid (3:1). Slides were prepared by forceful blowing of cell suspensions and were air dried. After 4 days at room temperature or, alternatively, an overnight incubation at $60^{\circ}C$, the sister chromatides were differentially stained with Hoechst for 30 min and Giemsa for 12 min. More than 14 metaphases per cell were scored for SCE frequency analysis.

RESULTS

Lymphocyte Proliferation

Lymphocytes were activated by HE-anti-CD3 in the absence or in the presence of various concentrations (between 0.3 pM and 0.5 mM) of $CoCl_2$, $NiCl_2$,

$FeC_6H_5O_7$ or MoO_3. Results showed that $CoCl_2$ was the metal salt causing higher inhibition; in fact concentrations as low as 18.8 μM were able to inhibit cell proliferation up to 90%, whereas similar inhibitions caused by $NiCl_2$ were obtained at concentrations higher than 300 μM (Fig.1). In contrast MoO_3 appeared to cause a lower effect since only less than 22% inhibition of lymphocyte proliferation was observed (Fig.1). In the case

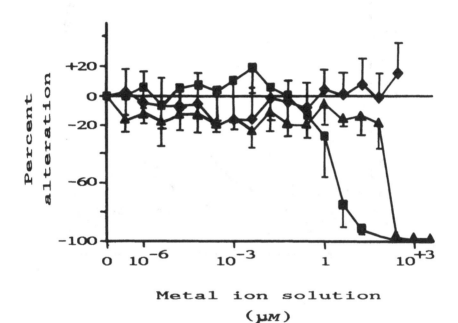

Figure 1. Effects of various concentrations of $CoCl_2$ (■—■), $NiCl_2$ (▲—▲) or MoO_3 (◆—◆) on lymphocyte proliferation. Results are expressed as mean ± SD of at least three separate experiments.

of lymphocytes cultured in the presence of $FeC_6H_5O_7$ divergent results were obtained according to the lymphocyte donor, and thus inhibitions as high as 70-80% or as low as 20-35% were observed (Fig.2 A and B, respectively).

The cell viability of lymphocytes cultured with metal solutions, either in the presence or in the absence of the stimulating agent (HE-anti-CD3), was very high as it is shown in table 1, indicating that cell toxicity was not occuring.

Sister Chromatid Exchange Analysis

Table 2 shows the frequencies of SCE/metaphase in peripheral blood

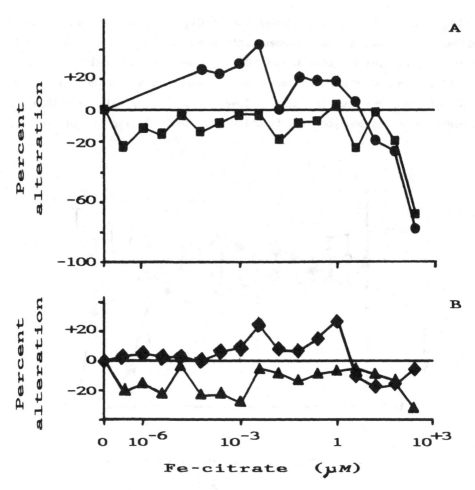

Figure 2. Effects of various concentrations of $FeC_6H_5O_7$ on lymphocyte proliferation. Inhibition is observed in two experiments (A), but not in the other ones (B).

lymphocytes stimulated with PHA in the presence of mitomycin C (positive control), and in the presence, or in the absence (negative control), of metal solutions. No significant increases were observed when lymphocytes were exposed to various concentrations of $CoCl_2$, $NiCl_2$ or $FeC_6H_5O_7$. At higher concentrations of nickel and cobalt no lymphocyte proliferation occured (NP in Table 2) and the SCE test could not be done. This blockade of PHA-stimulated lymphocytes in the presence of high concentrations of $CoCl_2$ or $NiCl_2$ is in perfect agreement with the inhibition of lymphocyte proliferation observed in lymphocytes stimulated via CD2 and CD3 (Fig.1).

TABLE 1
Cell viability of lymphocytes cultured for 3 days

Conc.[a]	$CoCl_2$		$NiCl_2$		MoO_3		$FeC_6H_5O_7$	
	No Act.[b]	Activ.[c]	No Act.	Activ.	No Act.	Activ.	No Activ.	Activ.
(mM)	%[d]	%	%	%	%	%	%	%
0.00	100.0	100.0	100.0	100.0	100.0	100.0	100.0	100.0
0.01	93.8	86.7	96.0	92.6	100.0	85.7	91.7	94.4
0.04	92.3	100.0	100.0	84.6	88.8	100.0	94.1	93.5
0.15	100.0	100.0	83.3	94.1	90.9	100.0	94.7	97.0
0.60	100.0	96.9	100.0	100.0	86.0	100.0	95.6	100.0

a) Metal salt concentration.
b) Non activated lymphocyte cultures.
c) Lymphocyte cultures activated with HE-anti-CD3.
d) Viable cells: percent of total cells.

TABLE 2
Sister chromatid exchange results

Treatment	Concentration (mM)	Mean (SCE/cell)	SD	n
Negative control	−	8.6	0.2	4
Positive control*	0.001	28.8	6.8	4
$CoCl_2$	0.078	8.9	1.5	2
	0.156	8.1	0.1	2
	1.000	NP	−	2
$NiCl_2$	0.078	8.4	1.6	2
	0.156	8.1	0.3	2
	0.313	8.6	2.1	2
	1.000	NP	−	2
$FeC_6H_5O_7$	0.313	8.3	0.2	4
	0.625	8.1	2.0	2

n: number of experiments in which more than 14 metaphases were scored.
*Mitomycin C.
NP: No proliferation.

DISCUSSION

In previous studies we have shown that the exposure of normal human lymphocytes to cobalt, nickel and iron causes a decrease in the proportion of lymphocytes expressing CD2 [8-10], nickel also affects the proportion of CD3 positive lymphocytes [9,10], and molybdenum causes no effect on either CD2 or CD3 positive cells [9,10]. The fact that CD2 and CD3 are particularly susceptible to these metals is of considerable interest since both molecules are involved in T cell activation: CD3 is a component of the T cell receptor acting in the antigen mediated pathway of T cell activation, and CD2 is involved in the alternative pathway of T cell activation; whether both pathways are linked or not has been a matter of recent discussions [reviewed in 12].

In the present work we stimulated normal human lymphocytes *via* CD2 and CD3, using an anti-CD3 monoclonal antibody coupled to human erythrocytes: anti-CD3 binds the lymphocyte surface molecule CD3, and LFA-3 (Lymphocyte Function-associated Antigen 3) on human erythrocytes binds to its natural ligand on T lymphocytes, the CD2 molecule [reviewed in 13]. The high inhibition of cell proliferation when lymphocytes were cultured in the presence of cobalt and nickel suggests that lymphocyte functions may be suppressed *in vivo* when concentrations similar to the ones used in the present *in vitro* tests occur. Such high levels of metal occur *in vivo* in soft tissues adjacent to metallic biomaterials, following corrosion and accumulation of metal ions [14].

In the case of iron, two sets of results were observed in four separate experiments. Further studies are now going on at the laboratory in order to clarify whether these differences are associated to differences among individuals, probably related to genetic specificities expressed in the context of the human histocompatibility complex, HLA.

Trypan blue dye is not able to penetrate the membrane of the cells culured in the presence (or in the absence, for control) of metallic solutions, indicating that the cells were alive. Similarly, the exposure of lymphocytes to cobalt, nickel and iron did not cause increase in the frequency of SCE suggesting that these metals, at the concentrations tested, are not cytotoxic nor mutagenic. The sister chromatid exchange test has been used as a sensitive index for both cytotoxicity of drugs and mutagenicity/carcinogenicity of environmental chemicals [reviewed in 15]: high positive correlation has been demonstrated between frequencies of

SCE and both cell killing, and mutagenicity and/or carcinogenicity of compounds.

Taken together our results show that cobalt, nickel and iron are not cytotoxic but they interfere in lymphocyte proliferation, so that compromising immune functions.

ACKNOWLEDGEMENTS

We thank Prof. Maria de Sousa for giving us the facilities for the development of this work as well as for her constructive criticism.

This work was carried out with support of grants 87.05 and 908.86.222 from the JNICT.

REFERENCES

1. Harmand, M.F., Bordenave, L., Duphil, R., Jeandot, R. and Ducassou, D., Human differentiated cell cultures: "In vitro" models for characterization of cell/biomaterial interface. In Biological and Biomechanical Performance of Biomaterials, ed. P. Christel and A.J.C. Lee, Elsevier Science Publishers B.V., Amsterdam, 1986, pp. 361-366.

2. Davies, J.E., Price, N.M. and Matsuda, T., In vitro biocompatibility assays which employ bone derived cells. In Oral Implantology and Biomaterials, ed. H. Kawahara, Elsevier Science Publishers B.V., Amsterdam, 1989, pp. 197-204.

3. Matsuda, T. and Davies, J.E., The in vitro response of osteoblasts to bioactive glass. Biomaterials, 1987, 8, 275-284.

4. Nicolas, V., Nefussi, J.R., Collin,P. and Forest, N., Effects of acidic fibroblast growth factor and epidermal growth factor on subconfluent fetal rat calvaria cell cultures: DNA synthesis and alkaline phosphatase activity. Bone Min., 1990, 8, 145-156.

5. Itakura, Y., Kosugi, A., Sudo, H. and Yamamoto, S., Development of a new system for evaluating the biocompatibility of implant materials using an osteogenic cell line (MC3T3-E1). J. Biomed. Mater. Res., 1988, 22, 613-622.

6. McAuslan, B.R., Johnson, G., Delamore, G.W., Gibson, M.A. and Steele, J.G., Cell growth on metallic glasses: The interaction of amorphous metal alloys with cultured neuronal, osteoblast, endothelial, and fibroblast cells. J. Biomed. Mater. Res., 1988, 22, 905-917.

7. Browne, R.M. and Tyas, M.J., Biological testing of dental restorative materials in vitro — a review. J. Oral Rehabil., 1979, 6, 365-374.

8. Carvalho, G.S. and DeSousa, M., Iron exerts a specific inhibitory effect on CD2 expression of human PBL. Immunol. Lett., 1988, 19, 163-168.

9. Bravo, I., Carvalho, G.S., Barbosa, M. and DeSousa, M.,   Differential effects of eight metal ions on lymphocyte differentiation antigens *in vitro*. J. Biomed. Mater. Res., 1990, **24** (in press).

10. Carvalho, G.S., Bravo, I., DeSousa, M. and Barbosa, M., Degradation of metallic biomaterials and the immune system: *in vitro* effects of iron, nickel and cobalt on the expression of human lymphocyte antigens. Proc. 8th European Conference on Biomaterials (in press).

11. Schad, V.C. and Greenstein, J.L., T cell accessory molecules mediating cell adhesion and signal transduction. Current Opinion in Immunology, 1989, **2**, 123-128.

12. Kabelitz, D., Do CD2 and CD3-TCR T-cell activation pathways function independently? Immunol. Today, 1990, **11**, 44-47.

13. Bierer, B.E. and Burakoff, S.J., T lymphocyte activation: the biology and function of CD2 and CD4. Immunol. Rev., 1989, **111**, 267-294.

14. Pohler, O.E.M., Degradation of metallic orthopedic implants. In Biomaterials in Reconstructive Surgery, ed. L.R. Rubin, C.V. Mosby, St. Louis, 1983, pp. 158-228.

15. Deen, D.F., Morgan, W.F., Tofilons, P.J. and Barcellos-Hoff, M.H., Measurement of sister chromatid exchanges and their relationship to DNA damage, repair and cell killing. Pharmac. Ther., 1989, **42**, 349-360.

DIFFERENTIAL RATE OF BONE GROWTH TO TITANIUM ALLOY AND STAINLESS STEEL IMPLANTS

RONI HAZAN AND URI ORON
Department of Zoology, The George S. Wise Faculty of Life Sciences,
Tel Aviv University, Ramat Aviv 69978, Israel

ABSTRACT

The kinetics of bone growth around titanium alloy (Ti-6Al-4V) and stainless steel 316L screw implants that were inserted into the medullary canal of the femur in rats were followed using mechanical and histological measures. The force needed to pull out the screw from the femur served as an indication for the extent of bone growth into the grooves between the ridges of the screw. A progressive and significant increase in the shear strength of the screw with time was observed from 4 until 35 days post operation for both metals. At all time intervals (4, 5, 6, 10 and 35 days) the shear strengths of the Ti-6Al-4V implants were significantly higher (1.4-6 fold) than those of the stainless steel implants. The experimental model in the rat can serve as a simple and reliable model for investigation of basic aspects of bone growth to metal implants. It is concluded that various factors (surface chemical characteristics, corrosion resistance, etc.) within the metal alloy itself may significantly affect the rate of bone growth to it. Ti-6Al-4V may be a preferable alloy over stainless steel for medical use in the case of cementless artificial joints.

INTRODUCTION

Problems with the use of polymethylmethacrylate cement in the fixation of total joint prosthesis have led to the development of new prosthetic designs that use the ingrowth of bone into a porous surface coating for the attachment between the bone and the implant [1]. Indeed, many studies have reported a progressive ingrowth of bone to porous implants in experimental animals and in humans [2,3]. The understanding of basic processes of bone growth to implants and factors controlling it, which may play a significant role in the success of cementless artificial joints, have become, therefore, of prime importance in the last decade. It is well known that the metals that are of clinical relevance (stainless steel, titanium alloys, Co-Cr alloys etc.) have different mechanical, biocompatible and corrosion resistant properties, as well as different chemical composition at their outermost layers [4-6]. It was also shown that the tissue response and the tissue in immediate contact with the metal may be different for the various alloys [7,8]. However, a quantitative comparative study on the rate of bone growth to the various metals is an _in vivo_ model has not been performed.

We describe here a simple and reliable model where the process of bone growth next to a metal implant inserted into the medullary canal of the femur can be followed by mechanical and histological methods. This model has been used to compare the rate of bone growth to stainless steel and Ti-6Al-4V implants.

MATERIALS AND METHODS

Animals and surgical procedure

Surgery was performed on mature Sprague-Dawly male rats (100 days old, 300-350 g) that were anaesthetized by intraperitoneal injection (1 ml per 100 g body weight) of Avertin under sterile conditions. The rats, at this age, are sexually mature and at a stable stage with respect to their bone formation [9]. Following longitudinal incision in the skin at the patellar region, the ligament of the quadriceps femoris muscle was shifted laterally and the distal part of the femur exposed. The patellar groove was widened using small conical drills of about 3 mm maximal diameter in order to create a precise space for the conical head of the screw. A series of dental screws with increasing diameter (0.8, 1.6 and 2.0 mm) and of 7 mm in length were then introduced (by hand drilling), in sequence, into the medullary canal through the symphysis to finally create an unobstructed pathway for subsequent insertion of the implanted screw. The final reaming was performed by a screw identical to the implanted screw. It should be noted that by histological methods it was revealed that the medullary canal is about 2 mm diameter, up to about 8 mm proximally to the knee joint and then narrows. Thus, the above procedure allowed a similar reaming process in each rat by which bone trabeculae in the medullary canal were broken but the cortical bone remained intact. In a preliminary study it was found that identical reaming process in each rat is extremely important in obtaining a constant kinetics of bone growth to the metal implant inserted to the canal. The screws used were self-made from medical grade rods of stainless steel 316L and Ti-6Al-4V alloy (2 mm diameter and 6 mm in length). All screws were cleaned, passivated and sterilized with ethylene oxide a few days before their insertion (non-screwed) into the medullary canal of the rats, with the head directed towards the distal head of the femur. Care was taken to ensure that the screws would fit tightly but could be pulled out with minimal force. The ligament of the quadriceps femoris was then cut distally to the patella and the cut muscles in the vicinity of the knee region were sutured around the head of the screw, as well as the skin in the patellar region. The rats were treated with penicillin (1000 u/g body weight) into the thigh muscles immediately post surgery, and with Bioxin (1 g/l) in drinking water for 3 days after surgery. No post-operative complications (e.g., acute infections or paresis) were observed in the animals. The animals were kept under constant environmental conditions, fed on purina and supplied with water _ad libitum_. They were usually inactive in the cage but did step on the operated leg. Since the patellar ligament was cut in this leg the implant was practically in an unloaded situation.

Mechanic measurements and statistics

Eighty five rats served for measurements of the force needed to dislodge
the screw from the medullary canal at various time intervals following
surgery. Six rats were used to measure the force immediately after
insertion (0 days time interval) and groups of 7-9 rats were used to
measure the force at each time interval (4, 5, 6, 10 and 35 days) post
operation. The device to measure this force consisted of a fixed part (a
small chuck of a drill) to which the head of the cortex screw was
anchored, while the proximal part of the femur was fixed in a special
device made of plexiglass and connected through flexible chains to a
steel basket (30 cm x 20 cm) into which weights were added. Measurements
of the force needed to dislodge the screw from the femur were made by
progressive addition of weights (50 g initially and then when a 1 kg
strength was reached the weights were of 200 g) to the basket until the
screw was completely pulled out. This force was defined as the total
weight in the basket at the moment the screw was completely pulled out
from the femur minus half the weight of the last added weight. The shear
strengths were calculated in MPa by dividing the dislodgement forces by
the surface area of a cylinder of 2 mm diameter (the outer diameter of the
screw ridges) and a height of 6 mm. The results were then statistically
evaluated using single classification analysis of variance and Student
Neuman-Keuls test for multiple comparison among means [10].

RESULTS

Bone growth into the grooves between the ridges of the screw implant that
was inserted into the medullary canal of the femur as revealed by the
measurements of the shear strengths is presented in Fig. 1. This force
was negligible (0.005 MPa) immediately following the insertion of the
screw (0 time interval) for both metals. A progressive and significant
(P < 0.05) increase in this force with time was observed up to 5 weeks
after surgery.

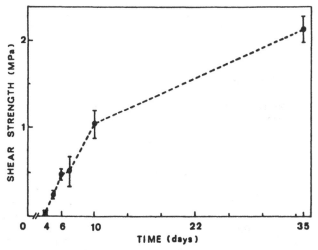

Figure 1. Shear strength of titanium alloy implants as a function of time
after insertion. Each point is mean ± s.e.m. of 7-9 rats.

At all time intervals after surgery the shear strengths of the Ti-6A1-4V implants were significantly higher (P < 0.05) than those of the stainless steel 316L implants comprising to 3.0, 6.0, 1.7, 1.8 and 1.35 fold at 4, 5, 6, 10 and 35 days respectively post surgery (Table 1).

TABLE 1

Shear strength of screw implants made of Ti-6A1-4Va and stainless steel 316L as functions of time after insertion

Time after insertion	Shear strength (MPa)	
(days)	stainless steel	Ti-6A1-4Va
4	0.01+0.001	0.03+0.010
5	0.04+0.010	0.24+0.050
6	0.28+0.060	0.47+0.160
10	0.58+0.090	1.04+0.160
35	1.58+0.180	2.13+0.160

Each number is mean $\pm$ s.e.m. of 7-9 rats.

DISCUSSION

A progressive and significant increase in the shear strength of the screw implant that was inserted into the medullary canal of the femur with time was observed both in stainless steel and Ti-6A1-4V implants in the present study. During the first few days after the insertion of the screw there was no increase in shear strength, probably indicating the initial inflammatory stage post surgery. Thereafter, there was a steep increase in the interfacial strength indicating accumulation of osteoblasts and collagen and initial maturation of bone around the screw. The progression in shear strength which correlates with bone maturity proceeded to a lesser extent until 35 days post surgery. Due to various animals, methodologies, biomaterials, implant design and implantation site, it is impossible to make a simple comparison of the kinetics of bone growth obtained in this study to other investigations. While in some studies progression in bone growth with time was noticed [11-14], in others bone growth could not be detected in the first few weeks after insertion and thereafter the increase was non-linear with time [15].

The present work extends the knowledge on bone growth around metal implants in rats but also presents a simple and reliable model for a further investigation of various aspects of bone growth and factors controlling it in adult animals. The advantages are several: The experimental animals used (rats) are inexpensive, easy to obtain and handle and possess a relatively low susceptibility to infection. The use of the screw as an implant allows us to obtain a simple general histological evaluation of the tissue around the implant by unscrewing the screw and without the need to use special, sophisticated equipment to do sections through metals.

The experimental model presented in this study enabled us to demonstrate a significantly different osteogenic response to implants made of two different alloys of clinical relevance. During the initial phase of bone formation (woven bone) the rate of bone growth was 3-6 fold higher around the Ti-6Al-4V alloy than the stainless steel implants but even after 35 days, when a compact bone is present next to the implant, the shear strength was 40% higher for the titanium alloy. It may be speculated that the initial response of tissue to a certain metal as shown, also has a long term effect on the bonding strength between bone and metal.

The mechanisms involved in the differential regulation of rate of bone growth to the different metals are not yet known. The course of tissue reaction to the implants, namely bone growth and development from stem cells, is a sequence of events each of which may be triggered by its predecessor and may also be influenced by environmental factors created by the implant. The activity of the osteoprogenitor cells may be mediated by a variety of biochemical substances which in turn are influenced by the physico-chemical interaction of the outermost layers of the metal with the molecules and cells in the adjacent milieu. Indeed, it has been previously shown that molecules of the extracelular matrix (proteoglycans) are the molecules in immediate contact with the metal oxide layer of the implants and also associated with collagen bundles next to the formed bone. Furthermore, it has been demonstrated that in the case of stainless steel the proteoglycan layer is wider than that observed next to titanium alloy. Sundgren et al. [5] have compared the interface between implant and human tissue in stainless steel and titanium implants using Auger electron spectroscopy. They found chemical changes due to the metal composition but, other than that, the oxide layer had similar thickness in both metals. However, in the case of titanium implants that were embedded in bone marrow in humans, the oxidation process occurred up to several years, a phenomenon that was not found for the stainless steel implants. Since Ca and P were embedded in this layer, it may be speculated that the higher rate of bone growth around the titanium implants, even at long-term intervals, may be associated with the difference in the kinetics of the oxide layer thickness in the two metals in a defined in vivo environment. Differences in the nature of the interface zone between bone and implants made of various metals have also been reported for other metals. Zirconium was found to be less "natural like" at the interface than titanium, and Ti-6Al-4V less than pure titanium [16]. The titanium alloys in general are also considered to be more biocompatible and corrosion resistant than stainless steel. Thus, the possibility that different metal ions that are shed from the metal to surrounding tissues affect the processes associated with bone growth and maturation cannot be ruled out.

The results of the present study also have clinical implications. The location of the implant in this model next to marrow and canceleous bone bone is very similar to the cell milieu in the case of cementless hip joint. The bone growth is measured in a practically unloaded situation since we intended to exclude the effect of load on this complex process. Similarly, during the initial period after insertion of cementless hip or knee joints in humans, the implants are subjected to a period of protected healing with minimal weight-bearing to minimize any movements of tissue-implant interface. Although it is difficult to extrapolate from an animal model to humans, it may be concluded that the Ti-6Al-4V alloy should be prefered to the stainless steel 316L in those cases where anchorage between bone and implant depends solely on bone growth to it. The mechanisms associated with the interesting phenomenon of regulation of bone growth to implants by their surface characteristics and/or metal ions released from them will have to be further elucidated.

ACKNOWLEDGEMENT

We wish to thank Mrs. L. Maltz for excellent technical assistance and Mrs. C. Shapiro for typing the manuscript.

REFERENCES

1. Haddad, R.J., Cook, S.D. and Thomas, K.A., Biological fixation of porous-coated implants. J. Bone Joint Surg., 1987, 69-A, 1459-1466.

2. Ronningen, H., Solheim, L.F., and Langeland, N., Invasion of bone into porous fiber metal implants in cats. Acta Orthop. Scand., 1984, 55, 352-358.

3. Landon, G.C., Galante, J.O. and Maley, M.M., Noncemented total knee arthroplasty. Clin. Orthop. Rel. Res., 1986, 205, 49-57.

4. Albrektsson, T. and Hansson, H.A., An ultra-structural characterization of the interface between bone and sputtered titanium or stainless steel surfaces. Biomaterials, 1986, 7, 201-205.

5. Sundgren, J.E., Bodo, P. and Lundstrom, I., Auger electron spectroscopic studies of the interface between human tissue and implants of titanium and stainless steel. J. Colloid Interface Sci., 1986, 110, 9-20.

6. Hanker, J.S. and Giammara, B.L., Biomaterials and biomedical devices. Science, 1988, 242, 885-892.

7. Albrektsson, T., and Jacobsson, M., Bone-metal interface in osseointegration. J. Prosthet. Dent., 1987, 57, 597-607.

8. Johansson, C.J., Lausmaa, M.A., Hansson, H.A. and Albrektsson, T., Ultrastructural differences of the interface zone between bone and Ti-6Al-4V or commercially pure titanium. J. Biomed. Eng., 1989, 11, 3-8.

9. Nishimoto, S.K., Chang, C.H., Gendler, E., Strykes, W.F. and Nimni, M.E., The effect of ageing on bone formation in rats: Biochemical and histological evidence for decrease bone formation capacity. Calcit. Tissue Int., 1985, 37, 617-624.

10. Sokal, R.R. and Rohlf, F.J., Biometry. London, W.H. Freeman 1969.

11. Galante, J.O., Rostoker, W., Lueck, R. and Ray, R.D.. Sintered fiber metal composites as a basis for attachment of implant to bone. J. Bone Joint Surg., 1971, 53-A, 101-114.

ENHANCEMENT OF BONY INGROWTH TO T TANIUM AND TITANIUM ALLOY ORTHOPAEDIC
IMPLANTS IN EXPERIMENTAL MODEL IN THE RAT

RONI HAZAN AND URI ORON
Department of Zoology, Tel Aviv University, Ramat Aviv 69978, Israel

ABSTRACT

Kinetics of bony ingrowth to control and heat-treated TI-6Al-4V screw
implants and pure titanium mesh implants inserted into the medullary canal
of the femur in rats was followed by mechanic and biochemical methods. A
progressive and significant increase in the bony ingrowth to the implants
as reflected by their interfacial shear strengths in the femur was
observed with time. These values were negligible at the time of insertion
and exceeded 2.13 MPa at 35 days post surgery for control Ti-6Al-4V
implants. The specific activity of alkaline phosphatase in tissue
extracts around the screw implant showed a peak at six days
postoperatively. At all time intervals following surgery the shear
strengths of the heat-treated Ti-6Al-4V screw implants and pure titanium
mesh cylinders were 2 to 5.5 fold significantly higher than the control
untreated implants. It is concluded that the heat treatment which most
probably modifies the surface chemical properties of the implants,
significantly affects the process of differentiation of osteoblasts from
osteoprogenitor cells in the marrow or the rate of bone deposition next to
the implants. The results of the present study are of direct clinical
relevance in improving the function of cementless artificial joints and
dental implants.

INTRODUCTION

Long term success of total joint replacement is the permanent fixation of
the prosthetic components to the host bone. Although acrylic cement is a
widely accepted method for implant stabilization, it is generally agreed
that the bone/cement interface is the usual point of failure of the
implants. Biological fixation has proven to be a viable alternative to
acrylic bone cement [1-3]. Indeed, many studies reported progressive bony
ingrowth to metal implants in experimental animals and humans [2,4,5]. It
became apparent that a faster initial stabilization of the prosthetic
components of cementless joints may shorten recovery time after surgery
but may also contribute significantly to the long term success of the
joints, especially in those patients with limited potential for bone
regeneration. Several attempts have been made in the past to enhance bony

ingrowth to metal implants in *in vivo* animal models. Berry *et al.* [6] have investigated the effect of electrical stimulation and impregnation of porous material with tricalcium phosphate on bone growth into porous implants. However, in a canine experimental model the interface bond strength between the implant and bone was not affected by these treatments. Liebrecht *et al.* [7] demonstrated that spherical hydroxyapatite particles that were packed around an intramedullary rod caused the enhancement of the initial stabilization of the implants. Hydroxyapatite coating of titanium implants resulted in a moderate but significant increase in the interfacial strength of these implants in an experimental model in dogs [8]. However, the possibility of augmenting bony ingrowth to implants by altering the surface chemical properties of the metals has not yet been investigated. We report here the possibility to enhance the bony ingrowth to metal implants by heat treatment in optimal temperature prior to their insertion to the medullary canal in rats using mechanical and biochemical methods.

MATERIALS AND METHODS

Experimental Procedure and Mechanic Measurements

Surgery was performed on mature (3-4 month old [9]) rats. Animal housing and the surgical procedure of insertion of the implants into the medullary canal of the femur in the rats was done exactly as described previously [10]. In some experiments the control or heat treated implants were inserted to one leg of each animal. For these experiments CR rats were used. In another experiment (see Results) the heat treated implants were inserted to one leg (randomly chosen) while the control implants were implanted to the contralateral leg. In these experiments Sprague-Dawley rats (Charles River Inc., England) were used.

At different time intervals after insertion of the implants measurement of the force needed to completely dislodge the implant from the femur were performed as described previously [10]. The shear strength of the implants served as a quantitative indication for the rate of bone growth into the implants.

Implants

The implants were either screw implants (2 mm diameter and 6 mm in length) self made from medical grade rods of Ti-6Al-4V or pure titanium mesh rods (2 x 2 x 7 mm, supplied by Zimmer Inc., USA) that were slightly shaped to fit the medullary canal of the femur. All the implants were extensively cleaned and passivated according to common standards for medical implants prior to their sterilization or heat treatment. A series of experiments was performed using the same experimental model in rats in order to find the optimal temperature, time of heating and the proper gaseous combination around the implants that would give the maximal enhancement phenomenon. All the implants (control and heat treated) were then sterilized with ethylene oxide in the standard procedure for medical devices and implanted to the rats usually not later than one week after the heating process.

Analytical Procedure and Statistics

The enzymatic activity of alkaline phosphatase (ALP) was determined in newly-formed tissue around the implants by collecting it immediately after it was pulled out from the femur and force was measured. The screw with the tissue around it was transferred to 1 ml ice-cold 10 mM Tris-Hcl pH, 7.2 buffer and the tissue adhered to it was carefully removed using a 16 gauge needle under a dissecting microscope. The extracted tissue was further homogenate in the same buffer by 10 strokes (about 20 sec) in glass-teflon homogenizer revolving at 480 rpm. The homogenate was then centrifuged at 6,700 g for 15 min at 4°C and the supernatant served for ALP and protein determination in fresh samples. In a preliminary study, it was found that addition of the detergent Triton X-100 to a final concentration of 0.1% to the homogenization medium had elevated the activity of ALP only in 10% in the soluble fraction and, therefore, it was omitted from the experimental procedure as it interfered with protein determination. The specific activity of ALP was also measured as described above on intact marrow of the femur and was found to be not more than 10% of the specific activity in the newly-formed extracted tissue around the screw at 5-10 days after implantation. Thus, the possibility that the ALP activity in the samples is partially derived from bone marrow rather than newly-formed bone (in particular in the early stages) can be ruled out. Alkaline phosphatase was assayed with 0.5 mM p-nitrophenyl phosphate as substrate. The reaction was carried out in 0.1 M sodium barbital buffer, pH 9.3, containing 2 mM $MgCl_2$ with 0.1-0.4 ml of the homogenate At 37°C. Protein was determined as described by Lowry *et al.* (1951) using bovine serum albumin as standard. One unit of phosphatase was defined as the enzyme activity that liberates 1 μmol p-nitrophenol per hour in 37°C.

Results were statistically evaluated using single classification analysis of variance and Student Neuman Keuls test for multiple comparison among means [11].

RESULTS

A progressive and significant ($P < 0.01$) increase in the bony ingrowth into the Ti-6Al-4V screw implants as reflected by the increase in their shear strength in the femurs with time was noticed (Fig. 1). The specific activity of alkaline phosphatase in the newly formed tissue extracts around the screw implants increased about 10 fold from 4 to 6 days post operatively, reaching its peak activity at the 6 day interval (Fig. 1). Thereafter a decrease and leveling off was noticed.

Figure 2 represents the kinetics of bone growth onto heat treated Ti-6Al-4V implants prior to implantation versus control untreated implants. The treated implants were inserted to one leg while the control implants were embedded in the contralateral leg of each rat. It can be seen that at the time intervals up to 10 days there was a 3-5 fold significant ($P < 0.05$) increase in the shear strength of the heat treated implants over the control ones. At 22 days post operatively there was also a 2.5 fold increase ($P < 0.05$) in the shear strength in the treated implants over the control ones.

A similar augmentation phenomenon of the bonding strength between the bone and the implants was also noticed when pure titanium mesh implants were inserted to the rats. There was 2.5 to 5.6 fold significant (P <

0.05) increase in the shear strength of the heat treated implants as compared to the control ones at 4 to 6 days following surgery (Table 1).

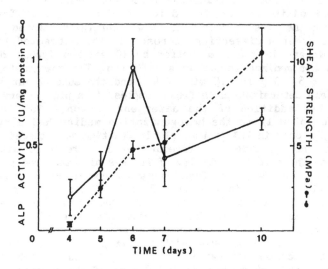

Figure 1. Shear strengths and alkaline phosphatase activity in tissue extracts around control Ti-6Al-4V screw implants as a function of time after insertion. Alkaline phosphatase activity was measured in the newly formed tissue between the ridges of the screw implants that was collected from the same implants that were pulled out from the femur during measurements of pull out forces. Each point is mean ± s.e.m. of 5-8 rats.

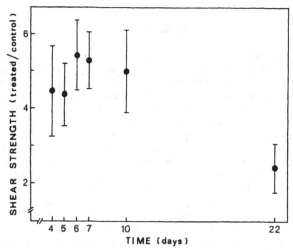

Figure 2. Shear strength of heat treated versus control Ti-6Al-4V implants as a function of time post insertion. The results are expressed as the ratio of the shear strength (MPa) of heat treated and control implants that were implanted in the same rat. Each point is mean ± s.e.m of 7-9 rats

TABLE 1

Shear strength of control and heat treated pure titanium mesh cylinders
inserted to the medullary canal of the femur in rats

Time after implant insertion (days)	Shear	strength	(MPa)
	control implants	heat treated implants	heat treated/control
4	-	-	5.55±1.36
5	2.37±1.10	5.88±0.65	-
6	-	-	5.14±1.13

At five day intervals results are from an experiment where the implants
(control or heat treated) were inserted separately to the femur of
different rats. At the 4 and 6 days time interval the control implants
were inserted to one leg and the heat treated ones to the contralateral
leg. The results are expressed in this case as the ratio between treated
and control implants. Each number is mean ± s.e.m. of 5-7 rats.

Another experiment was performed in order to evaluate the shelf life
of the heat treatment. Ten Ti-6Al-4V screw implants were heat treated and
then left under sterile conditions (ethylene oxide sterilization) for 14
months. They were then implanted to the femur in rats together with
another group of control (ethylene oxide sterilized) implants. The shear
strengths 7 days post surgery were 0.49±0.15 MPa for the heat treated
implants which was significantly higher (P < 0.05) than the value for the
control implants (0.11±0.04 MPa)

DISCUSSION

The present experimental model indicates a progressive increase of bone
deposition and maturation around the Ti-6Al-4V screw implants. The steep
increase in alkaline phosphatase activity, which is a good marker for
osteoblasts [12], in the newly formed tissue next to the implants
indicates the progressive differentiation of osteoblast from
osteoprogenitor cells in the marrow or the endosteum of the femur. The
peak activity at the 6-day interval is followed by a reduction in alkaline
phosphatase activity although bone formation and mineralization continues
until about two weeks following insertion, as indicated by the mechanic
measurements. Indeed, it was previously found that reduction in alkaline
phosphatase is associated with increase in calcium deposition in other
systems where bone growth was followed [13,14].
At all time intervals following surgery the rate of bone growth to
the preheated implants was several fold higher than control implants both
in the case of pure titanium and titanium alloy implants. The extent of
enhancement in the present study is higher than that previously reported
in hydroxyapatite coated implants [8].

The mechanism(s) by which the heat treatment of the implant induce the enhancement of bony ingrowth to it is not yet clearly understood. The course of tissue reaction, namely osteoprogenitor cells, to the implants is a complex sequence of events which may be mediated by a variety of biochemical substances. These substances are in turn affected by the physico-chemical interaction of the outermost layer of the metal with the molecules and cells in the adjacent milieu. Indeed, it has been previously shown that molecules of the extracellular matrix (proteoglycans) are the molecules in immediate contact with the metal oxide of the implants and also associated with collagen bundles next to the formed bone [1]. Since it was found that the heat treatment does not change the physical (roughness) properties of the surface of the implants [15], it may be postulated that this treatment alters the oxide layer characteristics (thickness, chemical nature, charge etc.). This in turn may cause different adsorbtion of these molecules onto the heat treated implants which may be different from their adsorbtion kinetics to the control implants. This may then change the response and growth rates of adjacent cells to the implants. It may also be suggested that change in the corrosion resistance due to the heat treatment, modifies the rate of the metal ions that are shed from the metal to surrounding tissue. The possibility that this phenomenon also affects the processes associated with osteoblast differentiation from osteoprogenitor cells and/or bone growth and maturation cannot be ruled out.

The process of augmentation of the bone growth to the implants, as described in this study has several significant advantages over other inventions for enhancement of bone growth. It is a low cost process, simple to perform and control without changing the core metal or the shape of the implants that have been developed for many years. Furthermore, it has an advantage over hydroxapatite coating both in the extent of enhancement and also the possibility that the point of failure of the implants may be between the metal and the hydroxyapatite coating. In addition, the heat treatment causes a change in the surface layer of the metal which has been found to be stable with a long shelf life.

The clinical relevance of the heat treatment is self-evident. It is well known that a rapid bony ingrowth during the initial period of minimal weight-bearing postoperatively in humans ensures a rapid recovery after cementless joints operations, and also contributes to a better long-term performance of the artificial joints [1-3]. Furthermore, it may be speculated that by altering heating temperatures, time or the gaseous environment during heating, it would be possible to create conditions where bony ingrowth to certain implants or regions within the implants could be predetermined according to specific clinical demands.

The present study also demonstrates a novel approach in the understanding of implant tissue interaction and may shed light on factors that affect bone growth in general. The exact mechanisms by which the chemical characteristics of the surface layer of the implant (most probably the oxide layer) affect the process of bony growth remain to be the subject of further study.

Acknowledgment

The authors wish to thank Mrs. L. Maltz for excellent technical assistance.

REFERENCES

1. Albrektsson, T. and Jackobsson, M., Bone-metal interface in osseointegration. J. Prosthet. Dent., 1987, 57, 597-607.

2. Collier, J.P., Mayor, M.B., Chae, J.C., Surprenant, V.A., Surprenant, H.P. and Dauphinais, L.A., Macroscopic and microscopic evidence for prosthetic fixation with porous coated materials. Clin. Orthop. Rel. res., 1988, 235, 172-180.

3. Pilliar, R.M., Porous-surface metallic implants for orthopedic applications. J. Biomed. Mater. Res., 1987, 21, 1-33.

4. Ronningen, H., Solheim, L.F. and Langeland, N., Bone formation enhanced by induction: bone growth in titanium implants in rats. Acta Ortop. Scand., 1985, 56, 67-71.

5. Landon, G.C., Galante, J.O. and Maley, M.M., Noncemented total knee arthroplasty. Clin. Orthop. Rel. Res., 1986, 205, 49-57.

6. Berry, J.L., Geiger, J.M., Moran, J.M., Skraba, J.S. and Greenwald, A.S., Use of tricalcium phosphate or electrical stimulation to enhance the bone-porous implant interface. J. Biomed. Mater. Res., 1986, 65-77.

7. Liebrecht, Pl, Ricci, J.L., Parsons, J.R., Salsbury, R. and Alexander, H., Enhanced stabilization of orthopaedic implants with spherical hydroxyapatite particulate. In 32nd Annual ORS, New Orleans, Louisiana, 1986, p.347.

8. Thomas, K.A., Kay, J.F., Cook, T. and Jarcho, S.D., The effect of surface macrotexture and hydroxylaptite coating on the mechanical strengths and histologic profiles of titanium implant materials. J. Biomed. Mater. Res., 1987, 21, 1395-1414.

9. Nishimoto, S.K., Chang, C.H., Gendler, E. and Nimni, E., The effect of ageing on bone formation in rats: biochemical and histological evidence for decrease bone formation capacity. Calcit. tissue Int., 1985, 37, 617-624.

10. Hazan, R. and Oron, U., Differential rate of bone growth to titanium alloy and stainless steel implants. 1990, this volume.

11. Sokal, R.R. and Rohlf, F.J., Biometry, W.H. Freeman, 1969.

12. Farley, J.R., and Baylink, D.J., Skeletal alkaline phosphatase activity as a bone formation index in vitro. Metabolism, 1986, 35, 563-571.

13. Genge, B., Sauer, G.R., Wu, L.N.Y., McLean, F.M. and Wuthier, R.E., Correlation between loss of alkaline phosphatase activity and accumulation of calcium during matrix vesicle-mediated mineralization. J. Biol. Chem., 1988, 263, 18513-18519.

14. Reddi, A.H. and Sullivan, N.E., Matrix-induced endochrondral bone differentiation: influence of hypophysectomy, growth hormone and thyroid-stimulating hormone. <u>Endocrinol.,</u> 1980, 107, 1291-1299.

15. Hazan, R., Bone ingrowth into porous surface metal implant in rat femur and factors that influence it. <u>M.Sc. thesis</u>, Tel Aviv University, 1985.

THE INFLUENCE OF THE PLATE STIFFNESS OF THE TIBIAL PLATEAU OF A KNEE PROSTHESIS

J. Vander Sloten[*], R. Van Audekercke, G. Van der Perre
Katholieke Universiteit Leuven
Division of Biomechanics and Engineering Design
Celestijnenlaan 200A
B-3030 Heverlee (Belgium)

[*] Research Assistant of the Belgian National Fund for Scientific Research

ABSTRACT

We investigated the influence of the plate characteristics of the tibial plateau on the stress distribution in the bone of the proximal tibia and on loosening at the anterior side. The proximal tibia was simulated by a balsa-wood model with fiber-reinforced outer layer and the tibial plateau by plates made of steel (3 mm), aluminium (1 mm) and polyethylene (6 mm). Stress analysis was made by strain gauge measurements and two-dimensional finite element analysis. From our experiments we have concluded that a compromise had to be found between a flexible and a stiff plateau. A plateau with anisotropic properties was designed and tested.

INTRODUCTION

Although clinical experience with knee prostheses started later than with hip prostheses, the results of the modern types of total knee replacement are good to very good. Clinical problems seem to concentrate most on the tibial part [1], [5]. The tibial parts of most modern types of knee prosthesis are based upon a plateau that fits the cut surface of the entire proximal tibia as good as possible. A short stem or studs may be present to provide rotational stability.

This research will be confined to the influence of the mechanical characteristics of the tibial plateau upon the stress distribution in the cortical and trabecular bone of the proximal tibia and eventual loosening underneath the plateau. A large medullary stem was eliminated from the beginning, because of the stress shielding it causes [6] and potential difficulties in the case of revision arthroplasty.

The stress distribution in the cancellous bone is important because of changes in bone geometry and structure that can be expected, in accordance with Wolff's Law. There is also the potential danger of bone resorption due to local overload. Loosening between tibial plateau and bone is important because it compromises bone ingrowth in a porous coating underneath a tibial plateau.

Finally, one should be aware that the factors we will investigate in this chapter are not the only causes of failure of a knee implant. Some investigators are even convinced that long term results are more determined by the quality of the operation than by the design of the implant [2], [4].

MATERIALS AND METHODS

The approach of the research

In the analysis of the stress distributions in the proximal tibia, our interest is focused upon the three dimensional stress situation in the cancellous bone. The application of strain gauges is limited to surface strains, though other investigators have embedded strain gauges into acrylic cement to investigate the three dimensional stress underneath a tibial plateau [7]. A finite element investigation gives a global view of the stress distribution in the trabecular bone. Both methods will be used together to investigate the influence of the plate stiffness of the tibial plateau upon bone stresses and loosening. A statical analysis will be made, though we are aware of the fact that the fatigue resistance of the trabeculae is of major importance.

The geometry

An idealized model for the proximal tibia was investigated (figure 1). This is justified because of the large dispersion in geometry and material properties within the proximal tibia and because we are dealing with a qualitative analysis.

The dimensions of the model are average values for a number of intact tibiae [9] and it was built of balsa-wood to simulate the trabecular bone and of fiber-reinforced epoxy-resin to simulate the thin cortex in the proximal tibia. The upper end was flat, as if it had been cut during surgery.

The tibial plateau was simulated by plates made of steel (3 mm thick), aluminium (1 mm) and High Density Polyethylene HDPE (6mm). The plate stiffness K for these three plateaus is given in table 1.

It was calculated as $K=EI/(1-\mu^2)$ [10]

with E : Young's modulus

I : planar moment of inertia per unit of length

μ : Poisson's ratio

The effect of a rim between tibial plateau and tibia and of anisotropic material properties for the proximal tibia was also investigated. The test models were not meant to be final designs ; their aim was to investigate the influence of some design concepts. Based upon this analysis, some of these concepts can be integrated in a new design for a tibial plateau.

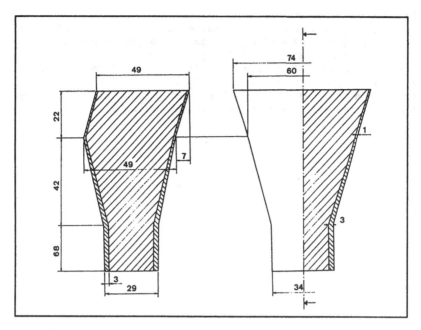

Figure 1. The model for the proximal tibia.

TABLE 1

Plateau	Plate stifness
Steel 3 mm	520 kN.mm
Aluminium 1 mm	6 kN.mm
HDPE 6 mm	12 kN.mm

In order to have a reference stress distribution, a part that simulated the proximal end that is normally removed at surgery was glued to the model of the proximal tibia. Of course, this was the last measurement that was made.

Loading
The load transfer in a natural knee is very complex : the structure is hyperstatic because of the presence of a multitude of ligaments and the menisci spread the contact forces over a wide area. We have focused our attention upon the situation of deep knee bending, where the contact zone shifts to posterior. From the observation of a number of knee surgical interventions, it was clear that the tendency of loosening anteriorly was strongest in this case. The magnitude and direction of the load on the tibia are however more dependent upon the relative position of the knee and the center of gravity of the body than upon the knee flexion [8]. Investigations have shown that the load is transfered more laterally than medially [3]. We have assumed a vertical load of 100 kg, 20 mm from the posterior edge of the tibial plateau. The effect of the patella tendon was not taken into account in the experimental set-up.

Strain gauge measurements

Strain gauges were glued at the entire proximal circumference of the model. Single gauges as well as rosette strain gauges were used.

The model was clamped in a heavy bench-screw and was loaded by a horizontal cylinder, simulating a medial and a lateral condyle. The whole system is placed on a force transducer to measure the force that is applied to the condyles by means of a screw and linear spring. Figure 2 shows an overview of the entire experimental set-up.

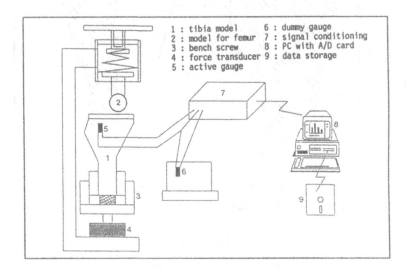

```
1 : tibia model        6 : dummy gauge
2 : model for femur    7 : signal conditioning
3 : bench screw        8 : PC with A/D card
4 : force transducer   9 : data storage
5 : active gauge
```

Figure 2. Overview of the experimental set-up.

Finite element models

Because of wavefront limitations, mainly a two-dimensional finite element study was made (sagittal plane). This approach was also used by Vasu et al. [11]. The finite element program ANSYS-PC (University Version), implemented on an IBM PS/2, model 80-111, was used for these calculations.

The two dimensional geometry reflected a mid-sagittal cross section of a tibia, with the same dimensions as the model for the strain gauge measurements. The mesh included two dimensional linear plane stress elements with 3 nodes (triangular) or 4 nodes (rectangular) each (figure 3).

The thickness h of the elements in the bone was calculated, based upon equal equivalent area and assuming trabecular bone at all places :

$$h = h_{trab} + h_{cort} \cdot E_{cort}/E_{trab}$$

where h_{trab} = local thickness of the trabecular bone,
h_{cort} = local thickness of the cortical bone,
E_{cort} and E_{trab} : Young's moduli of cortical and trabecular bone, respectively.

For the elements representing the plateau, the exact width was input as thickness of the elements.

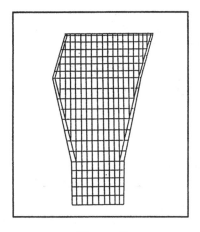
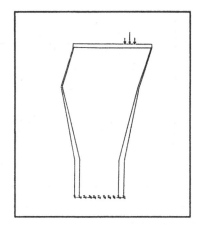

Figure 3   Figure 4

Figures 3 and 4. Two dimensional finite element mesh of a proximal tibia
(sagittal plane) and the loading and boundary conditions.

The equivalence of area is justified, because the effect of the
normal force is dominant in the proximal tibia. The plateau was
connected to the underlying cancellous bone by two-dimensional gap-
elements with no friction in the tangential direction. This simulated
the condition of a non-cemented plateau with no bone ingrowth. The
analysis is then based upon an automatic iterative non-linear algorithm.
The material properties were those of balsa-wood and fiber-
reinforced epoxy, to enable comparison of the analytical results with the
results of the strain gauge measurements. The elastic modulus of balsa-
wood was derived from literature, the elastic modulus of the epoxy layer
was determined by a tensile test. This gave E=3000 MPa for balsa and
E=8000 MPa for the own-made fiber-reinforced epoxy.
The nodes on the distal boundary were assumed to be fixed as
boundary condition. The loading is applied posteriorly as in the case of
knee bending (figure 4), and divided over three nodes to avoid unnatural
stress concentrations.

RESULTS

The results of the strain gauge measurements

The results of the longitudinal strain gauges will be presented
graphically by means of vectors along the circumference of the tibial
plateau, following the conventions shown in figure 5.

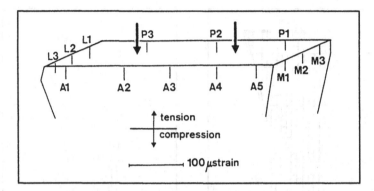

Figure 5. Conventions for the presentation of the strain gauge results :
A stands for Anterior; L for Lateral; P for Posterior and M for Medial.

The basic measurements

Figures 6, 7 and 8 present the results of the plateaus made of steel
(3mm), aluminium (1 mm) and HDPE (6mm) respectively.

The strains are average values for a number of measurements. Each
strain gauge was measured in three independent measurement sessions and
the experimental set-up was reconstructed each time. During each of
these sessions three measurements were made, between each of which the
load was applied and removed by turning the screw on top. Although the
dispersion for all measurements was large (e.g. -1.018 Volts with
standard deviation 0.446 Volts for the mid-posterior gauge), the
reproducibility within one measurement session was good (-1.448 Volts
with s.d. 0.05 Volts for the same gauge). In order not to overload the
illustrations, the error bars are not shown. The dispersion may be
explained by changes in the loading of the model, the clamping in the
bench-screw and the contact zones between plateau and tibia model.
Apparently the results are very sensitive to changes in the boundary
conditions of the experimental set-up and absolute strain values should
be interpreted with precaution.

This is undoubtedly a clinical problem as well.

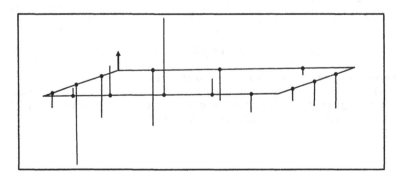

Figure 6. The longitudinal strains under a plateau in 3 mm thick steel.

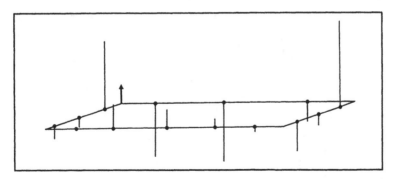

Figure 7. The longitudinal strains under a 1 mm thick aluminium plateau.

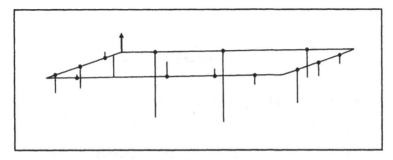

Figure 8 : The longitudinal strains under a plateau in 6 mm thick high density polyethylene.

The effect of a 'rim'

We have investigated whether we could enhance the load transfer to the cortex by means of a rim, made in aluminium of 1 mm and placed between the plateau and the tibia model (figure 9).

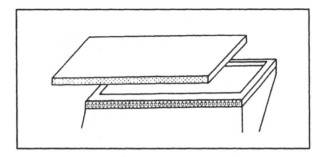

Figure 9. The aluminium rim, placed between plateau and tibia.

The result is shown in figure 10 for a plateau made in HDPE (6 mm). The entire cortex is now under compression, which will show to be a physiological situation.

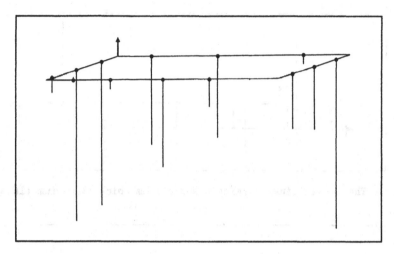

Figure 10. The longitudinal strains under a plateau in 6 mm thick
polyethylene with intermediate rim.

A plateau with anisotropic material properties

From the results, it appears that we will have to find a compromise
between a stiff and a flexible plateau (cfr. IV. Discussion). A plateau
with inhomogeneous and anisotropic material properties may be the
solution. In the posterior region, a stiffer material should be present
to avoid local overstressing of the cancellous bone, in view of the
fatigue resistance of the individual trabeculae. Furthermore the plateau
should be rather flexible in the sagittal plane, because this might
assist to avoid that the plateau behaves as an undeformable rotating
plate and by that means avoid or lower tensile strains anteriorly. We
have tried to realize this by inserting some steel wires of 1.5 mm
diameter medio-laterally into the posterior part of a 6 mm HDPE plateau
(figure 11). The results of the strain gauge measurements are shown in
figure 12.

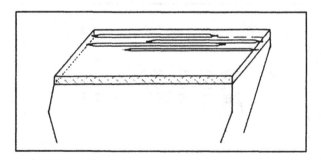

Figure 11. The plateau with anisotropic properties in the posterior part.

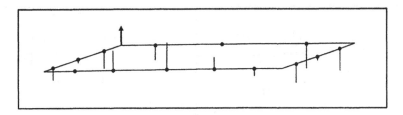

Figure 12. The longitudinal strains under a plateau with anisotropic
properties in the posterior part.

The reference model
The 'tibia' we have used so far was completed by a top, made of
balsa wood covered by fiber reinforced epoxy to represent the tibial
condyles. It was glued to the tibia (figure 13).

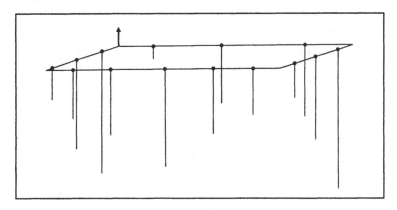

Figure 13. The longitudinal strains for the intact tibia.

Results of the two-dimensional finite element analysis
Calculations were made for plateaus made of steel (3 mm), aluminium (1
mm) and HDPE (6 mm). Besides, an intact tibia was simulated. The
results are shown in figure 14. The contour lines represent lines of
equal vertical stress, which is the dominant stress in the proximal
tibia.
The following code is used for the contour lines : (values in N/mm^2)

```
1 :    > 3          6 :  -1.5 .. -1
2 :    2 .. 3       7 :  -2.5 .. -1.5
3 :    1 .. 2       8 :  -3.5 .. -2.5
4 :    0 .. 1       9 :  -4.5 .. -3.5
5 :   -1 .. 0      10 :    < -4.5
```

52

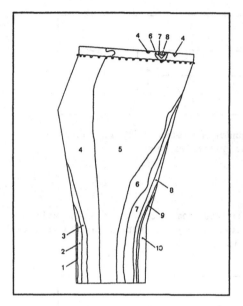

Plateau in 3 mm steel.

Plateau in 1 mm aluminium.

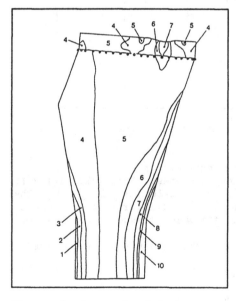

Plateau in 6 mm polyethylene.

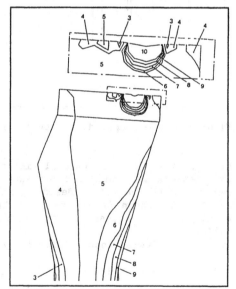

Intact tibia.

Figure 14. Results of the finite element calculations.

DISCUSSION

From figures 7 and 8 (strain gauge measurements on plateaus with low plate stiffness) it is clear that the load is transferred immediately and locally to the underlying material and the posterior cortex. The bending effect of the posterior load dominates more distally. The stiff plateau (figure 6) shows almost no out-of-plane deformation and rotates around a medio-lateral axis, hereby transferring most of the load to the cortical rim. The bending moment is already dominant very proximally, causing tensile strains at the anterior cortex and tendency of loosening. These findings will be confirmed by the finite element calculations.

The dispersion of the results appears to be larger in the case of the steel plateau than with the more flexible plateaus. These flexible plates have the ability to adapt more to the irregularities of the underlying soft material, whereas the stiff steel plateau has only little contact points, which depend upon the way the plateau is placed on the tibia. A porous surface may also assist in creating a larger contact surface between the plateau and the cancellous bone.

The presence of a rim simulates a perfect contact of the tibial plateau with the cortex. This creates strains in the proximal cortex that compare very well with the strains in an intact tibia.

When we compare figure 12 with figure 8, we see that the medio-lateral stiffening lowers the strains at the posterior side since the load is transferred to the medial and lateral cortex.

The connection of the top with the tibia (the situation of an intact tibia) has a very important influence upon the strain distribution : the entire cortex is now under compression. This situation was also measured in the case of a rim, placed between bone and plateau. The positive influence of a rim that assures contact of the circumference of the tibial plateau with the cortex might be realized in practice by the concept of a 'tailor made' tibial plateau, manufactured to the exact shape of the surface that was cut by the orthopaedic surgeon during the operation. The question arises however whether the stiffness of the top that was used in our experiments is a correct representation of the stiffness of the tibial plateau, removed during surgery.

From the results of the finite element analysis, it appears that the transition between the region of compression (posterior) and the region of tension (anterior) is located near the anterior cortex. The region of tension corresponds with the zone of loosening between bone and tibial plateau, as can be seen by comparison with the status of the interface elements. The effect of the plate stiffness upon the vertical stresses in the cancellous bone appears to be that the tensile region tends posteriorly in the case of the stiffer plateaus and tends anteriorly, thus minimizing loosening, in the case of the more flexible plateaus. The result of the intact tibia is situated between both the stiff and flexible plateaus.

Furthermore, the more flexible plateaus show a stress concentration in the trabecular bone underneath the load bearing region, which is not present in the intact tibia where the stress concentration is limited to the region that is removed during surgery. The stresses in the distal part are independent of the material used for the plateau, which confirms that the stress transfer in a total knee prosthesis is a local phenomenon.

CONCLUSIONS

The following conclusions are based upon insights that were brought together by the strain gauge measurements and the finite element analysis. The most accurate results were undoubtedly offered by the strain gauge measurements, though the large dispersion that was measured has shown that the stress distribution is very sensitive to changes in the boundary conditions of the problem. The finite element models have assisted in understanding the mechanics of load transfer from epiphysis to diaphysis.

1. A plateau with high plate stiffness $K=EI/(1-\mu^2)$ rotates around a medio-lateral axis, hereby loading the proximal region in bending and creating significant longitudinal tensile stresses anteriorly, leading to loosening. There is no stress concentration in the cancellous bone, which seems to be physiological.

2. A plateau with low plate stiffness transfers the load to the underlying cancellous bone and the posterior cortex. The stress concentration in the trabecular bone is unphysiological and further study will have to show whether these stresses do not exceed the fatigue resistance of the individual trabeculae. There is little load transfer to the other proximal cortices and there is little tendency of loosening.

3. A plateau with inhomogeneous and anisotropic material properties may combine the positive aspects of both extremes that were discussed so far. The positive influence of a rim at the plateau, assuring intimate contact of the circumference with the cortex was shown. The concept of a 'tailor made plateau' should be investigated further in this view. A tendency of anterior loosening seems to be present with all plateaus, but the tensile stresses can be transmitted by bone ingrowth into a porous coating, provided there is no loosening in the immediate postoperative period. Clamps, placed anteriorly over the interface and made of a bioresorptive material, could realize this by transmitting tensile stresses until porous ingrowth is complete.

4. The load transfer is truly three dimensional and should be studied by an efficient combination of strain gauge measurements and finite element models. These models and experiments should also take into account the asymmetries in geometry and loading that are present in reality.

REFERENCES

1. Ducheyne, P., Kagan, I.A., Lacey, J.L., 'Failure of total knee arthroplasty due to loosening and deformation of the tibial component.', J. Bone and Joint Surgery, Vol. 60-A, 1978, pp. 384-391.

2. Freeman, M.A.R., Samuelson, K.M. and Bertin, K.C., 'Freeman-Samuelson Arthroplasty of the knee.', Clinical Orthopaedics and Related Research, No. 192, Jan-Feb 1985, pp. 46-58.

3. Harrington, I.J., 'Static and dynamic loading patterns in the knee joint with deformities.', J. Bone and Joint Surgery, Vol. 65-A, 1983, pp. 247-259.

4. Hungerford, D.S. and Krackow, K.A., 'Total joint arthoplasty of the knee.', Clinical Orthopaedics and Related Research, No. 192, Jan-Feb 1985.

5. Insall, J.N., Ranawat, C.S., Aglietti, P. and Shine, J., 'A comparison of four models of total knee replacement prosthesis.', J. Bone and Joint Surgery, Vol. 58-A, 1976, pp. 754-765.

6. Klever, F.J., 'On the mechanics of failure of artificial knee joints.', Ph.D. Thesis, T.H. Twente, The Netherlands, 1984.

7. Little, E.G. and O'Keefe, D., 'An experimental technique for the investigation of three-dimensional stress in bone cement underlying a tibial plateau.', Proc. IMechE., Journal of Engineering in Medicine, Vol. 203, pp. 35-41.

8. Maquet, P.G.J., 'Biomechanics of the knee.', Springer Verlag, 1984.

9. Timmerman, P., 'Eindige elementen onderzoek van de knieprothese.', Internal report Div. Biomechanics, K.U.Leuven, 1988.

10. Van Gemert, D., 'Grafostatika en beginselen van sterkteleer.', ed. V.T.K. Leuven, 1983.

11. Vasu, R., Carter, D.R., Schurman, D.J. and Beaupré, G.S., 'Epiphyseal-based designs for tibial plateau components - I. Stress analysis in the frontal plane.', J. Biomechanics, Vol. 19, No. 8, 1986, pp. 647-662.

CHANGES OF ANTIBIOTIC SENSITIVITY AFTER CONTACT "IN VITRO" OF BACTERIA WITH METALS

ELISABETTA CENNI, LUCIA SAVARINO, DANIELA CAVEDAGNA, ARTURO PIZZOFERRATO
Laboratory for Biocompatibility Research on Implant Materials, Istituto Ortopedico Rizzoli, via di Barbiano 1/10, 40136 Bologna , Italy

ABSTRACT

Infection is one of the major complications in the use of prosthetic devices. Various Authors investigated the causes which favour such infection. We examined the changes in the resistance to antibiotics after contact with the metals used in the construction of prostheses, under the hypothesis that the persistence of the infection is due to a modification in the sensitivity of microorganisms. Compared with the results obtained from control tests, more or less evident variations of sensitivity to antibiotics were found in some strains which were put in contact with some metals.

INTRODUCTION

Prosthesis associated infection is one of the most feared complications in the use of prosthetic devices (1). Nosocomial infections, acquired during hospitalization, occur most frequently in prosthetic surgery. Prosthesis associated infection is characterized by its persistence, notwithstanding the host defence and the antibiotic therapy. In many cases, the only possible cure is the removal of the prosthesis (2). Often these infections are caused by antibiotic-resistant strains, which develop easily in the hospital environment. Also when caused by more sensible microorganisms, the therapy of prosthesis associated infection is very difficult, partly due to the problems arising in getting the antibiotics in the peri-prosthetic tissues, because of the intrinsec characteristics of biomaterials. Actually the latter can cause an increase in bacterial adherence (3) and modify the replication rhythm of microorganisms (4). Some Authors (5, 6) found that environmental pollution by some metals can increase the resistance to antibiotics. The aim of our study is to investigate whether some bacterial strains, which frequently cause peri-prosthetic infections, do modify their resistance to antibiotics after contact with metals usually employed in the construction of arthroprostheses and ostheosynthesis devices.

MATERIALS AND METHODS

The sensitivity variations to antibiotics were examined on 13 bacterial strains put in contact with various metals usually used in the construction of arthroprostheses and ostheosynthesis devices.

1. *Bacterial strains*

Among the 13 bacterial strains examined, 5 were Gram-positive and 8 Gram-negative. Three were standard cultures; five were isolated from post-surgical wounds and five from clinical specimens (blood, sputum, faeces). All the microorganisms, with the exception of the standard cultures, were collected from hospitalized patients. The clinical specimens from wounds and sputum were seeded on Blood agar, Mac Conkey agar, Mannitol-salt agar and Trypticase Soy Broth. Blood culture was performed according to the Bactec system. The faecal specimens were seeded on Selenite Broth and on Salmonella-Shigella agar. Staphylococci were identified on the basis of colony morphology, emolysis, the ability of producing acid from mannitol, coagulase and DNA-se production and other biochemical tests. The Gram-negative strains were identified on the basis of the oxidase test and biochemical tests of aminoacid hydrolysis and sugar fermentation. The bacterial strains assayed and their source are listed in Table 1.

2. *Drugs*

The following antibiotics with their concentrations were used for the sensitivity tests:

- Gram-positive and Gram-negative strains: ampicillin (10 μg), cephazolin (10 μg), co-trimoxazole (25 μg), chloramphenicol (10 μg), gentamicin (10 μg), tobramycin (10 μg), amikacin (30 μg), netilmycin (30 μg).

- Gram-positive strains: they were assayed also for penicillin (10 IU), erythromycin (15 μg), clindamycin (2 μg), cefamandol (30 μg), imipenem (10 μg), vancomycin (30 μg), fusidic acid (10 μg).

- Gram-negative strains: they were assayed also for carbenicillin (100 μg), rifampicin (30 μg), azlocillin (75 μg), cefotaxime (30 μg),ceftazidime (30 μg), aztreonam (30 μg), cefuroxim (30 μg).

TABLE 1

Bacterial strain	Source
Staphylococcus aureus	ATCC 25923
Staphylococcus aureus	Surgical wound
Staphylococcus aureus	Surgical wound
Staphylococcus aureus	Traumatic wound
Staphylococcus epidermidis	Blood
Escherichia coli	ATCC 25922
Escherichia coli	Sputum
Citrobacter freundii	Faeces
Proteus mirabilis	Faeces
Serratia marcescens	Surgical wound
Enterobacter cloacae	Surgical wound
Pseudomonas aeruginosa	ATCC 25823
Pseudomonas aeruginosa	Sputum

3. *Metals*

The following metals were used to assess whether they can modify antibiotic sensitivity:

- Stainless steel AISI 316 L (Hoganas AB, Sweden);

- Chromium, purity grade 99,0% (Goodfellow Metals, Cambridge, England);

- Cobalt, purity grade 99,6% (Goodfellow Metals, Cambridge, England);

- Titanium, purity grade 99,5% (Goodfellow Metals, Cambridge, England);

- Nickel, purity grade 99,8% (Goodfellow Metals, Cambridge, England).

4. *Tests*

For every test a single colony of each bacterial strain was used. It was seeded on Trypticase Soy Broth containing 0,01 g/ml of metal powder. The same strain seeded in Trypticase Soy Broth without powder was used as control. The cultures were incubated at 37°C for 24 hours; afterwards, the sensitivity tests to antibiotics were carried out, according to the Stokes method (7), on media containing the same metal at a concentration of 0,005 g/ml. After 24 hours, the zones of inhibition of growth around each antibiotic were measured.

RESULTS

The total number of tests carried out for the different antibiotics and for the various bacterial strains was 758. Only percentage differences higher than 20% were taken into consideration. The calculation of the variation in the resistance of the 13 strains used with respect to every assayed metal, shows that stainless steel caused most frequently an increase in resistance (19.4%), followed by nickel (16.8%), cobalt (12.8%), titanium (12.4%) and chromium (10.3%). A decrease in resistance can be noted in some tests, equal to 13.5% after contact with stainless steel and nickel, to 23.2% after contact with cobalt, to 14.4% after contact with titanium and to 11.0% after contact with chromium. In all other cases, only variations lower than 20% in reference to the control occurred. Tables 2 and 3 show the per cent frequency at which an increase or decrease in bacterial resistance occurred in respect to the single metals assayed, without considering the kind of antibiotic used.

TABLE 2

Per cent frequency of the cases in which an increase in bacterial resistance occurs in respect to the single assayed metals. The evaluation was done for each microorganism, without considering the kind of antibiotic used.

Strain	Metal				
	Inox	Cr	Co	Ti	Ni
Staphylococcus aureus ATCC 25923	0	0	6.7	0	0
Staphylococcus aureus (surgical wound)	93.3	-	-	-	-
Staphylococcus aureus (surgical wound)	7.1	0	7.1	0	0
Staphylococcus aureus (traumatic wound)	0	0	0	6.7	6.7
Staphylococcus epidermidis	20	0	0	20	20
Escherichia coli ATCC 25922	14.3	0	21.4	0	0
Escherichia coli (sputum)	7.1	4.3	35.7	21.4	21.4
Citrobacter freundii	21.4	35.7	21.4	21.4	35.7
Proteus mirabilis	14.3	28.6	-	35.7	71.4
Serratia marcescens	0	13.3	-	35.7	71.4
Enterobacter cloacae	8.3	8.3	8.3	0	0
Pseudomonas aeruginosa ATCC 25823	44.4	11.1	0	11.1	22.2
Pseudomonas aeruginosa (sputum)	33.3	11.1	33.3	22.2	11.1

Inox: stainless steel   Ti: titanium   Co: cobalt
Cr: chromium   Ni: nickel

TABLE 3

Per cent frequency of the cases in which a decrease in bacterial resistance occurs in relation to the single assayed metals. The evaluation was done for each microorganism, without considering the kind of antibiotic used.

Strain	Metal				
	Inox	Cr	Co	Ti	Ni
Staphylococcus aureus ATCC 25923	0	0	0	0	0
Staphylococcus aureus (surgical wound)	0	-	-	-	-
Staphylococcus aureus (surgical wound)	35.7	50	53.8	66.7	50
Staphylococcus aureus (traumatic wound)	26.7	0	53.3	20	13.3
Staphylococcus epidermidis	60	60	50	30	30
Escherichia coli ATCC 25922	0	7.1	21.4	0	14.3
Escherichia coli (sputum)	28.6	21.4	21.4	21.4	21.4
Citrobacter freundii	0	0	7.1	7.1	0
Proteus mirabilis	7.1	0	-	0	0
Serratia marcescens	6.7	0	-	6.7	6.7
Enterobacter cloacae	16.7	8.3	16.7	25	16.7
Pseudomonas aeruginosa ATCC 25823	0	0	0	0	11.1
Pseudomonas aeruginosa (sputum)	0	0	0	0	0

Inox: stainless steel   Ti: titanium
Cr: chromium   Ni: nickel
Co: cobalt

Indipendently from the assayed metal and strain used, the resistance to the various antibiotics increased by 30.2% in the tests performed for ampicillin, by 29.2% for chloramphenicol, by 26.3% for carbenicillin, by 20.4% for cephazolin, by 20% for penicillin. Frequencies of increased resistance lower than 20% were observed for gentamicin (18.5%), rifampicin (18.4%), amikacin (15.9%), tobramycin (14.8%), imipenem (14.3%), cefotaxime (12.5%), co-trimoxazole (10.2%), netilmycin (5.5%), cefuroxim (7.1%), fusidic acid (9.1%), azlocillin (7.9%), erythromycin, clindamycin and cefamandol (4.8%), ceftazidime (4.2%). No increase in resistance for aztreonam and vancomicin was observed.

In many tests also a decrease in antibiotic-resistance was noticed. The frequency at which such reduction was observed was 61.9% for clindamycin, 32.6% for co-trimoxazole, 29.5% for cephazolin, 28.6% for erythromycin, 25% for cefuroxim. Lower frequencies were observed for imipenem (19.0%), ampicillin (18.6%), rifampicin (18.4%), netilmycin (16.7%), amikacin (16.3%), penicillin (15.0%), chloramphenicol (12.5%). Rare cases of decreased resistance were observed in the strains assayed for cefamandol (9.5%), fusidic acid (9.1%), vancomycin (4.8%), aztreonam (4.3%), gentamicin and tobramycin (3.7%), for carbenicillin and azlocillin (2.6%). No resistance decrease was found in the strains assayed for cefotaxime and ceftazidime.

Tables 4 and 5 show the per cent frequencies of the cases in which respectively an increase or decrease in bacterial resistance occurs in regard to the single assayed antibiotics.

Considering the single assayed strains, a Staphylococcus aureus isolated from a surgical wound showed an increased resistance in 93.3% of the cultures in the presence of stainless steel. The strain of Proteus mirabilis showed a resistance increase in 37.5% of the cases; Citrobacter freundii in 27.1% of the cases. Pseudomonas aeruginosa and Escherichia coli isolated from sputum and Pseudomonas aeruginosa from standard cultures showed similar percentages of resistance increase (respectively in 22.2%, 20.0% and 17.8% of the cases). Less frequent resistance increase was observed for Staphylococcus epidermidis (12.0%). The lowest increase in frequency was observed for Serratia marcescens (8.3%), for Escherichia coli from standard culture (7.1%), for Enterobacter cloacae (5.0%) and for three strains of Staphylococcus aureus, the first isolated from a surgical wound (2.8%), the second isolated from a traumatic wound (2.7%) and the third coming from standard culture (1.3%).

No decrease in resistance was observed for the Staphylococcus aureus from standard culture, for the one isolated from a surgical wound and for Pseudomonas aeruginosa isolated from sputum. In other strains a decreased resistance was sometimes observed, not frequent in Proteus mirabilis (1.8% of the cases), in Pseudomonas aeruginosa from standard culture (2.2%), in Citrobacter freundii (2.8%), in Serratia marcescens (5.0%), in Escherichia coli from standard culture (8.6%). The frequence is higher in Enterobacter cloacae (16.7%), in Escherichia coli isolated from sputum (21.4%) and in Staphylococcus aureus isolated from a traumatic wound (22.7%). The highest frequency was observed for Staphylococcus epidermidis (46%) and for a Staphylococcus aureus isolated from a surgical wound (59.6%).

TABLE 4

Per cent frequency of the cases in which the bacterial resistance increases in regard to the single assayed metals. The evaluation was done for every single antibiotic, without considering the kind of antibiotic used.

Antibiotic	Strain	Metal				
		Inox	Cr	Co	Ti	Ni
Ampicillin	a,b	20	22.2	42.9	25	33.3
Cephazolin	a,b	30	22.2	14.3	0	11.1
Co-trimoxazole	a,b	9.1	10	0	20	30
Chloramphenicol	a,b	45.4	30	14.3	18.2	30
Gentamicin	a,b,c	25	9.1	11.1	18.2	27.3
Tobramycin	a,b,c	16.7	9.1	11.1	9.1	27.3
Amikacin	a,b,c	18.2	0	25	20	10
Netilmycin	a,b,c	8.3	0	11.1	0	0
Penicillin	b	40	0	0	33.3	25
Erythromycin	b	20	0	0	0	0
Clindamycin	b	20	0	0	0	0
Cefamandol	b	20	0	0	0	0
Imipenem	b	20	0	0	25	25
Vancomycin	b	0	0	0	0	0
Fusidic acid	b	33.3	0	0	0	0
Carbenicillin	a,c	25	25	33.3	25	25
Rifampicin	a,c	12.5	25	16.7	25	12.5
Azlocillin	a,c	0	12.5	16.7	0	12.5
Cefotaxime	a,c	25	12.5	16.7	12.5	0
Ceftazidime	a,c	14.3	0	0	0	0
Aztreonam	a	0	0	0	0	0
Cefuroxim	a	16.7	0	0	0	0

a) Gram-negative strains
b) Gram-positive strains
c) Pseudomonas

Inox: stainless steel   Ti: titanium
Cr: chromium   Ni: nickel
Co: cobalt

TABLE 5

Percentage frequency of the cases in which the bacterial resistance decreases in regard to the single assayed metals. The evaluation was done for every single antibiotic, without considering the kind of antibiotic used.

Antibiotic	Strain	Metal				
		Inox	Cr	Co	Ti	Ni
Ampicillin	a,b	20	22.2	28.6	12.5	11.1
Cephazolin	a,b	30	33.3	42.9	25	22.2
Co-trimoxazole	a,b	27.3	20	50	30	40
Chloramphenicol	a,b	9.1	0	14.3	30	10
Gentamicin	a,b,c	0	0	11.1	9.1	0
Tobramycin	a,b,c	0	9.1	9.1	0	9.1
Amikacin	a,b,c	0	20	37.5	10	20
Netilmycin	a,b,c	16.7	18.2	22.2	18.2	9.1
Penicillin	b	0	25	25	0	0
Erythromycin	b	40	25	25	25	25
Clindamycin	b	60	0	75	75	0
Cefamandol	b	20	0	25	0	0
Imipenem	b	20	25	0	25	25
Vancomycin	b	0	0	25	0	0
Fusidic acid	b	0	0	33.3	0	0
Carbenicillin	a,c	12.5	0	0	0	0
Rifampicin	a,c	25	0	16.7	12.5	25
Azlocillin	a,c	12.5	0	0	0	0
Cefotaxime	a,c	0	0	0	0	0
Ceftazidime	a,c	0	0	0	0	0
Aztreonam	a,c	0	0	0	20	0
Cefuroxim	a,c	16.7	0	75	33.3	16.7

a) Gram-negative strains
b) Gram-positive strains
c) Pseudomonas

Inox: stainless steel   Ti: titanium
Cr: chromium   Ni: nickel
Co: cobalt

DISCUSSION

The increasing incidence of infection observed in patients who had undergone prosthetic surgery is due, besides to causes common to every hospital infection, also to the influence exerted by biomaterials on the bacterial replication. In peri-prosthetic tissues, because of the wear process, metallic ions accumulate, and can induce an increase in bacterial adherence and changes in their replication and possibly in their sensitivity to antibiotics. While the biomaterial influence on the bacterial adhesion and replication were widely investigated (3, 4, 8), only few studies relate to the metal-induced antibiotic-resistance. Some Authors (5,6) tried to establish a link between the capability of growing in media containing mercury, cobalt or nickel and the antibiotic-resistance of hospital strains of Escherichia coli, Klebsiella pneumoniae, Pseudomonas aeruginosa and Staphylococcus aureus.

Our results show that metals mostly induced per cent variations in resistance lower than 20% in comparison to the control. However, a modification of the zone of inhibition of bacterial growth around the antibiotics was sometimes noticed. Even if the frequency of such variations is low, it can be observed that stainless steel and nickel determined more frequently an increase in resistance, while cobalt, titanium and chromium induced more often a reduction.

As for antibiotics, in most of the cases variations lower than 20% were observed, with the exception of the strains assayed for clindamycin, which showed a frequency of resistance reduction as high as 61.9%. The strains assayed for ampicillin, co-trimoxazole, chloramphenicol, gentamicin, tobramycin, penicillin, carbenicillin, azlocillin, cefotaxime and ceftazidime showed more frequently an increased resistance, while those assayed for cephazolin, chloramphenicol, clindamycin, netilmycin, erythromycin, cefamandol, imipenem, vancomycin, cefuroxim and aztreonam showed more often a decrease in resistance.

Among the examined strains, 69.5% did not show any major variation in the zone of inhibition of growth around the antibiotics. A reduced resistance was noticed in 17.2% of the cases, while 13.4% showed an increase. Gram-positive strains showed an increase of resistance in 8.9% of the cases, a decrease in 26.2% of the cases and they showed no variation in respect to the control in 64.9% of the cases. On the other hand, Gram-negative strains tended more often to increase their resistance in the presence of metals: 17.9% of them displayed an increased resistance, while 8.0% showed a reduction and 74.1% underwent no change in respect to the control.

These data show that in about 30% of the cases a modification in the resistance to antibiotics occurred after the contact with various types of metal. The resistance tended to increase in the Gram-negative strains, while it tended to decrease in the Gram-positive ones.

CONCLUSIONS

The resistance can undergo no change, increase or decrease according to the used antibiotics, the assayed microorganisms and metals. In most of the strains, no variations exceeding 20% were noticed. Among the microorganisms showing variations exceeding 20%, the Gram-negative strains, in particular Proteus mirabilis and Citrobacter freundii, show mostly an increase in resistance, while the Gram-positive strains display mainly a reduction. The metal causing more often an increased resistance is stainless steel, the one causing more often a reduction is cobalt. Chromium, titanium and nickel induce almost identical variations in both directions.

REFERENCES

1. Pizzoferrato A., Biomaterials and infection: clinical experience. In Biomaterials and Clinical Applications, ed. A. Pizzoferrato, P.G. Marchetti, A. Ravaglioli and A.J.C. Lee, Elsevier Science Publishers B.V., Amsterdam, 1987, pp. 1-12.

2. Canner G.C., Steinberg M.E., Heppenstall R.B., Balderston R., The infected hip after total hip arthroplasty. J. Bone Joint Surg., 1984, 66A (9), 1393-1399.

3. Arciola C., Versura P., Ciapetti G., Monti P., Pizzoferrato A., Caramazza R., Evaluation of adhesion capability of conjunctival Staphylococcus aureus on polymeric materials. In Implant Materials in Biofunction, ed. C. de Putter, G.J. de Lange, K. de Groot and A.J.C. Lee, Elsevier Science Publishers B.V., Amsterdam, 1988, pp. 349-354.

4. Stea S., Cavedagna D., Pirini W., Arciola C.R., Pizzoferrato A., Interferenza di alcuni biomateriali sulla replicazione batterica "in vitro". Biomateriali, 1989, 1/2, 11-14.

5. Nakahara H., Ishikawa T., Sari Y., Kondo I., Frequency of heavy -metal resistance in bacteria from inpatients in Japan. Nature (London), 1977, 266, 165-167.

6. Smith D.H., R factors mediate resistance to mercury, nickel, and cobalt. Science, 1967, 156, 1114- 1116.

7. Stokes E.J., Clinical Bacteriology, Edward Arnold Publishers, London, 1975, pp.203-225

8. Gristina A.G., Barth E., Webb L.X., Microbial adhesion and molecular mechanisms in biomaterial and compromised tissue centered infection. In Biomaterials and Clinical Applications, ed. A. Pizzoferrato, P.G. Marchetti, A. Ravaglioli, A.J.C. Lee, Elsevier Science Publishers B.V., Amsterdam, 1987, pp. 661-674.

A New Technique for Intact Interface Studies of Bone and Biomaterials using Light and Electron Microscopy.

M.V.Kayser, S.Downes, S.Y.Ali.
Institute of Orthopaedics(University of London), Royal National Orthopaedic
Hospital,Stanmore,Middlesex.

ABSTRACT

One of the main problems associated with cemented joint replacement is aseptic loosening, which occurs at the bone-cement interface; this often leads to failure of the implant.Standard techniques to date have been adequate for studying the histological changes at the interface. For ultrastructural studies,ultrathin sections of the edge of bone have always been difficult to obtain; phase problems exist due to the edge being composed of hard mineral banded by a seam of fibrous collagen and covered by a layer of lining cells, the osteoblasts. We present a new method for the fixation, processing and cutting of acrylic cements (poly-methylmethacrylate) and ceramics (*tri* -calcium phosphate, hydroxyapatite) to establish *in-vivo* response of bone to these materials.

INTRODUCTION

Standard histological techniques showing undecalcified bone-cement interfaces have proved inadequate for ultrastructural studies. Tissue samples, for histology at the light microscopy level, have to be decalcified, this generally requires the removal of calcium salts from the tissue [1]. Ultrathin sections of the edge of bone have always been difficult to obtain because of the phase problems that exists due to the edge being composed of hard mineral banded by a seam of fibrous collagen and covered by a layer of lining cells, the osteoblasts. Previous work processed for electron microscopy has shown us that there is a severe loss in ultrastructural morphology and the use of quantitative X- ray microanalysis is limited, preventing the detection of Ca: P ratios of newly laid down mineral. The use of acrylic bone cement as a biomaterial for the attachment of implanted prostheses has long been established [2], but very little histological information is known at the ultrastructural. level

Ceramic, being a new biomaterial, is used to coat prostheses prior to implantation, in order to achieve a more permanent bonding between bone and prosthesis [3]

A new method for the fixation, processing and cutting of acrylic cements (poly-methylmethacrylate) and ceramics (*tri* -calcium phosphate, hydroxyapatite) was used to study the *in-vivo* response of bone to these materials.

MATERIALS AND METHODS

In - Vivo Model

Sandy Lop rabbits of at least 3.5kg body weight were used. Access to the knee was gained through a medial parapatella capsulotomy. Using a 2mm diameter drill bit ,the medullary cavity of the femur was reamed to a depth of 200mm, starting at the intercondylar notch[4] . The cavity was filled with Poly-methylmethacrylate (PMMA) bone cement.

In the ceramic experiments pre-machined *tri*-calcium phosphate (TCP) pins (5.00mm in length and 2.00mm in diameter- supplied by C.KLEIN of Leiden) were inserted into the lateral cortex of rabbit femur. Post-operatively the rabbits were kept unrestrained in standard size cages.

Poly-methylmethacrylate samples .

One month after surgery the rabbits were sacrificed and femora removed. All adherent tissue was removed before the femora were cut into two halves longitudinally, starting at the intercondylar notch.

Rabbit Femur

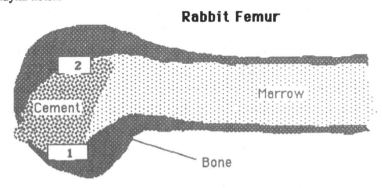

1 & 2= Sample areas.

Figure 1.

The halves were fixed in 2% Glutaraldehyde in 0.1M Sodium Cacodylate buffer at pH 7.2 for 2 hours. A small area, approximately 3mm x 4mm, was cut out(fig.1) and further fixed for 24 hours in fresh fixative at 4^0 C., followed by secondary fixation in 1% Osmium Tetroxide in 0.1 M Sodium Cacodylate buffer. The sample was washed in the same buffer, dehydrated through a graded series of alcohol (70%,90%,100%)and impregnated with a 1:1 alcohol/Spurrs' resin mixture for 6 hours, 2 hours of which were vacuum impregnated at 150 mbar (Anglia Scientific-Vacuum embedding chamber). This was followed by 4 changes of 12 hours each with Spurrs' resin alternating every 6 hours with vacuum infiltration.

The blocks were embedded and cured at 70^0 C for 18 hours. One micron sections were cut with a diamond knife for light microscopy (L.M.) and stained with methylene blue-azure II-basic fuchsin stain [6]. Selected areas for electron microscopy (E.M.) were cut between 60 -90 nm. Sections for E.M. were stained with 2% Uranyl Acetate (10 min.) and Reynold's Lead Citrate (10 min.) Sections for X-ray microanalysis were left unstained.

Tri - calcium phosphate samples .

One month after surgery the rabbits were sacrificed, the femora removed and all adherent tissue dissected off. The areas of femur containing the pins were cut out leaving sufficient bone on either side. The introduction of any unnecessary artifacts at the ceramic/bone interface was thus kept to a minimum.

This entire section was fixed in 2% Glutaraldehyde in 0.1M Sodium Cacodylate buffer at ph 7.2 for 48 hours at 4^0 C. No secondary fixation was used; samples were washed in buffer, dehydrated through a graded series of alcohols (70%,90%,100%) and impregnated with a 1:1 absolute alcohol/ Spurrs' resin mixture for 12 hours, 6 hours of which were vacuum impregnated at 150 mbar. The Spurrs' resin was made up to give the hardest formulation,with longest pot life [5] . Changes of resin were made every 5 days for a total of 21 days. Intermittent vacuum impregnation at 150 mbar was used for several hours every day. Whole sections were blocked and polymerised at 70^0 C for 18 hours. The blocks were then trimmed well into the block, exposing the bone / ceramic interface with a Reichert-Jung 2050 Super Cut microtome. Sections 0.5 mm thick were cut using a diamond cutter,washed in absolute alcohol and soaked in Spurrs' resin for a further 24 hours. The slices were placed into embedding moulds with fresh resin and polymerized at 70^0 C for 18 hours.

Selected areas were removed (fig.2) and the ceramic trimmed back, leaving just sufficient ceramic at the interface for our studies. The blocks were then re-embedded using flat silicone rubber embedding moulds.

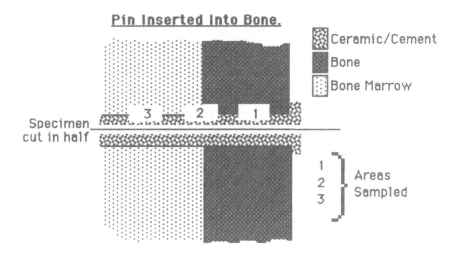

Figure 2.

polymerisation once again took place at 70^0 C for 18 hours. Sections for L.M. were cut on an L.K.B. Ultratome at 1 micron and stained with methylene blue-azure II-basic.fuchsin stain [6] .Selected areas were trimmed down and sectioned between 60- 90nm for E.M.,with a diamond knife .

Sections were picked up onto copper grids supported with 0.5% Pioloform in chloroform, stained with 2% Uranyl Acetate (10 min.) and Reynold's Lead Citrate (10 min.).Sections for X-ray microanalysis were left unstained. A Philips CM 12 Electron microscope with an EDAX PV9800 X-ray microanalysis system was used at 100KV for viewing the sections.

RESULTS

The relative densities of bone and cement lead to different cutting properties, resulting in poor sectioning; however,optimum results were obtained if the areas of cement adjacent to bone were kept to a minimum.

Problems with sectioning became more evident with the ceramic interfaces, but once the ceramic had been trimmed down to a 'thin' line and re-embedded ,sectioning improved.

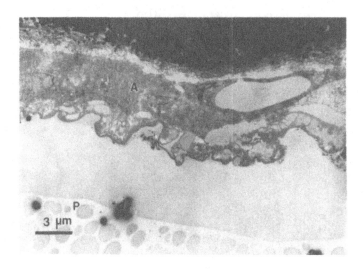

Figure 3.The results showed good contact between the bone and PMMA cement (P), with an area of tissue between the two layers (A).

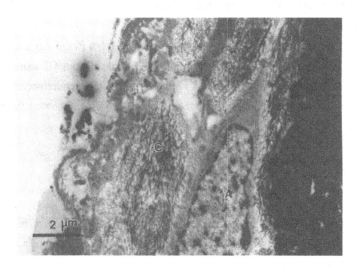

Figure 4. The cells at the interface with PMMA showed cells resembling fibroblasts(A). These fibroblast- or osteoblast-like cells had well-developed granular endoplasmic reticulum within cytoplasm and the cells were closely associated with collagen fibres (C), indicating metabolic activity.

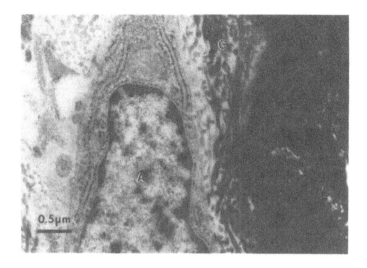

Figure 5. A higher magnification shows this close association with cells (A) and the unmineralised osteoid seam (C).

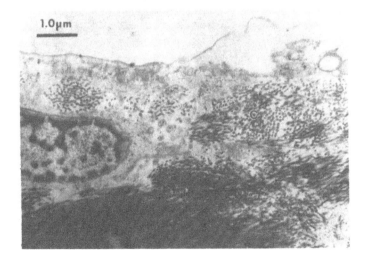

Figure 6. A similar area shows unmineralised matrix containing a mixture of broad and narrow collagen banding (C).

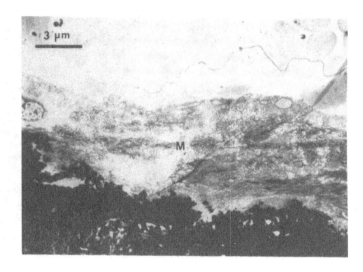

Figure 7 Other areas revealed a narrow zone of organic matrix(M), very poorly mineralized, with a few cells.

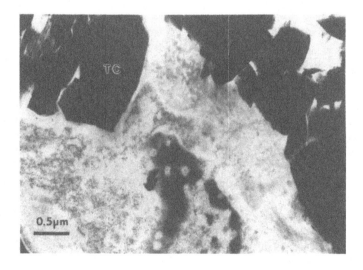

Figure 8 Morphology and cellular detail with the TCP interface (TC) was adequate, even though samples were not osmicated.

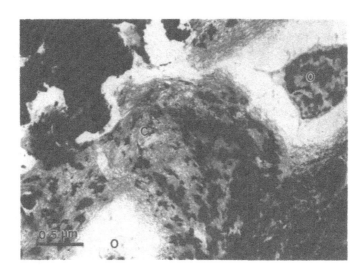

Figure 9.Shows an active collagen matrix indicating an advancing mineral front with osteoid(C) and newly formed osteocytes(O) along the interface

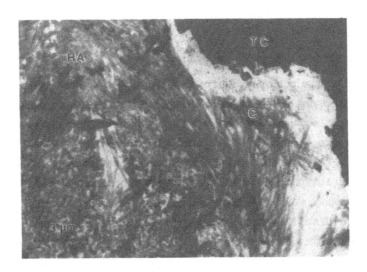

Figure 10. A higher magnification of the same area reveals the close proximity of broad collagen banding(C), with hydroxyapatite deposition(HA), to the ceramic interface(TC).

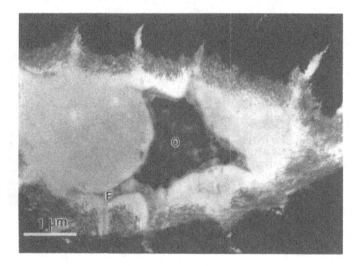

Figure 11.Further areas of osteocytes(O) with filipodial processes(F) and condensed nuclear chromatin were seen in close proximity to the interface, indicating active involvement in the maintenance of the bone matrix.

DISCUSSION.

Routine procedures for the processing of tissue showing the bone-cement interfaces can be done by decalcifying the mineral prior to dehydration and embedding in a paraffin wax, which in some cases can cause considerable shrinkage of cell components and 'break' at the cement interface. Paraffin sections are normally cut at 6 microns on a base sledge microtome. Another well-known method is embedding the un-decalcified bone in a methacrylate resin and sectioning on a motorized microtome. Alternatively sections are cut between 50-100 microns using the diamond slow saw technique and ground down until transparent. Preparations thus obtained are referred to as a ground section. In the latter case cellular structures are normally damaged due to the grinding process.

The first problem which arose in processing samples for Electron microscopy was the extreme density of the undecalcified mineral and trying to find a suitable resin to match it. In our early studies we sectioned undecalcified bone-cement embedded in Polymaster 1209 AC [7]. Morphology was poor, having been fixed in formalin . Araldite and Epon were subsequently used, because of their very high viscosities, but propylene oxide had to be used as an intermediate stage. This presented us with our second problem. Incorporated into the

PMMA are radiopaque markers i.e. Barium Sulphate (CMW-Densply Ltd) and Zirconium (Palacos-Kirby Warrick Pharmaceuticals Ltd). Which are used as X-ray markers. We could not afford to have the cement dissolving away in propylene oxide as suggested by previous workers [8], as these X- ray markers were useful in identifying the cement using X-ray microanalysis. In our laboratories, we overcame these difficulties by lengthening the infiltration times, using a harder Spurrs' resin formulation and, in the case of *tri* -calcium phosphate ceramic, double embedding. These procedures enabled studies of both PMMA and TCP interfaces to be carried out utilizing both the light and electron microscope, with X-ray microanalysis. In this report we were able to show intact interfaces and the response of cells and tissue components to the cements and ceramics. Acrylic cements are reported to be inert, so the lack of active remodelling was expected. Ceramic,on the other hand, is reported to be 'bioactive' [9] and active remodelling was seen at the interface.

REFERENCES.

1.Charnley J. (1970c) The reaction of bone to self- curing acrylic cement. A long term study in man. J Bone Joint Surg 52B:340-52.

2. Charnley J., (1970a) Acrylic cement in orthopaedic surgery. E.S.Livingstone, Edinburgh-London 1970.

3. Ducheyne P., McGuckin J F ., (1990) Handbook of Bioactive Ceramics : In press.

4. Downes S., Kayser M V., Ali S Y.,(1990) An Electron Microscopical Study of the interaction of bone with Growth Hormone loaded bone cement. Scanning Microscopy :In press.

5. Spurr A R. (1969) Low- Viscosity Epoxy Resin Embedding Medium for Electron Microscopy. J. Ultrastructure Research 26; 31-43.

6. Humphrey C.D. and Pittman F.E. (1974). A simple methylene blue -azure II-basic fuchsin stain for epoxy -embedded tissue sections. Stain Technol 49: 9-14.

7. Mawhinney W H B ., Ellis H A., (1983). A technique for plastic embedding of mineralized bone. J Clin Pathol 36: 1197-9.

8. Linder L., Hansson H A., (1983). Ultrastructural aspects of the interface between bone and cement in man. J Bone and Joint Surgery 65B: 646-649.

9. Ducheyne P., Hench L L., Kagan A., Martens M., Burssens A., Mulier J C., (1980) The effect of hydroxyapatite impregnation on skeletal bonding of porous coated implants. J Biomed Mater Res 14: 225.

PREFERRED ORIENTATION IN BONE

DAVID H. ISAAC [1] AND MICHAEL GREEN [2]

[1] Department of Materials Engineering, University College of Swansea, Singleton Park, SWANSEA SA2 8PP

[2] School of Engineering, University of Wales College of Cardiff, PO Box 917, Newport Road, CARDIFF CF2 1XH

ABSTRACT

Anisotropy in bone has long been recognised. In mammalian bone it may be observed at two distinct levels of organisation, the first being the arrangement of the osteons within the whole bone, and the second the ultrastructure of the osteons themselves. The preferred orientation of material within a single lamella of an osteon, (and in the whole bone of fish), has previously been explained in terms of the alignment of arrays of needle-like mineral crystallites within the collagen fibrils which are themselves arranged in parallel. Based on observations of bone from a variety of sources, this paper proposes an alternative hypothesis, in which the preferred orientation and ultimately the anisotropy have their origins in the existence of columns of mineral, the orientations of which are related to those of the collagen fibres. These columns are linked together to form a continuous mineral phase interpenetrated by a second continuous network of collagen fibres.

INTRODUCTION

The anisotropy of mammalian bone is a result of the preferred orientation which exists at two levels of structural organisation. The first of these levels is the arrangement of osteons within the bone. Osteons, or Haversian systems which are the basic structural units of mature mammalian bone, are roughly cylindrical and are arranged parallel to the long axis of the bone. The second important level of organisation is the ultrastructure of individual osteons. Each osteon is composed of a number of concentric layers of material, each of which contains collagen and the mineral, calcium hydroxyapatite. It has been demonstrated by Green et al (1), that the orientation of the collagen fibres of successive lamellae changes through approximately 90 degrees. It is the nature of the structure and spatial arrangement of the material of the individual lamellae which is the subject of this paper.

Bone is a composite material made up of collagen fibres and a chalk-like mineral, calcium hydroxyapatite. It has long been held that the mineral is in the form of needle or plate-like crystallites, a theory developed following the observation of ultrathin sections of undecalcified bone, using the transmission electron microscope (2,3). With the advent of the scanning electron microscope, some authors, for example Boyde et al. (4,5,6), have described spherical or spheroidal structures on the free surfaces of bone which had been treated with chemical etchants such as sodium hypochlorite. More recently, Turner and Jenkins (7) observed spheroidal particles ~ 100 nm across in both bovine and human bone which had been etched using the physical technique of ion bombardment. These particles were seen to be joined together in a vermiform structure constituting a continuous mineral phase.

Evidence of a continuous mineral phase in bovine bone has also been presented by Green *et al.* (8), using the technique of collagenase etching. Further applications of this technique to fish bone (9) and human trabecular bone (10, 11) revealed evidence that the mineral phase may exist in the form of columns which interconnect to produce a continuous mineral system which is interpenetrated by the collagen. This paper presents further evidence that the columnar structure of the mineral is a major factor in the generation of the observed preferred orientation.

MATERIALS AND METHODS

Materials

Transverse and longitudinal sections of the fin bones of the plaice, *Pleuronectes platessa*, were taken fresh to provide the fish bone samples. The human bone was from the trabecular volume of trans-iliac samples obtained post mortem from persons exhibiting no signs of bone pathology, while the rat bone was obtained from six day old rat femur.

Collagenase etching

The bone samples were polished with carbide papers and polishing alumina until the surfaces appeared under optical microscopy to be scratch free. In the case of the trabecular bone, when sections through the trabeculae were polished, some natural surfaces remained unpolished, and these were also examined. Following polishing, the specimens were first boiled in distilled water to denature the collagen and then incubated for 24 hours in 5 ml aliquots of Sigma type 1A collagenase in a calcium containing, Tris-buffer, pH 7.4 at 37° C. Each aliquot contained approximately 2100 units of collagenase.

Ultracryotomy

The femora were removed intact from six day old rats. They were carefully stripped of adherant soft tissue and plunged into liquid propane cooled by liquid nitrogen. Once frozen, the specimens were stored under liquid nitrogen until being sectioned at -105° C with a glass knife in a Reichert Ultracut FC4E ultramicrotome with cryo attachment. Sections were cut to a thickness of approximately 100 nm.

X-ray diffraction

The x-ray diffraction of whole fresh fish bone was carried out using CuKα radiation from a Philips PW 1820 x-ray generator, with a specimen to film distance of about 40 mm.

Electron microscopy and diffraction

The scanning electron microscopy was carried out following gold coating. A JEOL 35c scanning microscope, and for higher resolution a JEOL 120 Temscan operating in the scanning mode were used at accelerating voltages of 30 and 100 kV respectively. The transmission electron microscopy and electron diffraction of carbon coated thin sections were carried out on a JEOL 120 Temscan operating in the transmission mode at an accelerating voltage of 80 kV.

RESULTS

Figure 1 is a micrograph of a transverse section through the fin bone of a plaice. The specimen has been polished and etched with collagenase. Holes up to approximately 500 nm across may be seen, indicating the *in vivo* position of collagen fibres of up to that diameter. The mineral appears to be in the form of spheroids ~ 100 nm in diameter which are fused together forming units roughly circular in cross section and typically ~ 0.5-1.0 μm across. Figure 2 shows a longitudinal section of a similarly treated specimen. Here the mineral is seen to be in the form of columns, parallel with the long axis of the bone. This figure also reveals the small spheroidal units ~ 100 nm across and shows that they are fused together to form columns again typically 0.5-1.0 μm across. There is also evidence in this picture of columns ~ 100 nm across, suggesting that the 100 nm spheroids are more strongly bound along the bone axis than any other direction. Figure 3 is an x-ray diffraction pattern of a fresh fish bone, and it clearly demonstrates preferred orientation. The long axis of the bone was vertical and the sharp reflection at 0.344 nm, acred in the meridional direction, corresponds to the 002 reflection of hydroxyapatite, indicating preferred orientation of the c axis along the bone axis. The original diffraction pattern also shows that there is preferred orientation of the collagen fibres along the bone axis, but this is not so evident in the print.

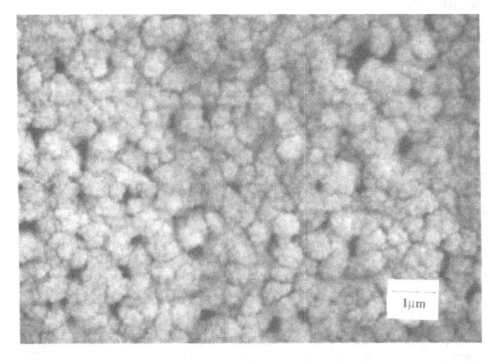

Figure 1   A scanning electron micrograph of a collagenase etched transverse section through a fish bone.

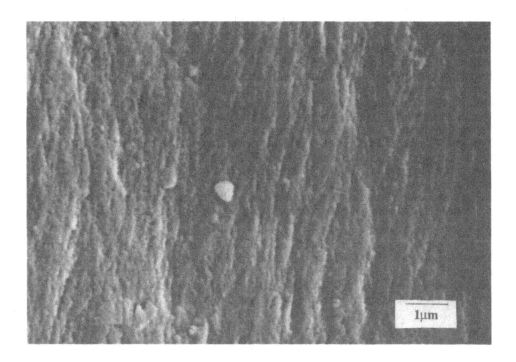

Figure 2   A longitudinal polished and collagenase etched section of a fish bone. The columns of mineral may clearly be seen.

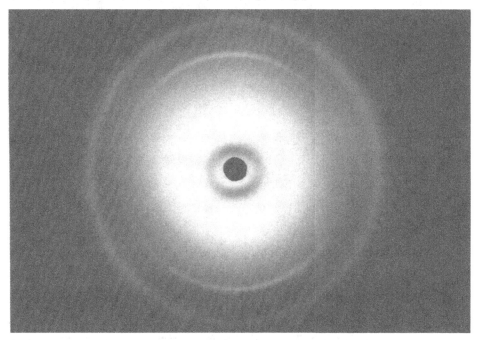

Figure 3   A flat plate x-ray diffraction pattern of fresh fish bone showing the preferred orientation of the mineral.

Figure 4 is an electron micrograph of a polished collagenase etched section through a human trans-iliac specimen. The polished surface can be seen to the bottom right and the rest of the micrograph shows unpolished, but collagenase etched, natural surfaces. There is some evidence on the polished surface of alternate light and dark regions resulting from the changing orientations of collagen fibres and mineral from one lamella to the next. Figure 5 is a much higher resolution picture of the polished surface and this reveals holes ~ 200 nm across from which collagen fibres have been removed by the collagenase. This figure also provides evidence of 100 nm spheroids (as described for fish bone above) which fuse together to form larger units up to ~ 500 nm across. Figures 6 and 7 show two different areas of the unpolished surface of Figure 4 at higher magnifications. Both these figures reveal the column-like structures similar to those seen in fish bone above. In Figure 6 these columns are seen sideways on and in Figure 7 the columns are running at an angle to the surface. The higher resolution picture of Figure 8 shows that columns to be ~ 100 nm across, but also indicates that not only do they link together to form larger diameter columns but also there are interconnections between these columns giving rise to a continuous mineral phase.

Figure 9 is a transmission electron micrograph of a thin section of rat femur. The columnar structure is again evident in this picture and the columns are typically up to ~ 1 μm across. Figure 10 shows part of one of these columns and Figure 11 is an electron diffraction pattern obtained from this column. The first sharp reflection, for which a preferred orientation is seen, indexes as the 002 of hydroxyapatite and after taking into account image rotation, it is found that this c axis corresponds to the long axis of the column.

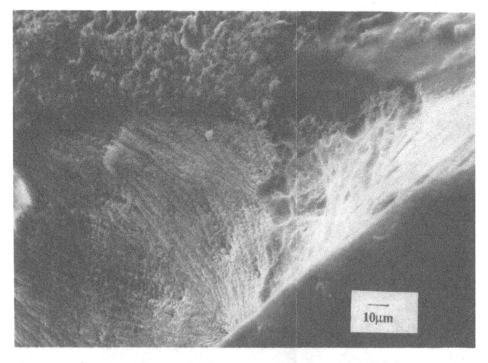

Figure 4   A polished and collagenase etched section through part of a human trabeculum. The polished surface is at the bottom right and the other areas are part of the natural surface which remained untouched during polishing, but which was subject to collagenase etching.

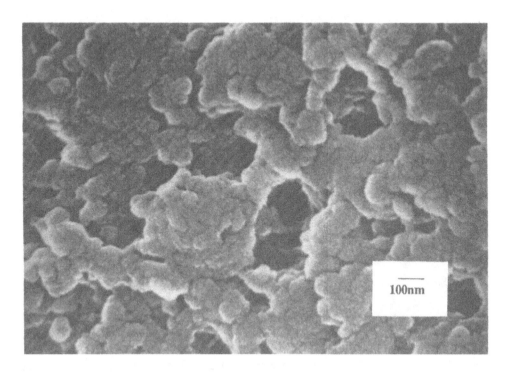

Figure 5   A higher magnification micrograph of the polished surface of Figure 4.

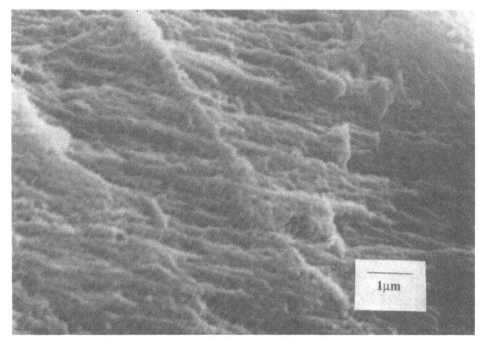

Figure 6   Part of the unpolished area of Figure 4 showing the parallel columnar arrangement.

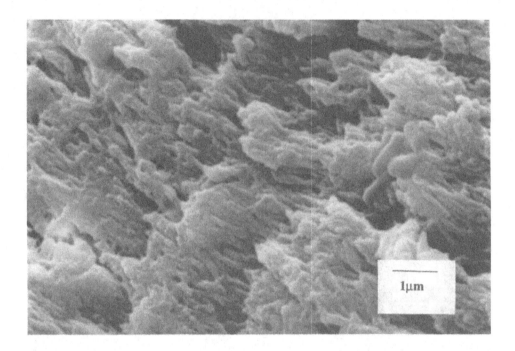

Figure 7   Another part of the unpolished surface of Figure 4, in which the columns are at a different angle to the natural surface.

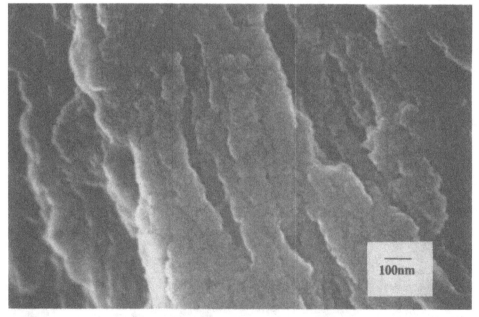

Figure 8   A high resolution micrograph of the unpolished surface of Figure 4. The columns, 100 nm in diameter are clearly visible.

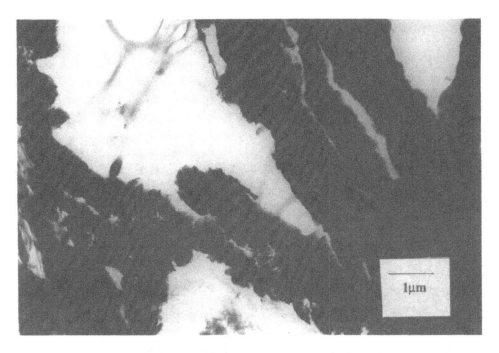

Figure 9   A transmission electron micrograph of an ultrathin section of rat femoral bone. The mineral can be seen to be in the form of columns up to ~ 1 μm across.

Figure 10   A higher magnification micrograph of one of the columns of Figure 9. The electron diffraction pattern of Figure 11 was taken from this part of the specimen.

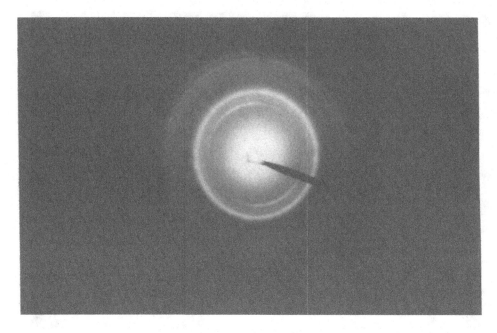

Figure 11   Electron diffraction pattern from part of the mineral column shown in
                   Figure 10.

DISCUSSION AND CONCLUSIONS

The micrographs shown in this paper are all consistent with a common model for the microstructure of bone. This model envisages the building units of mineral hydroxyapatite as 100 nm spheroids which fuse together to form columns ~ 100 nm across. The preferred orientation is incorporated at this level since, as the electron diffraction pattern of Figure 11 shows, there is a likelihood that the basal planes (001) of the hexagonal unit cells of adjacent spheroids will coalesce. This leads to the c-axis preferred orientation of the mineral along the column axis. The 100 nm columns also appear to aggregate and form columns on a larger scale typically ~ 0.5 to 1.0 µm across. Furthermore, these columns are linked together by additional mineral crosslinks, forming a continuous mineral phase. The collagen fibres are visualised as passing through this mineral phase with their axes approximately parallel to the predominant direction of the mineral columns.

For fish bone, the columns are seen in cross section (Figure 1) and longitudinal section (Figure 2) and the x-ray diffraction pattern of Figure 3 confirms that the c-axis of the hydroxyapatite is along the column axis. In human bone (Figures 4-8) the mineral columns are seen from a variety of angles, all of which are consistent with the above model. The transmission electron micrographs of rat bone (Figures 9,10) also support this columnar model, and the electron diffraction pattern of Figure 11 confirms that there is a preferred orientation of the hydroxyapatite c-axis along the columns.

Acknowledgements

We wish to thank the Arthritis and Rheumatism Council for Research for financial support of this project.

REFERENCES

1. Green, M., Isaac, D.H. and Jenkins, G.M., Collagen fibre orientation in bovine secondary osteons by collagenase etching. Biomaterials, 1987, **8**, 427-432.

2. Wolpers, C., Kollagenquerstreifung und Grundsubstanz. Klin. Wochschr., 1943, **22**, 624.

3. Jackson, S.A., Cartwright, A.G. and Lewis, D., The morphology of bone mineral crystals. Calcif. Tiss. Res., 1978, **25**, 217-222.

4. Boyde, A. and Hobdell, M.H., Scanning electron microscopy of lamellar bone. Zeitschrift fur Zellforschung und Mikroscopische Anatomie, 1969, **93**, 213-231.

5. Boyde, A., Scanning electron microscope studies of bone. In The Biochemistry and Physiology of Bone, 2nd. ed., ed. G.H. Bourne, Academic Press, 1972, Vol. 1 pp. 259-310.

6. Boyde, A. and Sela, J., Scanning electron microscope study of separated calcospherites from the matrices of different mineralising systems. Calcified Tissue Research, 1978, **26**, 47-49.

7. Turner, I.G. and Jenkins G.M., The spatial arrangement of bone mineral as revealed by ion bombardment. Biomaterials, 1981, **2**, 234-238.

8. Green, M., Isaac, D.H. and Jenkins, G.M., Bone microstructure by collagenase etching. Biomaterials, 1985, **6**, 150-152.

9. Green, M., Isaac, D.H. and Jenkins, G.M., Mineral structure and preferred orientation in the fin bones of the plaice, *Pleuronectes platessa*. Biomaterials, 1988, **9**, 319-323.

10. Mackie, I.G., Green, M., Clarke, H. and Isaac, D.H., Osteoporotic bone microstructure by collagenase etching. Annals of Rhuematic Disease, 1989, **48**, 464-469.

11. Mackie, I.G., Green, M., Clarke, H. and Isaac, D.H., Human bone microstructure by collagenase etching. J. Bone and Joint Surgery, 1989, **71B**, 509-513.

BONE MINERAL AS SEEN IN THE SCANNING AND TRANSMISSION ELECTRON MICROSCOPES

MICHAEL GREEN [1] AND DAVID H. ISAAC [2]

[1] School of Engineering, University of Wales College of Cardiff, P.O. Box 917, Newport Road, CARDIFF CF2 1XH

[2] Department of Materials Engineering, University of Wales Swansea, SWANSEA, SA2 8PP

ABSTRACT

Transverse sections of bovine tibial bone have been investigated in both the SEM and TEM. SEM photographs of collagenase etched surfaces reveal the mineral hydroxyapatite component as small spheroidal particles ~ 10 nm across which are fused together to form spheroidal units ~ 100 nm across. These 100 nm units aggregate to form a continuous mineral phase with holes or gaps (typically ~ 200 nm across) through which collagen fibres pass *in vivo*. TEM photographs of untreated thin sections confirm the existence of these spheroidal units but additionally reveal rod shaped mineral particles (typically ~ 10 nm by 100 nm). Both these morphological forms give electron diffraction patterns consistent with a poorly crystalline hydroxyapatite. Therefore, it is suggested that bone consists of a continuous fused spheroidal mineral component interpenetrated by collagen fibres which are closely associated with a second distinct morphological form in the shape of rod like mineral particles.

INTRODUCTION

It has been well established by x-ray diffraction (e.g. 1) that the mineral component of bone is a poorly crystalline hydrated form of hydroxyapatite, although some controversy still exists over whether there is additionally an amorphous mineral phase (e.g. 2).

The morphology of bone mineral has also been the subject of numerous investigations by various techniques the most prominent of which are electron microscopy and x-ray diffraction. The first transmission electron micrographs of undecalcified bone specimens prepared by thin sectioning were published by Wolpers (3). Subsequently many other authors (e.g. 4,5,6,7) using the same technique concluded that the mineral phase of bone consists of needle or plate-like particles with a variety of dimensions ranging from ~ 2 nm to ~ 150 nm. X ray diffraction experiments (e.g. 8) involving line broadening and low angle scattering have also been interpreted in terms of these regularly shaped particles. Weiner and Price (9) have recently studied the disaggregation of the mineral phase of bone by oxidising with sodium hypochlorite and sonicating. Their results suggested that all the crystals are tabular or plate-shaped, typically 40 nm long and 25 nm wide. Further recent studies based on the scanning electron microscope have however revealed spheroidal shaped mineral particles ~ 100 nm across in bone. Boyde (10) observed such "spheroidal particles" in fetal bone and similar particles have been isolated from solutions of NaOCl used to wash bone and dentine (11). Pautard (12) also observed spheroidal particles in 6 day old mouse calvarial bone after a treatment of freeze drying, enzymatic attack and density fractionation. Recently, a new SEM technique, developed in this laboratory, has been used to investigate the mineral component of bone by treating polished surfaces of bone with collagenase solution (13-16). The high resolution achieved by this method has demonstrated the presence of small rounded units (~ 10 - 20 nm across) which aggregate to form the larger contiguous spheroidal particles (~ 100 nm across) which coalesce to create a continuous mineral phase within the bone.

The present report looks at the spheroidal particles revealed in the SEM by collagenase etching and compares the results with TEM photographs of thinly sectioned bovine bone in an attempt to reconcile the differing views of bone mineral presented by these two techniques.

MATERIALS AND METHODS

Transverse sections approximately 2 mm thick were cut from the mid third of a fresh bovine tibial diaphysis using a hand hacksaw. For the scanning electron microscopy studies these sections were polished and etched as described below. Samples for the transmission electron microscope were prepared from blocks ~ 2 mm³, cut from areas immediately adjacent to the SEM specimens. Further blocks, approximately 5 mm³ were also cut from adjacent bone. These were ground to produce a fine powder of bone mineral. This material was treated as described below and examined using limited area electron diffraction.

a.   Scanning Electron Microscopy:

The transverse sections of fresh compact bovine bone were polished using carbide papers and polishing alumina. Segments, typically 3 mm by 5 mm, were cut from these sections and boiled in distilled water for 30 minutes to denature the collagen and so enhance the action of the collagenase. Aliquots (5 ml.) of collagenase solution each containing 2100 units were made up in 0.05 M Tris buffer, pH 7.4. These solutions also contained 0.1M calcium chloride, the Ca^{2+} ions being necessary for the activation of the collagenase. The solutions containing the specimens were incubated at 37°C for one week, each bone specimen being bathed in 5ml of enzyme solution which was agitated twice daily. Shorter etching times, down to 30 minutes, were also used to confirm that prolonged bathing did not introduce artifacts. After incubation the specimens were dehydrated in serial alcohols, dried in air from absolute alcohol, coated in gold and examined in a JEOL 120C Temscan operating in the scanning mode at an accelerating potential of 100 kV. Conducting coatings other than gold were also observed, to ensure that the features of interest were not artifacts of the coating.

b.   Transmission Electron Microscopy:

The 2 mm^3 blocks of undecalcified compact bone were mounted in LR white resin (London Resin Company). A Reichart OM U4 Ultra-microtome was used to produce transverse sections of the bone with thicknesses in the ranges 60 - 80 nm and 120 - 140 nm and it was found that a glass knife proved adequate for this purpose. The thicknesses were estimated by the colours of silver and gold respectively that were observed as the sections lay in the double distilled buffered water used as a flotation medium. The samples were examined without staining in a Philips EM400T transmission electron microscope operating at an accelerating potential of 100 kV.

c.   Limited Area Electron Diffraction:

Finely powdered bone was prepared for limited area electron diffraction in the following ways:

(i)   suspended in acetone and dispersed on carbon coated TEM grids,

(ii)   treated with collagenase etching solution under the same conditions as used to etch bone specimens.

Limited area electron diffraction was carried out on specimens treated as above and also on representative areas of the thin sections shown below. A JEOL 120C Temscan, operating in the transmission mode at an accelerating potential of 100 kV, was used with a field width of 20μm.

OBSERVATIONS AND RESULTS

Figure 1 is a scanning electron micrograph of polished and collagenase etched compact bovine bone. A continuous mineral phase is evident with holes or gaps (up to ~ 200 nm across) from which collagen has been removed by the action of the collagenase. The remaining mineral hydroxyapatite component is seen as spheroidal shaped units ~ 100 nm across which aggregate to form the continuous mineral phase. In the higher magnification scanning electron micrograph of Figure 2 a more detailed structure is revealed, indicating that the 100 nm units are composed of even smaller units ~ 10 - 20 nm across.

Figure 3 is a transmission electron micrograph of a transverse section through a fresh compact bovine bone. The section thickness is ~ 120 - 140 nm. The smallest units observed here are ~ 10 - 20 nm across and have a spheroidal shape, consistent with the 10 - 20 nm units of the SEM micrograph in Figure 2. There is also clear evidence that these units fuse to form larger units, ~ 100 nm across, which in turn appear to aggregate. Thus this TEM photograph may be interpreted in a manner consistent with the SEM photographs of Figures 1 and 2.

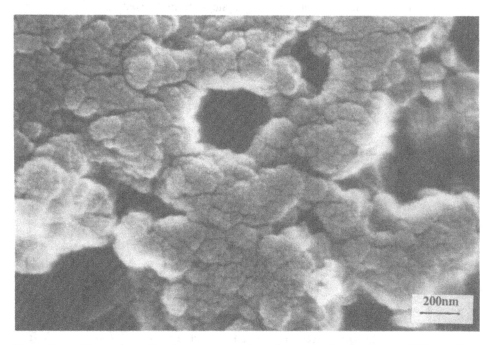

200nm

Figure 1   The surface of a transverse section of compact bovine bone which has been polished and collagenase etched. This scanning electron micrograph shows a continuous mineral phase of spheroidal units and holes (~ 200 nm across) from which collagen has been removed by the action of collagenase.

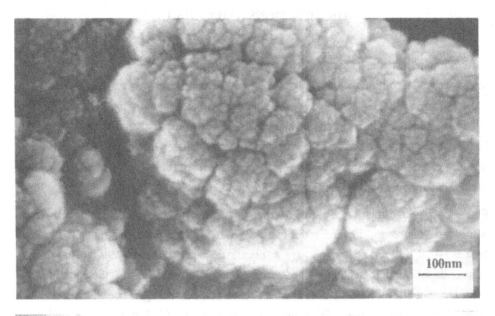

Figure 2   A higher magnification scanning electron micrograph of the surface shown in Figure 1. This indicates that spheroidal units ~ 10-20 nm across coalesce to form 100 nm units which also aggregate, to form the continuous mineral phase of macroscopic bone.

Figure 3   A transmission electron micrograph of a transverse section through fresh compact bovine bone of thickness ~ 120-140 nm. The 10-20 nm spheroidal units are clearly seen and they coalesce to form the 100 nm units which in turn appear to aggregate.

Figure 4 is a transmission electron micrograph of a transverse section cut to a thickness of 60 - 80 nm. In this micrograph there is still evidence of spheriodal units with dimensions ~ 10 - 20 nm which aggregate. However, they now appear more in the form of a background feature, with rod-like particles (typically 10 nm by 100 nm) becoming an important feature.

Limited area electron diffraction patterns were obtained from powdered bone which had not been exposed to an aqueous solution, powdered bone which had been treated in a collagenase etching solution under the same conditions as described above for the SEM bone samples and bone which had been thinly sectioned. All three categories of specimen yielded similar diffraction patterns and all were consistent with the material being a poorly crystalline form of hydroxyapatite.

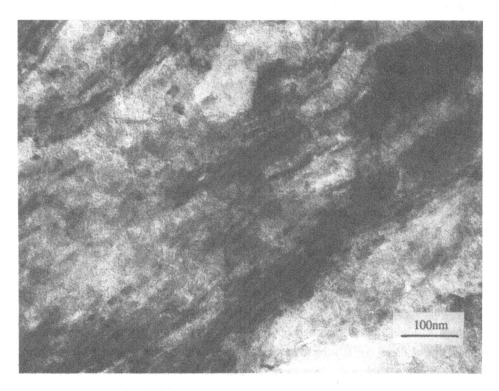

Figure 4   A transmission electron micrograph of a transverse section through fresh compact bovine bone of thickness ~ 60-80 nm. In addition to the background of spheroidal units there now appear to be rod like units ~ 10 nm by 100 nm.

DISCUSSION

The spheroidal shaped particles (~ 10 - 20 nm across and ~ 100 nm across) are evident in all the micrographs. They are clearly not an artifact of the preparation technique of collagenase etching, as they are present in the 120 - 140 nm thin sections. Also limited area electron diffraction, shows them to be a poorly crystalline form of hydroxyapatite and indistinguishable from both the sectioned material and anhydrously treated powdered bone. These spheroidal structures have previously been demonstrated to be the major morphological units in the mineral component of bone (eg. 16). The real difficulty in interpretation of the photographs presented is the presence in substantial quantities of needle - like particles (~ 10 nm by 100 nm) in Figure 4 which closely resemble results from other studies (e.g. 17). Such particles have not been observed in the scanning electron micrographs of collagenase etched polished surfaces or untreated fracture surfaces which still demonstrate the spheroidal units. Limited studies have not revealed these needle shaped particles in the supernatant of the collagenase solution that was centrifuged after collagenase treatment of the bone surface. Similar particles have been observed in transmission electron micrographs of the thicker sections but they are less widespread. There are various possible interpretations for these needle-shaped particles, but the crucial question that needs to be addressed is whether or not they constitute a genuine second distinct morphological form of mineral.

It seems unlikely that the needles could have originated in the spheroidal units and have been produced by disruption during sectioning. We certainly have great difficulty in reconciling needles (~ 10 nm by 100 nm) with the model of spheroids since the smallest spheroidal units are ~ 10 - 20 nm across. We have shown in a previous publication (13) that these units are indeed spheroidal shapes rather than rods or needles by also looking at longitudinal sections. For such spheroidal units to form the needle shapes observed in Figure 4 there would need to be substantial anisotropic forces maintaining a column of these units with 1 unit in cross section and about 10 units in length. This is an unlikely explanation but it cannot be totally ignored. Indeed in human trabecular bone (16) the 100 nm spheroids have been observed to aggregate to form columns (~ 100 nm across) which are approximately parallel and with numerous interconnections. However, we have not seen any detail within the rods of Figure 4 suggesting that they are composed of smaller integral units. Previous results have indicated that these smallest units always aggregate to form the 100 nm spheroids and it is the arrangement of these that is subject to variability. Thus in bovine bone (13) it has been suggested that these spheroids form larger conglomerates whereas in the fin bone of the plaice (15), which has been known for some time to have significant preferred orientation of the mineral, the 100 nm units form columnar shapes ~ 1 μm across and of significantly greater length. A similar pattern of these units forming columns of bone mineral (~ 100 nm across) has also been observed in both normal and porotic human trabecular bone (16, 18).

Assuming that the needles do not originate from disruption of the spheroidal units we must seek an alternative interpretation. First, they could be a genuine distinct second morphological form of mineral with dimensions approximately 10 nm by 100 nm. The main argument against this possibility is that we would expect to see such needles in the electron micrographs of the slightly thicker sections, although it could be argued that in the thicker sections these small units would be more difficult to see against the general background of spheroidal material (~ 120 nm thick). There have been many suggestions in the literature that there exist needle-like mineral units closely associated with the collagen fibrils (e.g. 19). A proposal based on this idea could be put forward for discussion. Of particular interest in Figure 4 is the feature that the needle shapes run approximately parallel. This would be consistent with their following the direction of collagen fibrils. Indeed it could be considered that in Figure 4 there is a collagen fibril ~ 100 nm across traversing the micrograph from middle left to top right. This dimension is consistent with the observed holes in Figure 1 from which the collagen fibrils have been removed. Collagen alone would not provide the absorption of electrons required to produce the strong images observed, but mineral needles closely associated with the collagen would have the required opacity. A further development of this is also illuminating. The lamellar structure of osteonic bovine bone has been shown to contain collagen fibrils running alternately along the axis of the osteon and circumferentially. However, the fibres are tilted from these ideal positions and in the case of the circumferential fibres the angle of tilt is typically 10 - 15 degrees (14). So if long rods of mineral were to be associated with the collagen, then a very thin transverse section, of the order of 80 nm thick could cut these rods to produce needles of the observed length (~ 100 nm).

The arguments presented here are by no means conclusive. However the weight of evidence does warrant serious consideration. For example, if the spheroids demonstrated here and elsewhere are indeed artifacts of the collagenase etching technique, why are they so clearly present in the 120 nm thick sections? It has been demonstrated elsewhere, both by the present authors (e.g. 13, 14, 15, 16, 18) and others (e.g. 20, 21) that the bulk of the mineral phase of bone exists in the form of a continuous network of contiguous spheroids ~ 100 nm in diameter and that this network contains gaps or holes up to ~ 200 nm in diameter which represent the *in vivo* position and dimensions of the collagen component. If these observations are valid then they are fully consistent with the hypothesis that two distinct morphologies of bone mineral exist. The contiguous spheroids constitute the major component, the second being the needle or plate-like particles which are intimately linked with the collagen residing within the holes in the continuous mineral matrix.

CONCLUSIONS

The interpretation of these results which appears to be the most reasonable is that there are two distinct morphological arrangements of the hydroxyapatite mineral in bovine bone. The bulk of the material is clearly in the form of small units (~ 10 nm) which coalesce to form spheroidal units ~ 100 nm across, which in turn aggregate to form a continuous mineral phase. The evidence presented suggests that there is a second distinct mineral component in the form of needles (~ 10 nm by 100 nm) or longer rods with a diameter of 10 nm which are cut by the sectioning knife to produce the needles. We should however qualify this interpretation by noting the scanty evidence for this second phase in transmission electron micrographs of the thicker sections and no evidence of it in the scanning electron micrographs of surfaces nor in the solutions used to etch the surfaces.

Acknowledgements

We wish to acknowledge the Arthritis and Rheumatism Council for Research for financial support of this project.

REFERENCES

1.   Carlstrom, D. and Finean, J.B., X-ray diffraction studies on the ultrastructure of bone, Biochim. Biophys. Acta, 1954, 13, 183-191.

2.   Bienstock, A. and Posner, A.S., Calculation of the x-ray intensities from arrays of small crystallites of hydroxyapatite, Archives of Biochemistry and Biophysics, 1968, 124, 604-615.

3.   Wolpers, C., Grenzgebiete Med., 1949, 2, 527.

4.   Robinson, R.A. and Watson, M.L., Collagen-crystal relationship in bone seen in the electron microscope, Anat. Rec., 1952, 114 (3), 383 - 409.

5.   Fernandez-Moran, H. and Engstrom, A., Electron microscopy and x-ray diffraction of bone, Biochim. Biophys. Acta., 1957, 23, 260-264.

6.   Cooper, R.R., Milgram, J.W. and Robinson, R.A., Morphology of the osteon - an electron microscope study, J. Bone Jt. Surg., 1966, 48A (7), 1239 - 1271.

7.   Bocciarelli, D.S., Morphology of crystallites in bone, Calcif. Tiss. Res., 1970, 5, 261 - 269.

8. Engstrom, A., Aspects of the molecular structure of bone, in The biochemistry and physiology of bone, 2nd ed., Ed. G.H. Bourne, Vol 1, Academic Press, 1972, 237-257.

9. Weiner, S. and Price, P.A., Disaggregation of bone into crystals, Calcif. Tiss. Int., 1986, 39, 365-375.

10. Boyde, A., Scanning electron microscope studies of bone, in The biochemistry and physiology of bone, 2nd ed., Ed. G.H. Bourne, Vol. 1, Academic Press, 1972, 259 - 310.

11. Boyde, A. and Sela, J., Scanning electron microscope study of separated calco-spherites from the matrices of different mineralising systems, Calcif. Tiss. Res., 1978, 26, 47-49.

12. Pautard, F.G.E., Phosphorous and bone, in New trends in bioinorganic chemistry, Eds. R.J.P. Williams and J.F.R. Du Silva, Academic Press, 1978, 261-354.

13. Green, M., Isaac, D.H. and Jenkins, G.M., Bone microstructure by collagenase etching, Biomaterials, 1985, 6, 150-152.

14. Green, M., Isaac, D.H. and Jenkins, G.M., Collagen fibre orientation in bovine secondary osteons, Biomaterials, 1987, 8, 427-432.

15. Green, M., Isaac, D.H. and Jenkins, G.M., Mineral structure and preferred orienta-tion in the fin bones of the plaice, *Pleuronectes platessa*, Biomaterials, 1988, 9, 319-323.

16. Mackie, I.G., Green, M., Clarke, H. and Isaac, D.H., Human bone microstructure by collagenase etching, J. Bone Jnt. Surg., 1989, 71-B, 509-513.

17. Jackson, S.A., Cartwright, A.G. and Lewis, D., The morphology of bone mineral crystals, Calcif. Tiss. Res., 1978, 25, 217-222.

18. Mackie, I.G., Green, M., Clarke, H. and Isaac, D.H., Osteoporotic bone microstruc-ture by collagenase etching, Annals of Rheumatic Diseases, 1989, 48, 464-469.

19. Lees, S., A model for the distribution of HAP crystallites in bone - an hypothesis, Calcif. Tiss. Int., 1979, 27, 53-56.

20. Turner, I.G. and Jenkins, G.M., The spatial arrangement of bone mineral as revealed by ion bombardment, Biomaterials, 1981, 2, 234-238.

21. Ralis, Z.A. and Turner, I.G., Two phases of the bone mineral as revealed by the high resolution SEM on etched bone surfaces and as seen on surfaces untreated and chemically etched, Microscopica Acta, 1981, 84 (4), 385-400.

EXPERIMENTAL AND THEORETICAL THERMAL EFFECTS DURING CEMENTATION OF AN ENDOMEDULLARY INFIBULUM

G. Paganetto, S. Mazzullo
HIMONT Italia. Research Centre "G. Natta"
44100 Ferrara Italy

ABSTRACT

A 1-dimensional mathematical model of a femoral prosthesis implant has been developed. The actual complex geometry has been idealized as a composite cylinder of infinite length having a multishell bone/cement/stem structure. Attention is focused on the cement, which is a polymer that polymerizes "in situ", inside the medullar channel and generates heat. The "cementation" phenomenon is described by the heat diffusion Fourier equation coupled with the polymerization kinetics. The numerical solution of the model has been accomplished by a finite difference explicit scheme. The experimental kinetic data has been obtained through isothermal polymerization tests, carried out by using a DSC calorimeter and a commercially available Howmedica Simplex P cement. The two most relevant results are:
1. The conversion of monomer into polymer is never 100% under the imposed initial and boundary conditions.
2. Bone/cement interface temperature is a function of the interface heat transfer coefficient.
Depending on such a parameter, ranging from 10. to 1000. (Wm^{-2} $°K^{-1}$), the temperature may reach the maximum values of 80 °C and 40°C, respectively. In other words, the lower the heat transfer coefficient, the higher the interface temperature. Therefore, the model predicts relatively high temperature values at the bone/cement inteface. Such values confirm the criticality of this type of implant. The mathematical model developed is capable of taking into account both the geometry of the implant and the chemical-physical and kinetic properties of the cement. It represents a useful tool for setting-up the optimal conditions for the new materials developed in this orthopaedic field.
Heat generated within the polymerizing mixture partly diffuses through the bone, causing damages. We have therefore tested bone behaviour before and after thermal treatment in order to find possible mechanical effects.

INTRODUCTION

Cementation technique of hip prostheses by
polymethylmethacrylate (PMMA) and its copolymers started less
than 30 years ago [6].
Total hip prostheses generally consist of two pieces: the
femoral component and the acetabular component. We shall deal
with the cementation thermal effects on the femoral component:
the cement is, in fact, a polymer which polymerizes "in situ",
inside the medullar channel, and generates heat. It is pretty
natural to idealize the complex geometry of a femoral implant
as a bone/cement/stem composite cylinder. Assuming that the
stem length/medullar channel ratio is generally greater than
5:1,the composite cylinder can be considered of infinite length
Thus the study shall be limited to the heat flow in the radial
direction only.

MATHEMATICAL MODEL

The cementation phenomenon is described by the heat diffusion
Fourier equation coupled with the polymerization kinetics.
If $T(r,t)$ is temperature and $X(t)$ the dimensionless extent of
polymerization, we obtain in radial coordinates:

$$\varrho \, c_p \frac{\partial T}{\partial t} = K\left(\frac{\partial^2 T}{\partial r^2} + \frac{1}{r} \, \frac{\partial T}{\partial r}\right) + \varrho \lambda \frac{dX}{dt} \tag{1}$$

$$\frac{dX}{dt} = k \, (T) \, f \, (T,X) \tag{2}$$

where ϱ_m, c_p, K, are density, specific heat, thermal
conducticity and enthalpy of polymerization, respectively. For
simplicity all these quantities are assumed independent of
temperature and are given a value in Tab. 1 and 2.

TABLE 1
Thermal properties (Huiskes, [3])

	Density Kg m^{-3}	Specific heat J Kg^{-1} °K^{-1}	Thermal conductivity W m^{-1} °K^{-1}
PMMA	1.19 10^3	1.6 10^3	0.17
Cr-Ni Steel	7.85 10^3	0.46 10^3	14.0
Cortical bone	2.10 10^3	1.26 10^3	0.3-0.5
Spongeous bone	2.1-2.3 10^3	1.15-1.73 10^3	0.5

Table 2
Kinetic contrants of a commercial cement (eq.(2))

Cement	λ J Kg^{-1}	k_o sec^{-1}	E J mol^{-1}
SIMPLEX P Howmedica	193. 10^3	2.640 10^8	62.966 10^3

Table 2
continue

Cement	α -	Tg °K	R J mol^{-1} °K^{-1}
SIMPLEX P Howmedica	9.20	378	8.314

Although polymerization of methyl-methacrylate (MMA) is qualitatively well known, not all kinetic parameters are always available for the more common cements in use [3], [4]. Therefore we have decided to carry out isothermal polymerization experiments (Fig.1) in order to identify all the actual kinetic parameters of a commercial cement.

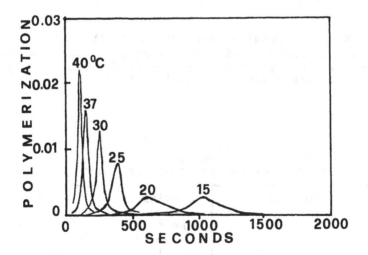

Figure 1. Isotherms of polymerization
of a cement by DSC

The acrylic cement is composed of polymer powder (PMMA) and monomer liquid (MMA). The composition of the mixture is expressed in the ratio gr polymer/ml monomer, which is usually around 2. The main problem in these experiments is to avoid the early polymerization of the mixture before the beginning of the isotherms. To this purpose the mixture has been prepared at the temperature of liquid nitrogen. Experimental data suggest an Arrhenius behaviour for the kinetic constant:

$$k(T) = k_o \exp - E/RT \tag{3}$$

and the following functional form for the extent of polymerization:

$$f(T,X) = \begin{cases} \dfrac{\alpha}{\eta(T)} X^{1-\frac{1}{\alpha}} (\eta(T)-X)^{1+\frac{1}{\alpha}} & ; \quad X < \eta(T) \\ \\ 0 & ; \quad X \geqslant \eta(T). \end{cases} \tag{4}$$

where >1 is an empirical parameter. The function $\eta(T)$ is the equilibrium conversion of monomer to polymer. It behaves as follows [1]:

$$\eta(T) = \begin{cases} \dfrac{T}{Tg} & ; \quad 0 \leq T \leq Tg \\ \\ 1 & ; \quad T > Tg \end{cases} \tag{5}$$

where Tg is the glass transition temperature of cement. The actual values of these kinetic constants are summarized in Tab.2 To complete the model, appropriate initial and boundary conditions are required. The temperature of cement, before its introduction into the femoral cavity, is either known or can be determined experimentally. The same applies also to the initial polymerization extent. Therefore, as initial conditions we have:

$$T(r,0) = T_o \tag{6}$$
$$X(0) = X_o \tag{7}$$

At the stem/cement boundary, $r = a$, we assume perfect contact between the cement and a perfect heat conductor (the metal infibulum) [2]:

$$\pi a^2 \rho_m Cp_m \frac{\partial T}{\partial t} = 2 \pi a K \frac{\partial T}{\partial r} \tag{8}$$

where $\rho_m, Cp_m,$ refer to the metal. At the cement/bone boundary, $r = b$, we assume linear heat transfer between the cement and the sorrounding medium (the bone) at body temperature $T_B = 37[C$:

$$K \frac{\partial T}{\partial r} = - U(T - T_B) \tag{9}$$

where $U(W\ m^{-2}\ °K^{-1})$ is the surface heat transfer coefficient. The actual value of U strongly depends on the preparation of the femoral cavity by the surgeon [3].

NUMERICAL SOLUTION

On physical grounds, the model formulated previously seems well posed. From a mathematical view point the model belongs to the class of Stefan problems. More formal sufficient conditions of existence and uniqueness of the solution can be found in [5]. The numerical solution of the model has been obtained in three steps as follows: i) transformation of the problem to dimensionless form, ii) change of the coordinate system from hollow cylindrical "r" to rectangular "x", by setting r=exp x, iii) numerical solution of the cartesian problem by explicit finite difference schemes. Use has been made of forward difference operators to approximate time derivatives, and central difference operators to approximate space derivatives. Stability restrictions on time step can be easily obtained.

NUMERICAL RESULTS

Numerical simulation shall be limited to a sensitivity analysis with respect to variations of the heat transfer coefficient, at the cement/bone interface. The U parameter takes values in the range 10 - 1000 $(W\ m^{-2}\ °K^{-1})$. The metallic infibulum has a 16 mm diameter and the cement a 6 mm thickness. Temperature at the cement/bone interface, as a function of time, is shown in fig.2.

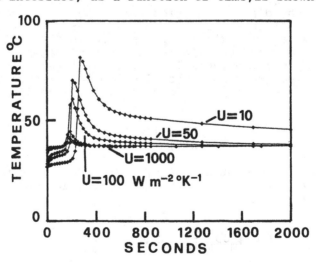

Figure 2. Cement/bone interface temperature

The temperature profile always reaches a maximum, T max, whose intensity increases when U decreases. The maximum conversion, X max, shows a similar trend. The results mentioned above are summarized in Tab.3. It should be mentioned that the temperature behaviour is basically in agreement with the results of Huiskes' model [3] which is numerically solved by finite elements. An important and so far unique feature emphasized by the numerical simulations, is that in no case a 100% conversion is achieved. Therefore, presence of residual unreacted monomer is always noticed in the cement. Fig.3 provides the radial conversion profile at fixed times, in case the U parameter assumes the value of 1000. The maximum radial conversion is 82.1%, while maximum cement/bone interface temperature is 41.7°, (Tab.3).

TABLE 3
Effects of variation of bone-cement
heat transfer coefficient U

U	10.	50.	100.	500.	1000	W m^{-2} °K^{-1}
T max	81.6	70.0	60.6	44.5	41.7	°C
t max	275.2	211.7	211.7	190.5	175.8	sec
X max	94.0	91.2	88.7	82.7	82.1	%

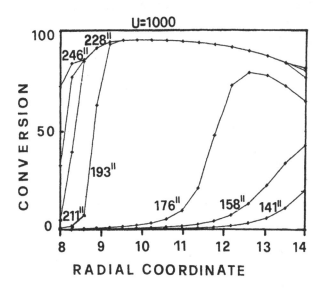

Figure 3. Radial conversions at fixed times, (sec)
U = 1000   W m^{-2} °K^{-1}

THERMOMECHANICAL BEHAVIOUR OF BOVINE CORTICAL BONE

Heat generated within the polymerizing mixture partly diffuses through the bone, causing damages. We have therefore tested bone behaviour before and after thermal treatment in order to study, if any, mechanical effects. For this purpose we have used Dynamical Mechanical Thermal Analyzer (DMTA) equipment, at the frequency of 1 Hz and a scanning rate of 2 °C/min in the temperature range from -140 to 160 °C.

The lower part of Fig.4 shows the real part of the complex dynamic bending modulus (E') as a function of temperature. The lower curve refers to fresh bone. The upper one to the same sample after the thermal treatment due to the first scansion. An increase of the elastic modulus is clearly shown.

The upper part of fig.4 gives the loss factor (tanδ) as a function of temperature. Three relative maxima are clearly distinguishable. We have labeled these peaks as α,β,γ transictions. The γ-transition occurs in the range from -140°C to -100°C and may be attributed to the glass transition of the aliphatic chains of the fatty component of the bone. The β-transition occurs around 0°C and can be explained in terms of water interaction with bone tissue.

Finally, the α-transition occurs in the broad range from 80°C to 160°C and can be associated to removal of bound water or to a solid-solid transition induced by water interaction with collagen [7].

The intensity of the β-transition increases after the thermal treatment due to the first scansion.

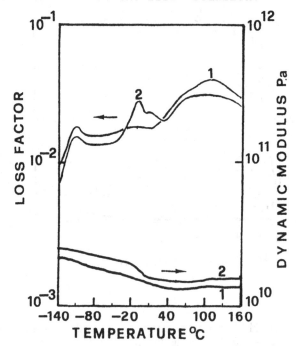

Figure 4. DMTA spectrum of bovine cortical bone
1) before and 2) after thermal treatment

REFERENCES

1. Burnett, G.M., Duncan, G.L., High conversion polymerization of vynyl system. I Methyl-methacrylate. Die Makromoleculare Chem. **51** (1962) 154-170.

2. Carslaw, H.S., Jaeger, J.C., Conduction of heat in solids. Oxford Univ. Press, Oxford (1959), Cap 1.

3. Huiskes, R.. Some fundamental aspects of human joint replacement. Acta Orthop. Scand.; Supp. 185 - Copenhagen (1979) Sect.2 Cap. 7.

4. Jefferis, C.D., Lee, A.J.C., Ling, R.S.M., Thermal aspects of self curing PMMA. J. Bone Joint Surg. **578** (1985) 511-518

5. Verdi, C., Visintin, A., Numerical analysis of the multidimensional Stefan problem with supercooling and superheating. Boll. U.M.I. **1-B** (1987) 795-814.

6. Pipino, F., Il punto sulla cementazione degli impianti protesici. OIC Medical Press, Firenze (1987)

7. Civjan, S., Selting, W.J., De Simon, L.B., Battistone, G.C. and Grower, M.F.. Characterization of Osseus Tissues by Thermogravimetric and Physical Techniques. J. Dent. Res. **51**, (1972) 539-542.

PIN-BONE INTERFACE: A GEOMETRIC AND MATERIAL NON-LINEAR ANALYSIS

A.N. NATALI, E.A. MEROI
University of Padova
Istituto di Scienza e Tecnica delle Costruzioni
via F. Marzolo, 9, 35131 Padova, Italy

ABSTRACT

The aim of this study is to investigate the interaction phenomena between cortical femoral bone and pins used in external fixation. Particular attention is paid to the evaluation of stress and strain fields at the pin-bone interface, analysing this problem according to a contact surface strategy. This geometric non linear aspect is coupled with material non-linearity by introducing an elasto-plastic material constitutive law for bone. A general finite element model is preliminarily developed to investigate bone-fixator interaction and to offer input data for subsequent detailed model for the analysis of interface phenomena, where both pin and bone are discretized by solid elements and with interface elements for the zone in which unilateral contact may arise. Theoretical and operational aspects of this analysis are described, with special attention to numerical formulation.

INTRODUCTION

Bone fracture treated by external fixation is studied from the biomechanical viewpoint, with evaluation of the morphological evolution of callus, also in comparison with results obtained using internal fixators. Because of its stiffness characteristics, fixator configuration deeply influences healing trend, both in static and intermittent loading [1 to 7]. Experimental data on fracture repair refer to both static and intermittent loading conditions [8 to 18]. Numerical modeling of pin-bone interaction phenomena is faced with definition of bone constitutive laws, which are studied using different experimental approaches in relation with induced loading conditions and tissue configuration [19 to 27]. The external and internal remodeling process may play a significant role [28 to 35]. A relevant feature is fracture monitoring. The interpretation of these experimental data in vivo is deeply influenced by stress transfer processes between bone and pins [36 to 42]. Investigations cover numerical modeling and experimental analyses of these interaction phenomena in the evaluation of stress-strain levels at the contact surfaces [43 to 45].

The present work underlines some characteristic aspects of the problem treated. It seems necessary to supply a definition of the interaction, in a general sense, between femoral bone and fixator device, performed using a global model. This allows the correct evaluation of loading induced in pins used in the detailed model and of the interaction zone where the phenomenon is significant. Moreover, the specific configuration of non-linear contact interaction between bone and pin is defined, together with a bone constitutive law in elastic and elastoplastic phases. As a consequence, the numerical model, performed using three dimensional finite elements, considers a coupled geometric and material non-linear problem.

NUMERICAL MODEL

The finite element analysis is developed in two different phases. The global model of a fractured human femur with a monolateral external fixation device is preliminarily investigated. Subsequently a more detailed model of a portion of bone with its pertaining pin is provided, in order to investigate stress-strain transfer mechanisms. Regarding preliminary analysis, the geometric configuration of the model is defined by processing data obtained via nuclear magnetic resonance of a human femur in vivo. The images are digitized and resulting data are used to generate the three-dimensional finite element model shown in fig. 1. Discretization is performed using first-order brick elements. The monolateral external fixator is modeled by beam elements. The loading condition assumed is in accordance with [7]. The reduction process of the fracture is simulated by using a non-linear material law. Both stress-strain patterns in this zone and the distribution and intensity of the forces loading the pins are determined at every stage of healing.

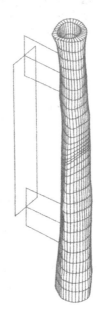

Figure 1. Femoral bone and fixation device: finite element model.

In the light of the results obtained, a more detailed model is adopted regarding the pin and the boundary region of bone, in which localised effects are redistributed (fig. 2).

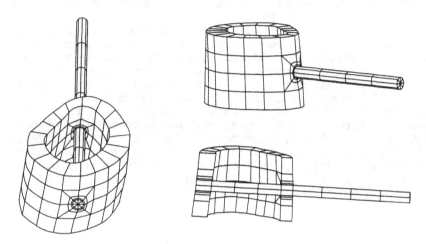

Figure 2. Finite element model for investigation of pin-bone interaction phenomena.

Load intensity is deduced by preliminary analysis and the values assumed can be compared with those reported in [43,44]. The bone layer nearest the pin is differentiated in the model in order to offer the possibility of a specific assumption of a material law and for better evaluation of results. Both pin and bone are now modeled by second-order isoparametric solid elements. Particular attention is devoted to the modeling of the zone in which unilateral contact may arise, defined by interface elements with nine nodes per face, with a mid-face node to give satisfactory modeling in cases of partial contact at the element surface. It is therefore necessary to use brick elements with at least one central node on the face where contact may arise and one node at the centroid, to allow compatibility with these interface elements. The problem is investigated assuming first an orthotropic linear elastic law for cortical bone and an isotropic one for the pin [21,25]. According to the stress-strain levels reached in this way, it seems reasonable to assume post-elastic behaviour for that portion of bone immediately around the pin, where the stress field is directly influenced by contact phenomena. The constitutive law assumed, that considers an isotropic hardening effect, is in accordance with [20,21,25,26]. A coupled geometric and material non-linear approach is thus induced.

THEORETICAL ASPECTS

Particular attention is paid to the theoretical formulation of unilateral contact phenomena, as a strategy to model the

geometric non-linear contact problem.
The numerical technique used here evaluates the relative
displacements at each integration point of the surface
elements. Lagrange multipliers are introduced at the
integration points where contact takes place to impose the
constraint to avoid surface interpenetration. The integration
scheme is Simpson's rule, in order to have accurate solutions
for the contact area and pressure distribution between
surfaces.
Surface contact problems are considered, as first approach, by
assuming small displacement and small displacement gradient,
with only negligible sliding between the two facing surfaces,
like the rotation of the normal to the surfaces. Further
conditions describe the contact between surfaces in tha case
of finite relative displacements, for both local deformation
and relative sliding possibly associated with a friction law.
Contacting bodies may be considered as one deformable and the
other rigid. It is also possible to configurate the
interaction between two deforming surfaces with finite
displacements and strains. The latter is the most general
case and requires modeling of contacts between deformable
bodies with possible finite slidings, and involving throughout
the analysis the automatic application and removal of
kinematic constraints. In order to define the problem it must
be specified if the contacting surfaces remain almost the same
or if they can vary during the analysis, which is of peculiar
interest in dynamic cases.
The first approach described here is suitable for this pin-
bone contact analysis.
The virtual work variation associated to the contact surfaces
may be defined, distingushing the contribution of contact
pressure from that of shear stress, as:

$$\delta W = \delta W_{sn} + \delta W_{st} = \int_S (p\delta h + t\delta r)\ dS \tag{1}$$

where:

δW_{sn}	virtual work variation associated with normal contact
δW_{st}	virtual work variation associated with shear action
p	pressure between contact surfaces
h	relative distance between contact surfaces
t	shear stress
r	relative tangential shear displacement
S	area of contact surface

Contact pressure p between two surfaces is a function of the
relative distance of the surfaces, that is p=p(h), and their
configuration. In place of a step function to define the
contact condition for rigid surfaces, pressure may be assumed
to follow an exponential trend in the contact phase. This
approach can also help convergence of computational procedure,
so as to allow partial overclosures.
In the case of a rigid surface, any positive pressure can be
transmitted when the surfaces come in contact at a point. The
value of p is null if no contact arises, otherwise, as a

function of relative distance, it becomes indeterminate. In order to use the rigid surface for the definition of contact pressure, independent variable p_m may be introduced and constraint $p=p_m$ imposed by a Lagrange multiplier. The term of virtual work variation related to normal pressure may be expressed as:

$$\delta W_{sn,m} = \int_S (p_m\delta h+\delta m(p-p_m)) \, dS \tag{2}$$

where m is the Lagrange multiplier.
Equation (1) can be expressed in the form:

$$\delta W_m = k\delta W_{sn}+(1-k)\delta W_{sn,m}+\delta W_{st} \tag{3}$$

where k is a scalar factor defined as the ratio between the evolutive stiffness of the surface in contact, that is $k=s_0/s$. In the first-order expansion about the approximate solution, the Newton scheme leads to discretized equilibrium equations for the contact potential in the form:

$$d\delta W_m = -\delta W_m \tag{4}$$

This may be expressed as:

$$\int_S [(kdp+(1-k)dp_m)\delta h+dt\delta r+(1-k)(dp-dp_m)\delta m+(kp+(1-k)p_m)d\delta h+$$

$$+td\delta r+[(kp+(1-k)p_m)\delta h+t\delta r]dS_0] \, dS =$$

$$-\int_S [(kp+(1-k)p_m)\delta h+t\delta r+(1-k)(p-p_m)\delta m] \, dS \tag{5}$$

where dS_0 is the variation in the dimension of surfaces in contact. Assuming $s\delta m=\delta p_m$ and $p_n=kp+(1-k)p_m$, the contribution to the Newton scheme becomes:

$$\int_S [(ksdh+(1-k)dp_m)\delta h+dt\delta r+(1-k)(dh-dp_m/s)\delta p_m+p_nd\delta h+$$

$$+td\delta r+(p_n\delta h+t\delta r)dS_0] \, dS =$$

$$-\int_S [p_n\delta h+t\delta r+(1-k)(p-p_m)\delta p_m/s] \, dS \tag{6}$$

and, recalling that $k=s_0/s$, equation (6) may be written as:

$$\int_S [s_0dh+(1-s_0/s)dp_m]\delta h+dt\delta r+(1-s_0/s)(dh-dp_m/s)\delta p_m+p_nd\delta h+$$

$$+td\delta r+(p_n\delta h+t\delta r)dS_0] \, dS =$$

$$-\int_S [(p_n\delta h+t\delta r+(1-s_0/s)(p-p_n)\delta p_m/s] \, dS \tag{7}$$

Since s, that reaches infinity for the hard surface, appears only at the denominator, this approach is suitable for both finite and infnite values of surface rigidity. In the present formulation, also in accordance with the assumptions made, no friction phenomena are taken into account.

Plastic material behaviour implies an improvement of mesh adopted for better results in stress distribution in plastic field. This implies in particular mesh redefinition in the interface and neighbouring zone so as to have a finite element mesh capable of interpreting the plastic trend in the non-linear response. The multipoint constraint technique is adopted for mesh refinement only in the zone where plastic behaviour can be found, to minimize further computational effort.

ANALYSIS DEVELOPED

This analysis leads to results describing pin-bone interaction phenomena. The stress fields induced at the contact surface are shown at different locations along the contact zone at subsequent load levels. In the hypothesis of having linear elastic behaviour, material characteristics are homogeneous in the whole region around the pin. Values are assumed according to Ashman et al. and reported in [7]. Furthermore, a reduced elastic modulus of about sixty per cent is assumed in the first layer. This is a second approach to simulate degeneration phenomena that may occur in the zone in contact with the pin.

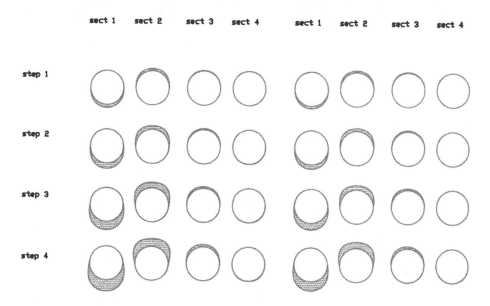

a)   b)

Figure 3. Pressure at interface level for elastic material a) and with the reduced elasticity modulus zone b).

Results are reported in fig. 3 for the first case a) and for the second b), for subsequent loading steps on the pin (32 N, 56 N, 88 N, 120 N respectively; pin diameter 3 mm) at four pin-bone intersections. Maximum values at subsequent steps are: 90.1 MPa, 157.7 MPa, 247.6 MPa., 338.2 MPa for case a), and 74.7 MPa, 130.9 MPa, 205.7 MPa, 280.5 MPa for case b). Bone behaviour in the plastic phase is subsequently investigated, according to the response of the previous analysis in which, the elastic limit of the material is overcome under the assumed loads. A plastic material law according to Cezayirlioglu et al. [20] is adopted. Results are reported in fig. 4 at the previously defined loading steps and locations. Maximum values at subsequent steps are 145.3 MPa, 196.1 MPa and 238.2 MPa.

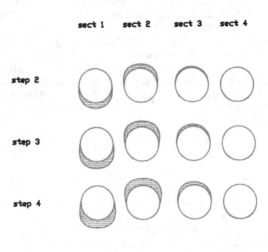

Figure 4. Pressure at interface level for plastic material.

Figure 5 shows radial relative displacements at interface level, as effective clearance between bone and pin, for elastic a) and elastoplastic material law b) respectively. Maximum values are found at the fourth step as 0.07 mm (case a) and 0.08 mm (case b).

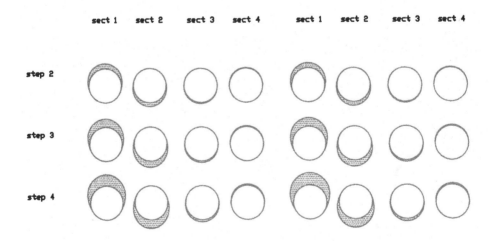

a)   b)
Figure 5. Clearance at interface level for elastic a) and elastoplastic material b).

It is interesting to point out that pin vertical displacements, evaluated at a distance of 35 mm from bone border, present significant differences if continuity is assumed between bone and pin instead of using interface elements. In the comparison of results obtained adopting different models, a variation of about twentyfive per cent is found and this fact confirms the necessity to use a contact strategy formulation.

Figure 6 reports material stress status in terms of Von Mises equivalent stress in the region near the pin, for elastic material a) and elasto-plastic material b). Maximum values for subsequent steps are 158.4 MPa, 244.4 MPa and 333.3 MPa for elastic material, and 155.5 MPa, 176.8 MPa and 180.0 MPa for elastoplastic material.

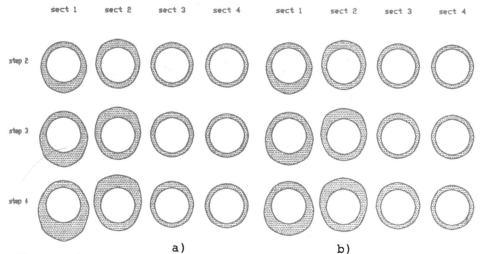

a)   b)

Figure 6. Stress in bone region near pin for elastic a) and elastoplastic material b).

CONCLUSIONS

It is necessary to point out that numerical formulation presents particular complexities, due to the specific configuration of a coupled geometric and non-linear problem. In the preliminary phase, it is difficult to define the geometric configuration to be used in a three-dimensional model for the orthotropic material law in elastic phase while following the precision required for the contact problem approach. The assumption of material characteristics in the post-elastic response of bone is not easy to define, as it depends on several factors and also directly on the experimental procedure adopted by different authors. The numerical approach implies considerable effort in defining both preliminary and detailed models. In the latter, particular attention must be paid to coupled non-linearities that also involve significant work, since several cases must be considered to report representative results for the problem trated.

The theoretical approach is complex even when partial simplification is assumed, as no friction phenomena are taken into account. The consideration of contact phenomena remains fundamental for the interpretation of pin-bone interaction. In fact, the approximations that may be induced must be evaluated for their influence on final results, but a reasonable tolerance margin should be considered in this kind of problem, in relation to the indetermination of both material properties and geometry characteristics and because of possible clinical aspects.

Experimental tests developed with the aim of evaluating deformational behaviour of fractured bone treated with external fixators are certainly influenced by pin-bone interaction phenomena, to such an extent that no correct evaluation can be performed neglecting this aspect. In this sense, another factor must be evaluated: creep response of bone. Short-term creep, if present, may have a relevant effect on experimental data.

All these aspects are necessary for the characterization of external fixators and for correct evaluation of the action that they exert on the healing of bone fractures.

REFERENCES

1. Aro HT, Hein TY, Chao EY. MECHANICAL PERFORMANCE OF PIN CLAMPS IN EXTERNAL FIXATORS. Clin Orthop 1989; 248: 246-253.
2. Chao EY, Kasman RA, An KR. RIGIDITY AND STRESS ANALYSES OF EXTERNAL FRACTURE FIXATION DEVICES - A THEORETICAL APPROACH. J Biomech 1982; 15: 971-983.
3. Chao EYS, An K-N. BIOMECHANICAL ANALYSIS OF EXTERNAL FIXATION DEVICES FOR THE TREATMENT OF OPEN BONE FRACTURES. In Finite Elements in Biomechanics, ed. by Gallagher RH, Simon BR, Johnson PC, Gross JF; John Wiley & Sons, New York; 1982: 195-222.
4. Kasman RA, Chao EYS. FATIGUE PERFORMANCE OF EXTERNAL FIXATOR PINS. J Orthop Res 1984; 2: 377-384.
5. Kempson GE, Campbell D. THE COMPARATIVE STIFFNESS OF EXTERNAL FIXATION FRAMES. Injury 1981; 12: 297-304.
6. McCoy MT, Chao EY, Kasman RA. COMPARISON OF MECHANICAL PERFORMANCE IN FOUR TYPES OF EXTERNAL FIXATORS. Clin Orthop 1983; 180: 23-33.
7. Meroi EA, Natali AN. A NUMERICAL APPROACH TO THE BIOMECHANICAL ANALYSIS OF BONE FRACTURE HEALING. J Biomed Eng 1989; 11: 390-397.
8. Aalto K, Holmstrom T, Karaharju E, Joukainen J, Paavolainen P, Slatis P. FRACTURE REPAIR DURING EXTERNAL FIXATION. Acta Orthop Scand 1987; 58: 66-70.
9. Carter DR. MECHANICAL LOADING HISTORY AND SKELETAL

BIOLOGY. J Biomech 1987; 20: 1095-1109.

10. Chao EY, Aro HT, Lewallen DG, Kelly PJ. THE EFFECT OF RIGIDITY ON FRACTURE HEALING IN EXTERNAL FIXATION. Clin Orthop 1989; 241: 24-35.

11. Davy DT, Connolly JF. THE BIOMECHANICAL BEHAVIOUR OF HEALING CANINE RADII AND RIBS. J Biomech 1982; 15: 235-247.

12. Gilbert JA, Dahners LE, Atkinson MA. EFFECT OF EXTERNAL FIXATION STIFFNESS ON EARLY HEALING OF TRANSVERSE OSTEOTOMIES. J Orthop Res 1989; 7:389-397.

13. Lewallen DG, Chao EY, Kasman RA, Kelly PJ. COMPARISON OF THE EFFECTS OF COMPRESSION PLATES AND EXTERNAL FIXATORS ON EARLY BONE-HEALING. J Bone Jt Surg 1984; 66A: 1084-1091.

14. Paavolainen P, Slatis P, Karaharju E, Holmstrom T. THE HEALING OF EXPERIMENTAL FRACTURES BY COMPRESSION OSTEOSYNTHESIS - I. TORSIONAL STRENGTH. Acta Orthop Scand 1979; 50: 369-374.

15. Perren SM. PHYSICAL AND BIOLOGICAL ASPECTS OF FRACTURE HEALING WITH SPECIAL REFERENCE TO INTERNAL FIXATION. Clin Orthop 1979; 138: 175-196.

16. Richard J. STIFFNESS IN HEALING FRACTURES. CRC Critical Reviews in Biomedical Engineering 1987; 15: 145-185.

17. White III AA, Panjabi MM, Southwick WO. THE FOUR BIOMECHANICAL STAGES OF FRACTURE REPAIR. J Bone Jt Surg 1977; 59A: 188-192.

18. Wu JJ, Shyr HS, Chao EY, Kelly PJ. COMPARISON OF OSTEOTOMY HEALING UNDER EXTERNAL FIXATION DEVICES WITH DIFFERENT STIFFNESS CHARACTERISTICS. J Bone Jt Surg 1984; 66A: 1258-1264.

19. Ashman RB, Cowin SC, Van Buskirk WC, Rice JC. A CONTINUOS WAVE TECHNIQUE FOR THE MEASUREMENT OF THE ELASTIC PROPERTIES OF CORTICAL BONE. J Biomech 1984; 17: 349-361.

20. Cezayirlioglu H, Bahniuk E, Davy DT, Heiple KG. ANISOTROPIC YIELD BEHAVIOR OF BONE UNDER COMBINED AXIAL FORCE AND TORQUE. J Biomech 1985; 18: 61-69.

21. Cowin SC. THE MECHANICAL PROPERTIES OF BONE. In: Proc. Int. Symp. Mechanical Behaviour of Structured Media, part A; Ottawa, May 18-21, 1981: 151-184.

22. Heiple KG, Torzilli PA, Takebe K, Burstein AH, Zika JM. THE MATERIAL PROPERTIES OF IMMATURE BONE. J Biomech Eng1982; 104: 12-20.

23. Katz JL, Meunier A. THE ELASTIC ANISOTROPY OF BONE. J Biomech 1987; 20: 1063-1070.

24. Knets IV, Pfafrod GO, Saulgozis JZ. DEFORMATION AND FRACTURE OF HARD BIOLOGICAL TISSUE. Zinatne, Riga, 1980, 319.

25. Natali AN, Meroi EA. A REVIEW OF BIOMECHANICAL PROPERTIES OF BONE AS A MATERIAL. J Biomed Eng 1989; 11: 266-276.

26. Reilly DT, Burstein AH. THE ELASTIC AND ULTIMATE PROPERTIES OF COMPACT BONE TISSUE. J Biomech 1975; 8: 393-405.

27. Van Buskirk WC, Cowin SC, Ward RN. ULTRASONIC MEASUREMENT OF ORTHOTROPIC ELASTIC CONSTANTS OF BOVINE FEMORAL BONE. J Biomech Eng 1981; 103: 67-71.

28. Cowin SC, Firoozbakhsh K. BONE REMODELING OF DIAPHYSIAL SURFACES UNDER COSTANT LOAD: THEORETICAL PREDICTIONS. J

Biomech 1981; 14: 471-484.

29. Cowin SC. MECHANICAL MODELING OF THE STRESS ADAPTATION PROCESS IN BONE. Calcif Tissue Int 1984; 36: 98-103.

30. Cowin SC, Hart RT, Balser JR, Kohn DH. FUNCTIONAL ADAPTATION IN LONG BONES: ESTABLISHING IN VIVO VALUES FOR SURFACE REMODELING COEFFICIENTS. J Biomech 1985; 18: 665-684.

31. Hart RT, Davy DT, Heiple KG. A COMPUTATIONAL METHOD FOR STRESS ANALYSIS OF ADAPTIVE ELASTIC MATERIALS WITH A VIEW TOWARD APPLICATIONS IN STRAIN-INDUCED BONE REMODELING. J Biomech Eng 1984; 106: 342-350.

32. Hart RT, Davy DT, Heiple KG. MATHEMATICAL MODELING AND NUMERICAL SOLUTION FOR FUNCTIONALLY DEPENDENT BONE REMODELING. Calcif Tissue Int 1984; 36: 104-109.

33. Hassler CR, Rybicki EF, Cummings KD, Clark LC. QUANTIFICATION OF BONE STRESSES DURING REMODELING. J Biomech 1980; 13: 185-190.

34. Meade JB, Cowin SC, Klawitter JJ, Van Buskirk WC, Skinner HB. BONE REMODELING DUE TO CONTINUOSLY APPLIED LOADS. Calcif Tissue Int 1984; 36: 25-30.

35. Misra JC, Samanta S. EFFECT OF MATERIAL DAMPING ON BONE REMODELING. J Biomech 1987; 20: 241-249.

36. Beaupre GS, Hayes WC, Jofe MH, White AA. MONITORING FRACTURE SITE PROPERTIES WITH EXTERNAL FIXATION. J Biomech Eng 1983; 105: 120-126.

37. Bourgois R, Burny F. MEASUREMENTS OF THE STIFFNESS OF FRACTURE CALLUS IN VIVO. A THEORETICAL STUDY. J Biomech 1972; 5: 85-91.

38. Evans M,, Kenwright J, Cunningham JL. DESIGN ANF PERFORMANCE OF A FRACTURE MONITORING TRANSDUCER. J Biomed Eng 1988; 10: 64-69.

39. Jorgesen TE. MEASUREMENTS OF STABILITY OF CRURAL FRACTURES TREATED WITH THE HOFFMANN OSTEOTAXIS. Acta Orth Scand 1972; 43: 188-218.

40. Karaharju EO, Aalto K. THE DEFORMATION OF EXTERNAL FIXATION DEVICES DURING LOADING. International Orthopaedics 1983; 7: 179-183.

41. Pelker RR, Saha S. WAVE PROPAGATION ACROSS A BONY DISCONTINUITY SIMULATING A HEALING FRACTURE. J Biomech 1985; 18: 745-753.

42. Van der Perre G, Borgwardt-Christensen A. MONITORING OF FRACTURE HEALING BY VIBRATION ANALYSIS AND OTHER MECHANICAL METHODS. In: Proc. of the specialists consensus meeting, Leuven, 1985: 195.

43. Huiskes R, Chao EYS, Crippen TE. PARAMETRIC ANALYSES OF PIN-BONE STRESSES IN EXTERNAL FRACTURE FIXATION DEVICES. J Orthop Res 1985; 3: 341-349.

44. Huiskes R, Chao EYS. GUIDELINES FOR EXTERNAL FIXATION FRAME RIGIDITY AND STRESSES. J Orthop Res 1986; 4: 68-75.

45. Manley MT, Hurst L, Hindes R, Dee R, Chiang FP. EFFECTS OF LOW-MODULUS COATINGS ON PIN-BONE CONTACT STRESSES IN EXTERNAL FIXATION. J Orthop Res 1984; 2: 385-392.

THE ELLIPSE OF ELASTICITY APPLIED TO A DENTAL PROSTHESIS

K. R. Williams* and G. Pallotti**
*Department of Basic Dental Science,
University of Wales College of Medicine
Heath Park, Cardiff.
**Department of Medical Physics, University of Bologna
Via Irnerio, 46–40126 Bologna, Italy

ABSTRACT

The early elastic theories of Culmann and associated workers has been reviewed with particular reference to beam bending. The theory allows an analytical and geometric description of the elastic deformation of beams of various geometries under a variety of loading conditions.

The calculated strains have been compared with numerically computed data on the same structures as a means of validation of implant and prosthesis behaviour. Results indicate that the FEM needs to be carefully examined particularly at large elastic strains even for structures of simple geometry.

INTRODUCTION

Many modern engineering structures are designed and optimised using the Finite Element Method (FEM). FEM analysis can deal with linear and non–linear materials both statically and dynamically loaded. A range of fluid and heat flow problems are also readily handled using this methodology [1]. Validation of the numerical results is available to the Engineer either by simulated laboratory testing or by strain gauging the loaded structure.

However, Clinicians, Surgeons and Scientists are not able to validate the FEM analysis of implants in vivo for very obvious reasons. If an analytical description of the loads and displacements in and around the implant were possible for a few simple cases, then it would be possible to validate the FEM results and possibly extend the analysis to more complex situations.

It has occurred to the authors that the early geometric formulations for the design of structures based on elasticity theory may help in FEM validation. Examples can include

(i)   elastic deformation of clasp arms in partial dentures
(ii)   tooth cusp movement following restoration by modern adhesive techniques
(iii)   orthodontic forces delivered by various fixed and removeable appliances.

The early elastic theory of Culmann [2], developed by Ritter [3] and Guidi [4] can be readily applied to the above problems. As proposed by Belluzzi [5] "the theory of the Ellipse of Elasticity [2] constitutes a most elegant method for the study of the deformation of beams with rectilinear or curvilinear geometry and with sections constant or variable".

The theory allows the determination of the rotation and the movement of a section from a given force which in turn can serve to validate FEM analysis of certain clinical situations

THEORY OF THE ELLIPSE OF ELASTICITY

Consider a beam (Figure 1) fixed at one end B and loaded at the other ;   then the point A will tend to move around a centre of rotation C.   In addition, all the elastic displacements of points connected to A provided by all possible loads and lines of action will form an ellipse of elasticity (Figure 1).   This is the ellipse of elasticity relative to the section A, [2].

Rotation Caused by a Couple   (Figure 2)

Point A rotates around the centre of the ellipse, O, through an angle φ proportional to the moment M

$$\varphi = M \; \mathrm{G} \qquad\qquad\qquad (1)$$

The constant G is called the elastic weight of the beam.

Movement Caused by a Couple

Section A rotates around O, and the movement δ is in a direction normal to $Oa = d_a$.

Since φ is small

$$\delta = \varphi \, d_a = M.\underset{\cdot}{G} \, d_a \tag{2}$$

To project δ on a direction x then

$$\delta_x = \delta \cos \alpha = M \, \underset{\cdot}{G} \, d_a \cos \alpha$$

but $\qquad d_a \cos \alpha = d_x$

thus $\qquad \delta_x = M.\underset{\cdot}{G} \, d_x \tag{3}$

Rotation Caused by a Force   (Figure 3)

The rotation of A and every point connected with it, occurs around the antipole C of r. In order to determine φ, we transport P parallel to itself and passing through the elastic centre O and add a couple $M = Pd_r$. The force P passing through O produces only a translation and no rotation. The couple M produces a rotation

$$\varphi = P.\underset{\cdot}{G} \, d_r \tag{4}$$

Movement Caused by a Force

A general point 'a' undergoes a movement δ normal to $Ca = d'_a$

$$\delta = \varphi \, d'_a = P. \, \underset{\cdot}{G} \, d_r \, d'_a$$

The component δ_x of the movement δ is

$$\delta_x = \delta \cos \alpha = P \, \underset{\cdot}{G} \, d_r \, d'_a \cos \alpha$$

Since $d'_a \cos \alpha$ is the distance d'_x then

$$\delta_x = P. \, \underset{\cdot}{G}. \, d_r \, d'_x \tag{5}$$

DECOMPOSITION OF A BEAM INTO SECTIONS

For beams of varying sections and particularly as a means of geometric solution, a process of division can be utilised and the theory applied to each section thus.

Rotation Provided by a Couple

Figure 4 shows a beam decomposed into sections Δs, which rotate around a point O of the axis, through an angle $\Delta\varphi$ proportional to M. If ΔG or w is now the constant of proportionality, we have

$$\Delta\varphi = M.w$$

The rotation of the section A from all elements Δs of the beam is

$$\varphi = \Sigma\Delta\varphi = \Sigma Mw = M.\Sigma w \tag{6}$$

Movement Caused by Couple

The point 'a' connected rigidly to A undergoes a movement

$$\Delta\delta_x = \Delta\varphi.\ d_x = M\ w\ d_x$$

The total movement from all elements,

$$\delta_x = \Sigma\Delta\delta_x = \Sigma Mwd_x = M.\ \Sigma\ wd_x \tag{7}$$

Rotation Caused by a Force

A force P acting at 'A' through a line of action r (Figure 4); the bending moment on element Δs is $M = Pd_r$

The rotation is

$$\Delta\varphi = Mw = Pd_r w$$

The total rotation is

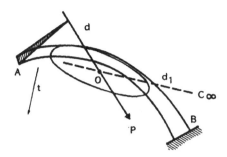

Figure 1. The ellipse of elasticity
for the beam illustrated.

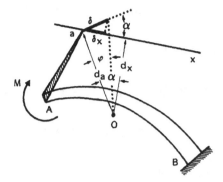

Figure 2. Rotation and movement
caused by a couple
applied to a beam.

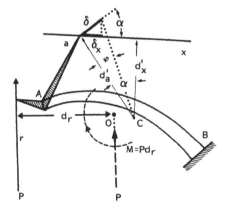
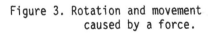

Figure 3. Rotation and movement
caused by a force.

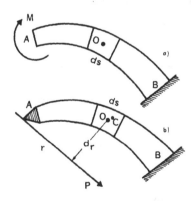

Figure 4. Decomposition of a beam
into sections.

$$\varphi = \Sigma\Delta\varphi = \Sigma \ P \ d_r \ w = P.\Sigma \ wd_r \qquad\qquad (8)$$

Movement Caused by a Force

The point 'a' undergoes a movement in the direction x of

$$\Delta\delta_x = \Delta\varphi.d_x = Pd_r \ w \ d'_x$$

Total movement

$$\delta_x = \Sigma\Delta\delta_x = \Sigma \ P \ d_r \ w \ d'_x$$

$$= P. \ \Sigma \ w \ d_r \ d'_x \qquad\qquad (9)$$

ELASTIC CONSTANTS

The elastic constants necessary for inclusion in the derived equations are most easily explained by reference to a simple cantilever beam under a variety of loading conditions see Figure 5.

(a) Elastic centre of gravity

Beam of length ℓ, with a couple applied at A. The centre of rotation along the beam is

$$d = f/\varphi = \frac{M\ell^2/2EJ}{M\ell/EJ} = \frac{\ell}{2}$$

(b) Elastic weight

If G is the elastic weight of the beam and J the moment of inertia of the section with respect to the axis of bending and E the modulus of elasticity, then

$$\varphi = M.G \quad ; \quad \varphi = \frac{M\ell}{EJ}$$

thus $G = \ell/EJ$.

For the case of division of the beam into sections ds or Δs then $d\varsigma = \dfrac{ds}{EJ}$ and $\Delta\varsigma = w = \dfrac{\Delta s}{EJ}$

(c) <u>Transverse semi-axis of the ellipse</u>

If ρ_{x_o} is the semiaxis of the ellipse the movement of A is

$$\delta_{x_o} = P.\ J_{x_o} = P.\ \varsigma\ \rho_{x_o}^2$$

but $\qquad \delta_{x_o} = \dfrac{P\ell}{EA}$

thus $\qquad \varsigma\rho_{x_o}^2 = \dfrac{\ell}{EA} \qquad$ or $\qquad \dfrac{\ell}{EJ}\ \rho_{x_o}^2 = \dfrac{\ell}{EA}$

from which $\rho_{x_o} = \left[\dfrac{J}{A}\right]^{\frac{1}{2}}$

(d) <u>Longitudinal semi-axis of the ellipse</u>

$$\delta_y = P.J_y = P\left[Jy_o + \varsigma\left[\ell/2\right]^2\right]$$

$$= P\left[\varsigma\rho_{y_o}^2 + \varsigma\dfrac{\ell^2}{4}\right]$$

$$\delta_y = \dfrac{P\ell^3}{3EJ} + \chi\ \dfrac{P\ell}{\varsigma A}$$

Hence $\qquad \dfrac{\ell}{EJ}\left[\rho_{y_o}^2 + \dfrac{\ell^2}{4}\right] = \dfrac{\ell^3}{3EJ} + \chi\ \dfrac{\ell}{\varsigma A}$

$$\rho_{y_o} = \left[\dfrac{\ell^2}{12} + \chi\ \dfrac{E}{\varsigma}\ \rho_{x_o}^2\right]^{\frac{1}{2}}$$

if $\ell \gg \rho_{x_o}$ then $\rho_{y_o} = \dfrac{\ell}{\sqrt{12}}$

APPLICATION TO CLASP DEFORMATION

The theory can readily deal with elastic deformation of variously shaped clasp designs. Figure 6 shows a variety of loading conditions for a full 360° clasp. These are summarised as follows:

a) The pole for rotation is found on the diameter, normal to the force.

$$\overline{GC} = r^2/2d_r \quad ; \quad \varphi = P.\varsigma\,d_r$$

The movement is normal to CA.

b) and c) The pole of rotation is at infinity along the diameter normal to P. Thus 'A' undergoes a translation parallel to the force.

$$\delta = P.J_r = P.\varsigma r^2/2$$

$$= P.\pi r^3/EJ$$

d) Vertical force through A

The centre of rotation C is found on the horizontal diameter distance r/2 from G

$$\delta = P.J_r = P.\varsigma r \left[\frac{3r}{2}\right]$$

$$= P.\frac{3\pi r^3}{EJ}$$

e) Force at 45° to A

The centre of rotation C is diametrically opposite the point of contact at a distance r/2 on the horizontal from A and 3r/2 on the vertical through A. Hence the vertical component of movement is three times the horizontal.

f) Couple applied at A

The section 'A' rotates around G

$$\varphi = M.\varsigma = M.\frac{2\pi r}{EJ}$$

Semi-circular Clasp

In a similar way the deflection of a semicircular clasp arm can be analysed, thus, the elastic centre of gravity is defined as; (Figure 7)

$$\overline{OG} = 2r/\pi = 0.6366r$$

$$\rho_{y_o} = \frac{r}{\sqrt{2}} = 0.707r$$

$$\rho_{x_o} = \frac{\sqrt{\pi^2 - 8}}{\sqrt{2\pi}} \cdot r = 0.308r$$

$$G = \frac{\pi r}{EJ} \; ; \; Jy_o = \frac{\pi r^3}{2EJ} \; ; \; Jx_o = \frac{\pi^2 - 8}{2\pi} \cdot \frac{r^3}{EJ}$$

In order to determine the rotation and movement of section A for various couples and forces we have

a) Couple M

Section A rotates by φ around G and the movement is normal to GA hence

$$\varphi = \pi \frac{Mr}{EJ}$$

$$\delta v = \pi \cdot \frac{Mr^3}{EJ} \; ; \qquad \delta_h = \frac{2Mr^3}{EJ}$$

b) Vertical force P_V

This causes a rotation φ of A around the pole C' where GC' = r/2 and the movement is normal to C'A.

Hence

$$\varphi = \pi \frac{P_V r^2}{EJ}$$

$$\delta r = \frac{3\pi}{2} \cdot \frac{P_V r^3}{EJ} \; ; \qquad \delta_h = 2 \frac{P_V r^3}{EJ}$$

c) Horizontal force P_h

This causes a rotation φ around C"

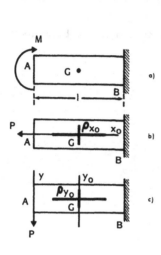

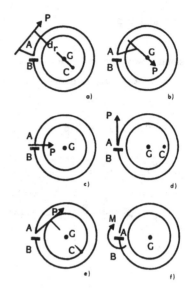

Figure 5. Definition of the elastic constants.

Figure 6. Theory of the ellipse applied to 360° clasps.

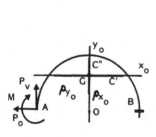

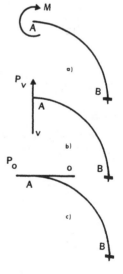

Figure 7. Movement of a 180° clasp caused by forces in the vertical and horizontal plane.

Figure 8. Couple and forces applied to a 90° clasp.

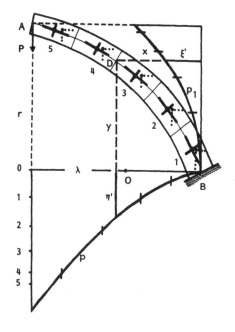

Figure 9. Calculation graphically of the vertical and horizontal movement of a generalised beam under point loading.

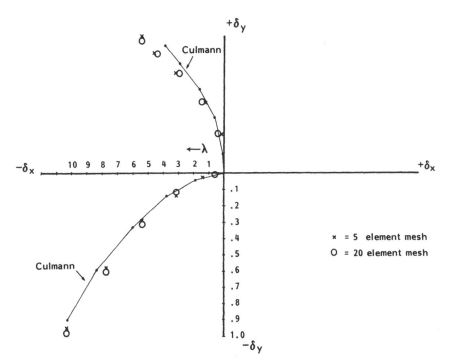

Figure 10. Comparison of the Culmann beam deflection (see Figure 9) with the finite element analysis.

$$\overline{GC''} = \rho_{x_o}^2 / \overline{OG} = \left[\pi^2 - 8\right]r/4\pi$$

$$\overline{OC''} = \overline{OG} + \overline{GC''} = \pi r/4$$

and the movement is normal to C''A.

Hence

$$\varphi = \frac{2P_h r^2}{EJ}$$

$$\delta v = \frac{2P_h r^3}{EJ} \quad ; \quad \delta_h = \frac{\pi}{2} \cdot \frac{P_h r^3}{EJ}$$

Quarter Circle Clasp

Finally for a quarter circle clasp and by similar reasoning we obtain, (see Figure 8)

a) $\qquad \varphi = \frac{\pi}{2} \cdot \frac{Mr}{EJ}$

$$\delta v = \frac{Mr^2}{EJ} \quad ; \quad \delta_h = \frac{\pi - 2}{2} \cdot \frac{Mr^2}{EJ}$$

b) $\qquad \varphi = \frac{P_v r^2}{EJ}$

$$\delta v = \frac{\pi}{4} \cdot \frac{P_v r^3}{EJ} \quad ; \quad \delta h = \frac{1}{2} \cdot \frac{P_v r^3}{EJ}$$

c) $\qquad \varphi = \frac{\pi-2}{2} \cdot \frac{P_h r^2}{EJ}$

$$\delta v = \frac{1}{2} \cdot \frac{P_h r^3}{EJ} \quad ; \quad \delta_h = \frac{3\pi - 8}{4} \cdot \frac{P_h r^3}{EJ}$$

DISCUSSION

These formulae of Culmann [2] should be particularly useful in designing clasp structures of various shapes and materials. It is important in such designs that the material does not exceed its elastic limit when sprung under the bulbous part of the tooth in order to secure the appliance. Bates [6] measured experimentally the tip displacements of a clasp material after casting into various shapes. Thus a straight Wipla 16, 30 mm long clasp has a deflection at the tip at the elastic limit of 0.92 mm. When bent around a tooth (270°) the deflection was reduced to 0.25 mm. Thus the change in geometry has effectively increased the rigidity by approximately four times.

Using the theory of Culmann it is possible to calculate the relative increases in rigidity for a clasp of constant length as:

	Straight	$\frac{1}{4}$ circle	$\frac{1}{2}$ circle	full circle
$\delta =$	1/3	$2/\pi^2$	$3/2.1/\pi^2$	$3/8.1/\pi^2$
	0.33	0.2	0.15	.04
relative rigidity	1	1.75	2.2	8.2

which is in good agreement with experiment [6]. However the main purpose of this work is to examine the analytical approach to elastic deformation as a means of validating the concepts and results of finite element analysis as applied to beam bending.

This has been attempted using the construction shown in Figure 9 in which the encastrated beam has been decomposed into five sections as a means of using the ellipse theory according to equations 6, 7, 8, 9. The δ_x and δ_y displacements have been extracted from Figure 9 and replotted in Figure 10 together with the results of a finite element programme. Initially the finite element analysis was carried out using only 5 isoparametric elements of exactly similar shape to the decomposition used for the elasticity theory and shown in Figure 9. As a means of checking the accuracy of the FEA results the beam was further divided in 20, 40 and 80 elements with no significant change in the displacements (see Figure 10).

The accuracy of the Culmann [2] displacement is claimed to be better than 1%, while the FEA results although good at small displacements $\delta \leqslant .04$ mm, become increasingly poor at larger displacements. This is probably a consequence of geometric instabilities in the mesh at larger elastic strains, suggesting a linearly large deformation analysis using an updated Lagrangian formulation is necessary.

CONCLUSIONS

An elegant theory of elastic deformation developed by Culmann [2] has been reviewed with a view to its use in describing the deformation behaviour of relatively simple geometric dental appliances. The theory may be particularly useful in clasp and orthodontic appliance design as described.

Furthermore, the theory provides a means of validating the FEA for certain prosthesis again of reasonably simple geometry. In particular the deformation of a curved beam has been examined analytically [2] and numerically, with certain errors at large elastic deformations in the FEM results.

It is suggested that linearly large FEM deformation analysis be applied to this particular application.,

REFERENCES

1. Desai, C. S. and Abel, J. F., Introduction to the finite element method, Van Nostrand Reinhold Co., 1972.

2. Culmann, K., Die graphische Statik, Meyer–Zeller, Zurigo, 1875.

3. Ritter, W., Der elastische Bogen, Meyer–Zeller, Zurigo, 1886.

4. Guidi, C., L'ellisse di elsticità nella Scienza delle costruzioni, Torino, 1904.

5. Belluzzi, O., La teoria dell'ellisse di elasticità Vol. II, Nicola Zanichichelli, Bologna, 1956.

6. Bates, J. F., Retention of partial dentures, Brit. Dent. J., 1980, 149, 171–176.

ACKNOWLEDGEMENTS

This work was made possible by financial support to on eof the authors (KRW) by the Royal Society, London and the Accademia Nazionale dei Lincei, Roma at the University of Bologna in May and June 1990.

THREE DIMENSIONAL FINITE ELEMENT ANALYSIS OF A SAGGITAL SPLIT OSTEOTOMY

G. WHITHAM*, J. MIDDLETON*, G. R. BARKER**
*Department of Civil Engineering, University College of Swansea,
Singleton Park, Swansea
**Dental School, University of Wales College of Medicine,
The Heath, Cardiff

ABSTRACT

A three dimensional finite element model was used to investigate the response of a human mandible subject to sectioning as for a Saggital Split Osteotomy. The material consists of relative proportions of cortical, cancellous and cavitational cancellous bone and the shape of the mandible was modelled using 8 noded linear elements. Displacements and stresses of the mandible are calculated for stability aided by zero, one or two Champy plates. Loading was transmitted through muscular forces which were nodally applied to provide a bite force on the first pre-molar.

INTRODUCTION

The finite element method (FEM) is now well established as an accurate technique for modelling the physical behaviour of complex three dimensional bodies. Recently various researchers have applied three dimensional FEM modelling to large scale biomechanical problems and the results have provided clinicians with information which was hitherto unavailable. In this paper the FEM is applied to model the physical behaviour of a mandible which has undergone a Saggital Split Osteotomy. In particular the strength of the jaw associated with the split and the use of Champy plates will be investigated.

The mandible is formed from a framework of mineralized bone material. This consists of alveolar bone and teeth, an outer cortex of dense lamellae material and an inner cancellous bone which is porous and trabeculated. The mandible is unique in that it is the only moving bone in the body. This movement is provided by the bilateral specialised condylar articulation and the associated coronoid muscle attachment process. Because of the nature of the oral cavity, teeth and the surrounding structure, which provides support, the mandible is subject to thermal and mechanical loading. These loadings can affect the way in which the mandible functions and repairs after trauma and surgery.

Experienced clinicians have developed empirical means of avoiding some of the potential problems associated with mechanical stresses. However it would be of considerable advantage to be able to provide the stress–strain behaviour of the system and hence predict the possibility of failure. The mathematical complexity of calculating the relationships of the interacting forces and material of the mandible are considerable but developments in numerical modelling together with advances in computer capabilities now permits accurate and realistic physical models to be analysed.

In this investigation the mandible was modelled using three–dimensional eight–noded linear isoparametric elements. These elements have been used extensively in many applications and provide versatility in mesh generation and also behave well for relatively high aspect ratios. Quantitative results are given for displacements and stresses for a mandible subject to a first molar bite force. The mandible was also considered to be cut as for a saggital split osteotomy and analyses were considered for the immediate post–operative condition when little time had been allowed for bone repair and for the condition after repair whilst immobilised in fixation.

FINITE ELEMENT DISCRETISATION

A mandible was sub–divided and cut into fourteen sections as shown in Figure 1. Measurements were then taken of the relative proportions of cortical, cancellous bone and the central cavitation, together with the external shape of the bone. This allowed the three dimensional mesh to be formed as shown in Figure 2. The mesh consists of 840 nodes, each with three degrees of freedom, and 623 elements and this forms the half symmetrical section of the mandible. The full mandible is shown reconstructed from the symmetrical mesh in Figure 3.

Loading of the mandible is usually taken to be a maximum in the premolar and molar region. Published literature has produced many different methods for assessing bite force, however a consensus view gives a value of 445 N, Mansour[1], Gupta[2], Knoell[3],

131

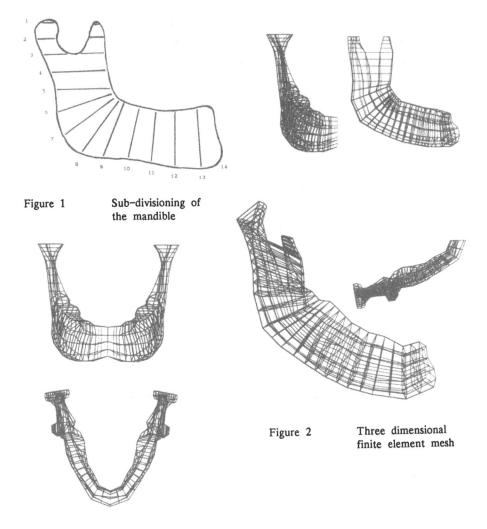

Figure 1   Sub–divisioning of the mandible

Figure 2   Three dimensional finite element mesh

Figure 3   Mesh of complete mandible

Haskell[4]. This bite force is applied in the first molar region and is opposed by the masseter and temporalis muscles which are attached to the mandible at the coronoid and over the angle between the body and ramus.

The mandible will pivot about the temporomandibular joint at the condylar head articulation and this will be considered and modelled as a pivot or hinge. Restraint will also be applied at a set of nodal points in the first molar region and this will be modelled as a sliding force which is restrained by the maxilla in the vertical direction. Haskell et al [4] suggest that muscle forces opposing a bite force of 445N produce forces of 800 N in the masseter and 500N in the temporalis muscles. To obtain these tractions the masseter force was sub–divided into fifteen point loads and spread evenly over the surface nodes of the mesh. Similarly the temporalis force was sub–divided into twelve point loads and spread onto both sides of the coronoid process. The restraints and applied muscular forces applied to the mandible are shown in Figure 4. Due to the restraint at the first premolar this allows the bite force to be calculated for subsequent use in the numerical solution process.

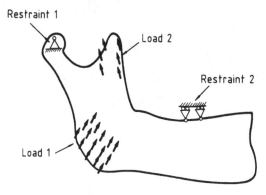

Figure 4   Loadings and restraints

MATERIAL PROPERTIES OF THE MANDIBLE

Gupta et al [2], Knoell [3] and Haskell [4] have discussed the finite element analyses of the mandible and from these investigations relatively small displacements have been predicted, hence the model used here will be restricted to elastic behaviour. Published work on the material properties of the bone forming the mandible can also be found in the references of McElhanay [5], Bonfield [6], Haskell [4] and Natali [7]. Scaled values have also been published by Gupta [2], Knoell [3] and Tanne [8]. Using this experimental data the following material properties were derived for the Elastic Modulus and Poissons ratio of bone (Table 1).

TABLE 1

	$E(N/mm^2)$	ν
Cortical bone	15000	0.3
Cancellous bone	9000	0.3
Cavitational Cancellous bone	1800	0.3

To model the saggital split osteotomy (SSO) it is necessary to postulate that the mandible is cut in three ways. These are a cut through the ramus which involves the cortex and cancellous bone, a horizontal cut along the length of the mandible and a cut through the body which consists of cancellous bone carrying the nerves and blood supply. This sectioning is illustrated in Figure 5.

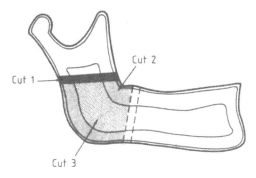

Figure 5   Sectioning for SSO

Two post operative situations were investigated these being shortly after the operation when the bone has had little time to repair and also some time after repair whilst immobilised in fixation. The material properties used to describe these two cases are given in Table 2.

TABLE 2

		E(N/mm²)	Poisson's ratio ν
ANALYSIS A (early release)	CUT 1	75	0.3
	CUT 2	900	0.3
	CUT 3	75	0.3
ANALYSIS B	CUT 1	1800	0.3
	CUT 2	3000	0.3
	CUT 3	450	0.3

For the conditions described above separate analyses were performed with both a single and two Champy plates used to restrain the mandible. The material properties prescribed for the Champy plates were $E = 2.04 \times 10^5$ N/mm² and $\nu = 0.3$ and the position of fixture is shown in Figure 6.

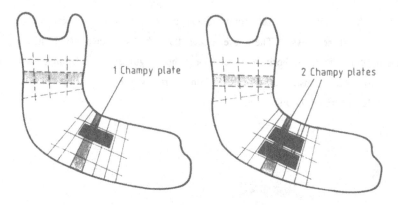

Figure 6   Champy plating of mandible

NUMERICAL RESULTS

Seven different analyses were performed using the three dimensional mandible mesh together with loading and fixity conditions previous described. Analysis 'O' represents a normal mandible under first molar bite force. Analysis type 1, 2 and 3 considers the conditions of no restraint, a single Champy plate and two Champy plates respectively. Each of these three cases were examined for the two following conditions. Condition 'A' which represents early release of maxillomandibular fixation (MMF) while analysis B represents much later release of the MMF.

Displacements of the pogonion point (ΔP_0) and the gonial angle (ΔG_0), as shown in Figure 7, were sampled for each of the seven cases considered and the results are given in Table 3.

TABLE 3

	ΔG_0 (°)	ΔP_0 (mm.)		
		X	Y	Total (Maximum)
Analysis O	0.48	0.032	-0.183	0.186
Analysis A 1	1.20	0.877	-0.315	0.932
2	0.92	0.428	-0.294	0.519
3	0.90	0.401	-0.292	0.496
Analysis B 1	0.64	0.260	-0.237	0.352
2	0.62	0.208	-0.239	0.316
3	0.61	0.194	-0.240	0.309

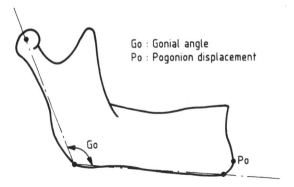

Go : Gonial angle
Po : Pogonion displacement

Go

Po

Figure 7   Reference points for displacement/rotation

The results from the normal mandible under premolar bite force show a gonial angle change of 0.48° and a pogonion displacement of 0.186 mm. It may be considered that any increase in these magnitudes when compared with the other analyses is a result of the sectioning and the mandibular restraint applied. This is expressed in percentage terms in Table 4 where no change in ΔG_0 or ΔP_0 would be equivalent to 100%.

TABLE 4

		$\dfrac{\Delta Go}{\Delta Go(An.0)} * 100\%$	$\dfrac{\Delta Po}{\Delta Po(An.0)} * 100\%$
Analysis A	1	250	502
	2	192	280
	3	188	267
Analysis B	1	133	189
	2	129	171
	3	127	166

The results from Analysis A show large increases in both ΔP_0 and ΔG_0 when compared with Analysis 'O'. This indicates that the mandible is under-restrained which can be seen from the comparative displacement plots shown in Figure 8(a) and (b). Analysis A represents the mandible with early release of MMF and no Champy plates. The angle change of 250% and displacement change of 502% indicates that the mandible is most prone to relapse when subject to these conditions.

When a Champy plate system is applied to restrain the mandible a single plate reduces the gonial angle movement by 58% and the pogonian displacement by 222%. A second plate, analysis 3A, adds further to stability but reduces the gonial angle and pogonion displacement by only a further 4% and 13% respectively.

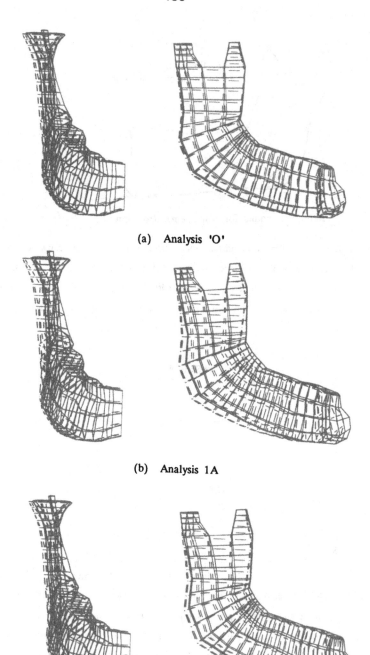

(a) Analysis 'O'

(b) Analysis 1A

(c) Analysis 2A

Figure 8

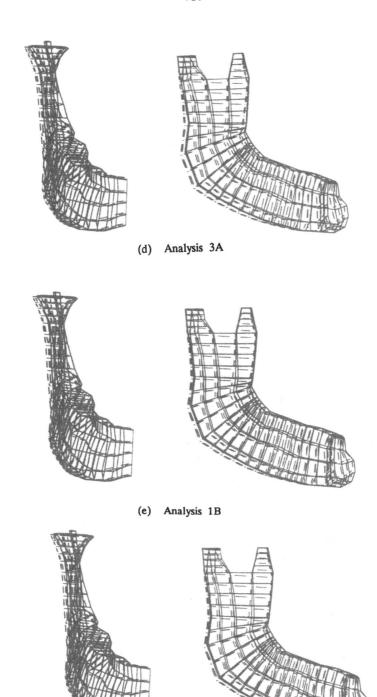

(d) Analysis 3A

(e) Analysis 1B

(f) Analysis 2B

Figure 8

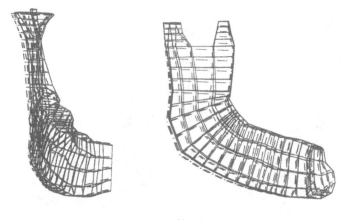

(g)   Analysis 3B

Figure 8   Displaced mesh plots

Analysis B which represents much later release of MMF shows smaller movements than those of analysis A.   This is due to progressive healing and the associated increase in the bone modulus which provides greater stability.   Analysis 1B shows a gonial angle change of 133% and a pogonion displacement of 189% when compared with analysis O. The addition of one Champy plate reduces the gonial angle and pogonion displacement by 4% and 38% respectively.   The addition of a second Champy plate gives further reductions although these are considerabley smaller at 2% for ΔG_O and 55% for ΔP_O. Figure 8 shows the displacement plots for Analysis B and as can be seen the displacements are considerably smaller than the earlier post operative results of Analysis A.

CONCLUSIONS

A three dimensional FE model has been used to analyse a human mandible which has been subject to a saggital split osteotomy.   The analysis shows the effects of plating under a first molar bite force and the consequence of releasing intra–maxillary fixation.   The maximum stress of 80 N/mm^2 and the calculated biteforce agreed with published values by Haskell [4].

The results show the importance of restraining the mandible by a Champy plate although the application of a second plate only provides limited benefit. Early release of intra maxillar fixation is advisable to prevent other complications such as muscle wasting and restriction of temporal mandibular movement, however this must be weighed against the increase in movement brought about by first molar loading. It is not known at present if the stresses at the location of sectioning and plating are an adjunct to healing and this will be the subject of further research.

REFERENCES

1. Mansour, R. M. and Reynik, R. J.   Forces and moments generated during maximum bite in centric occlusion. Int. Assoc. Dent. Res. Abs., 1972.

2. Gupta, K. K. et al.   Mathematical modelling and structural analysis of the mandible. Biomat., Med. Dev., ARt. Org. 1, 469-79, 1973.

3. Knoell, A. C.   A mathematical model of an in-vitro human mandible. J. Biomech. 10, 159-66, 1977.

4. Haskell, B. et al.   Computer-aided modelling in the assessment of the biomechanical determinants of diverse skeletal patterns. Am. J. Orthod. 89, 363-82, 1986.

5. McElhaney, J. H.   Dynamic response of bone and muscle tissue. J. Appl. Physiol. 21, 1231-6, 1966.

6. Bonfield, W. and Datta, P. K.   Youngs modulus of compact bone. J. Biomech., 7, 147-9, 1974.

7. Natali, A. N. and Meroi, E. A.   A review of the biomechanical properties of bone as a material. J. Biomed. Eng., 11, 266-76, 1989.

8. Tanne, K. et al.   Biomechanical effect of anteriorly directed extraoral forces on the craniofacial complex: A study using the finite element method. Am. J. Orthod. Dento. Orthop., 95, 00-7, 1989.

DEBONDING

R.G. OLIVER, M.Sc.D.,F.D.S.
Senior Lecturer/Consultant,
Department of Child Dental Health,
Dental School, Heath Park, Cardiff, CF4 4XY, UK

K.R. WILLIAMS, B.Sc., Ph.D.
Senior Lecturer,
Department of Basic Dental Science,
Dental School, Heath Park, Cardiff, CF4 4XY, UK

ABSTRACT

This paper gives an extensive review of factors which may influence the location of the weak link in the bonding/debonding chain. The engineering principles of finite element stress analysis have been applied to three different methods of orthodontic bracket removal. This method identifies the area of maximum principal stress which is the likely location for the initiation of the crack within the enamel/composite/complex.

INTRODUCTION

'Bonding' is a loose term used to describe the various processes involved in placing an orthodontic bracket on the enamel surface of a tooth. The term is also used in restorative dentistry when placing composite retained bridges, veneers or building up diminutive or fractured teeth using composite. The aims of the restorative dentist are to produce a 'bond' which will not break. The orthodontist, however, would like a 'bond' which is strong enough to withstand the various forces placed upon it during treatment, but which, at the end of treatment, may be easily broken with minimal risk of damage to the enamel.

There has been an abundance of research directed at improving and enhancing the strength of the initial bond, and some work on the best method of composite removal subsequent to bracket removal which is least harmful to the enamel surface. However, relatively little attention has been paid to the mechanisms involved in the removal of orthodontic brackets, which is the first stage of the process given the equally loose term 'debonding'.

In order to remove an orthodontic bracket a line of cleavage in the enamel/composite/bracket base complex must be created. Bond strength trials record tensile or shear bond strength. Some also record the location of failure, which in turn, can suggest the site of the weakest link in the enamel/composite/bracket base complex. The point at which the initiation of this line of cleavage occurs, together with its subsequent path, may be influenced by several variables :

1. Tooth Enamel

Nordenvall et al.[1] found that 'young' permanent tooth enamel reacted in a different manner to 'old' permanent tooth enamel when acid etched for 60 seconds, with a greater surface irregularity found on 'old' enamel. Oliver[2,3] could find little difference in the etch score and type of enamel from unerupted vs. erupted teeth, and both Jacobs et al.[4] and Oliver[5] could find no difference in bond strength between unerupted and erupted teeth; Jacobs et al. found no consistent pattern in the site of bond failure, Oliver used the Adhesive Remnant Index (ARI) devised by Artun and Bergland[6] and found identical mean ARI values for unerupted and erupted teeth, with predominantly adhesive/bracket base failure.

Prior exposure of the enamel to fluoride has been shown to influence the etch characteristics of the outermost 2 - 4 microns of enamel[7], however, Brannstrom et al.[8] could find no difference in etch score between enamel which had been pretreated with fluoride, and enamel with no fluoride exposure, however the inner composite surface was rougher on fluoride pretreated enamel indicating better retentive tags. The bond strength of enamel which had been pretreated with fluoride has been shown in vitro to be not significantly different to that of untreated enamel[9]. Although the site of failure was recorded in this experiment no meaningful results were given. Incorporation of 2% sodium fluoride in the etching acid has been shown to lead to a significant reduction of both bond strength and area of composite remaining on the enamel[10], whereas Thornton et al.[11], although producing a range of mean tensile bond strengths from 15.3 MNm^{-2} to 11.8 MNm^{-2} with varying concentrations of sodium fluoride incorporated into their etchant, found that this produced no statistically significant reduction of bond strength. Nor could they support other work which showed the addition of fluoride to the etching solution produced enamel/composite interface failure, instead of failure within the composite[12,13].

Recent work has shown statistically higher bond strengths for brackets bonded on the lingual enamel surface compared with the labial enamel[14]. These increased bond strengths have also been found to lead to significantly greater incidence of enamel damage when debonding using a shear force.

2. Acid Concentration

The effect of reducing the concentration of phosphoric acid has been shown to alter the morphology of the etched enamel surface[3,15,16]. One paper has shown that a concentration of 10% phosphoric acid gave maximum bond strength, with 40% phosphoric acid second; 70% phosphoric acid gave statistically significantly weaker bond strengths[17]. Later work from the same centre showed that 40% phosphoric acid produced the greatest etch depth and dissolution of calcium, with 70% and 10% phosphoric acid solutions both producing the shallowest etch depth and least calcium dissolution[18], suggesting therefore that etch depth has little relationship to bond strength.

An in vitro study using commercially available bonding systems showed that 15% phosphoric acid gave the highest bond strength when used for 30 seconds, whereas 5% phosphoric acid for 10 seconds gave the weakest bond strength[19]. With few exceptions the majority of the composite remained on the tooth following bracket removal.

3. Acid Vehicle

Brannstrom et al.[20] showed no difference in etch scores (representing surface roughness) when 50% acid gel was compared with 37% acid liquid. Gwinnett[21] suggests that the washing time for a gel should be double that for a liquid, otherwise the cellulose vehicle may remain and prevent penetration of the etched surface by the composite resin. Sheykholeslam and Brandt[22] suggest that a liquid is preferred to a gel because the liquid may be agitated and thereby avoid neutralization of the acid by by-products of the etching process which are concentrated at the enamel surface.

4. Length of Etch

There are now several references in the literature with the results of both in vitro[23,24,25] and in vivo[26,27,28] studies which suggest that a reduced etch time of 15 seconds, is satisfactory for bonding. In fact, it is suggested in the second paper by Kinch et al.[28] that a 15 second etch time gives a stronger enamel/composite bond than a 60 second etch time.

5. Agitation of Acid

As mentioned in section 3 above, agitation of the acid by gentle dabbing with the applicator is supposed to overcome problems of re-precipitation of calcium and phosphate salts which will build-up in solution during the etch procedure. It has been shown by some authors[3,29,30,31] that agitation of the acid will lead to a different etch pattern, but this has yet to be related to bond strength.

6. Wash Time

Williams and Fraunhofer[32] found that etch time and wash time were intimately related to bond strength, with a short etch/long wash or long etch/short wash producing the highest bond strengths. They suggest that a longer etch leads to less surface ortho-phosphates (a by-product of the etch process) which act as a contaminant of the etched surface. Gwinnett[33] suggests doubling wash times when an acid gel is used. Gottlieb et al. [17] link acid concentration to wash time. An acid concentration greater than 27% produces monocalcium phosphate monohydrate, a readily soluable salt. Acid concentrations less than 27% produce dicalcium phosphate dihydrate which is much less soluable and therefore more difficult to wash away. It is entirely possible that there is further interaction between acid concentration and etch time. It is clear, however, that relatively insoluable salts may be precipitated which will require thorough debridement.

7. Contamination

It is recommended that the newly etched surface should be washed and dried in an oil free air/water delivery system, and that subsequently the etched surface should not be touched by saliva, gingival fluid, blood, or even the operator's fingers. Contamination of the etched surface will lead to a reduced bond strength. Thomson et al.[34] have shown, however, that immediately washing the contaminated surface with water will restore the enamel surface without the need for re-etching.

8. Intermediate Unfilled Resin Layer

Early work by Zachrisson[35] suggested that the use of an intermediate unfilled resin layer on the newly etched enamel surface would provide a weak link to facilitate debonding. Other workers [36,37] suggest that an intermediate layer of unfilled resin has no effect on bond strength. However, McCabe and Storer[38] and Asmussen[39] claim enhanced bond strength when using a layer of unfilled resin. Gwinnett[33] states that the unfilled resin encapsulates the enamel crystallites which then become the 'filler' particles for the resin and the unfilled resin does increase bond strength. Faust et al.[40] have shown a linear relationship between tensile bond strength and penetration coefficient (a measure of the flow of the material along a capillary tube). In contrast Jergensen and Shimokobe[41] demonstrated that an unfilled resin (Nuva-Seal) would not penetrate etched enamel any further than a highly filled composite resin such as Concise. Moin and Dogon [42] produced slightly ambivalent results when testing bond strength with and without an unfilled resin.

9. Composite Type and Mix

In the early years of bonding, the composite used was predominantly a two paste chemical cure system. More recently the 'no-mix' or

'one-step' chemical cure adhesives have become popular, and there is
a growth in the number of light cure systems specifically for orthodontic bonding. Scanning of the literature on bond strength trials
would suggest that Concise has survived from the early two paste
systems and become almost a 'bench standard' for evaluation of other
bonding adhesives. Recently Delport and Grobler[43] compared
the in vitro tensile bond strengths of four 'no mix' composites and
three two paste composites. They drew the distinction between
bench testing of an enamel-composite-bracket complex, and an enamel-
composite arrangement. The former being a test of the weakest link
in the bonding chain, the latter being a true test of the enamel/
composite bond. Delport and Grobler used an enamel-composite
experimental model and reported no significant differences between
any of the composites tested.

Pender et al.[44] tested the bond strength of the enamel-
composite-bracket base complex comparing a light cure adhesive
against a 'no-mix' and a two paste composite adhesive. They concluded that the light cure system produced the weakest bonds, but
pointed out that failures with this material occurred at the
composite/bracket base interface and therefore the true enamel-
composite bond strength remained unknown. Wright and Powers[45]
also found light cure composite to give weaker bonds.

Moin and Dogon[42] used a two paste system (Concise) and
compared various combinations of diluted composite mix. Unfilled
resin on its own produced the weakest bond, and the use of an
unfilled resin layer together with a heavily filled (restorative)
composite produced the strongest bond strength.

Rezich et al.[46] compared the bond strengths of a paste-paste
composite, with a 'no-mix' composite, a fluoride releasing composite
and a glass ionomer cement. Bond strengths reduced in the given
order of the materials tested.

10. Thickness of Composite

Early work by Eden et al.[47] suggested that bond strength was
adversely affected by increasing composite thickness. Schechter et
al.[48] found that tensile bond strength remained unaffected by
increased thickness, shear bond strength was reduced with increased

thickness and peel bond strength was slightly increased with increased thickness and varied according to material. Evans and Powers[49] showed that for Concise in very thin section, tensile bond strength was weak, but reached a maximum when 0.25 m.m. thick. Other composites (Mono-Lok and System 1+) exhibited a critical thickness of 0.3 m.m., above which bond strength was effectively reduced to zero. Unite maintained reasonable bond strength up to a thickness of 0.38 m.m. In almost all cases the break occurred at the composite/bracket base interface.

In vitro and in vivo work by Perry[50] showed that rebonding a bracket without first removing adhesive remaining on the enamel surface, (and thereby increasing composite thickness) produced a significantly weaker bond which tended to fail along the same interface as the initial failure, other workers however, have shown that rebonding a bracket without further tooth preparation produced a stronger bond than the original[51].

11. Bracket Base

The use of perforated metal bases has been shown to produce weak bonds under tension[52], but stronger bonds in shear[53], however, mesh size or bracket base area would not seem to significantly affect bond strength[53,54]. The weld 'nuggetts' produced when attaching the base to the foil mesh would seem to have some effect. Ferguson et al.[55] compared a mesh base with two undercut base types, using a paste/paste composite and a no-mix composite. They found the highest bond strengths with a 'photo etch' base, in direct contrast to Maijer and Smith[56].

Hanson et al.[57] used a stainless steel powder sintered to the bracket base for comparison with a foil mesh base, and found a significant increase in bond strength for the powdered bases. Treatment of the bracket base by etching, silanation or surface activation was examined by Siomka and Powers[58]. They found significant increases in bond strength for all forms of bracket base conditioning, although the effect was also dependant on the composition and morphology of the bracket base. Peutzfeldt and Asmussen[59] have confirmed that silocoating increases bond strengths. Recycling of brackets produces a reduction in bond strength[45,60,61].

12. Tooth Position

There is a considerable body of evidence to show that the position of the tooth within the dental arch has an influence on the location of failure of the enamel-composite-bracket base complex[28,44,62, 63,64]. Reasons why posterior teeth should demonstrate a higher incidence of enamel-composite failure have been discussed[28, 44]. It would seem that both the ultra structural properties of the surface enamel as well as the gross morphology of the labial/ buccal surface and the bracket base adaptation are more likely to be influencing factors than previous suggestions of moisture con- tamination[35,63].

13. Method of Debond

Three different methods of bracket removal were examined in vivo by Oliver[64]. He concluded that the use of debonding pliers with the blades applied at the enamel-composite or composite-bracket interface resulted in more adhesive being removed with the bracket. The two other methods used were essentially applying a tensile force to the enamel-composite-bracket base complex via the wings of the bracket.

Thanos et al.[53] tested the bond strength in vitro of an enamel/composite/bracket system using three principle directions of force application. They found that torsion required the least force, followed by tension and maximum bond strength was found for shear testing. Unfortunately torsional forces lead to problems of distortion and fracture of the wings of the bracket before the bond fractured.

Sheridan et al.[66] described an electrothermal device for removing brackets. This instrument relies on local heat to disrupt the composite-bracket base interface, and, unlike other mechanical methods of debonding, does not rely on the formation and propagation of a cleavage somewhere within the enamel-composite-bracket base complex.

Zidan et al.[67] examined the fractured surfaces of broken bonds and reported that although a failure may appear to occur at the enamel-composite interface on macroscopic inspection, closer examination showed that a very thin layer of resin still remained on the enamel surface. They concluded that true interfacial failure at the enamel-composite boundary never occurred.

Bennett et al.[88] produced a photoelastic stress model to illustrate the size, location, and distribution of forces generated within the enamel-composite-bracket base complex when applying a load to either the tie wings of an edgewise bracket, the bracket base or the composite. High concentrations of stress in the enamel surface were found with a compressive load applied at the level of the composite layer.

Mulville and Vaishnav[69] examined interfacial crack propagation of epoxy resin bonded to aluminium. They found that a roughened aluminium surface produced a higher strain energy release rate and that the crack line occurred close to the aluminium surface but, similar to the findings of Ziden et al.[67] who used composite resin on etched enamel, they found a residue of adhesive remaining on the aluminium.

From the foregoing account it can be seen that the weak link in the bonding chain can be influenced by several different factors. Some of these are universally accepted to be true, whereas some factors remain controversial in their effect.

Many of the variables which affect bracket bonding already discussed can be readily examined using the finite element method. Although, basically formulated for engineering design purposes, its use is becoming widespread in Biomechanics. It has been used by previous workers[70,71] to examine orthodontic tooth movement and also the stress on the periodontium and alveolar bone generated by orthodontic loading on a tooth[72]. They state that finite element stress analysis is superior to photoelastic techniques which are both crude and difficult to interpret. This technique of mathematical modelling of stress and crack initiation would seem to be ideally suited for the study of the mechanism of debonding.

The following discussion examines some initial findings of the localised principal and effective stresses generated at the bracket/composite/enamel interfaces and within the body of each material under three conditions of loading corresponding to three different methods of bracket removal described earlier[64].

This preliminary investigation examines the use of the finite element method as an indicator of the initial crack nucleation and fracture path. The bracket, composite layer and enamel are shown diagrammatically in Figures 1 - 4, based on travelling microscope measurements of a sectioned tooth and bonded bracket. Because of the symmetrical configuration of the geometry, only one half of the bracket is illustrated. The finite element modelling techniques and procedures will not be fully explained here, (for a full explanation see references 73 and 74) only the essential points for an understanding. A number of assumptions are made in this initial work.

Firstly; linear elastic behaviour is assumed for all materials, i.e. the resultant displacements (ϵ) within the material are directly proportional to the imposed stress (σ_o).

$$\sigma_o = \epsilon E \quad \ldots \quad (1)$$

where E is a constant, the elastic modulus. This behaviour is true for the stainless steel metal and tooth enamel, but less exact for the composite. Indeed it is probable that the composite has a strain dependent modulus, E, where

$$E = f (\epsilon) \quad \ldots \quad (2).$$

However, within the accuracies of this initial investigation, we have assumed linear elastic behaviour for the three materials, with appropriate modulus values.

Secondly; perfect bonding is assumed between the metal/composite/ enamel interfaces. In reality, little if any chemical bonding occurs across these interfaces. The bond strength relies on micromechanical locking with some degree of secondary bonding along these interfaces. In this work we have assumed the maximum composite/enamel bond strength is 20 MPa[39].

A range of debonding techniques have been analysed using the finite element method. These debonding techniques include Method A : whereby compressive forces are applied close to and parallel to the enamel surface, the point of application being either the composite surface, the composite/bracket interface, or the edge of the bracket (Figures 1 & 2). A compression technique (Method B) where the wings of the bracket are forced together (Figure 3), and finally a tensile force at right angles

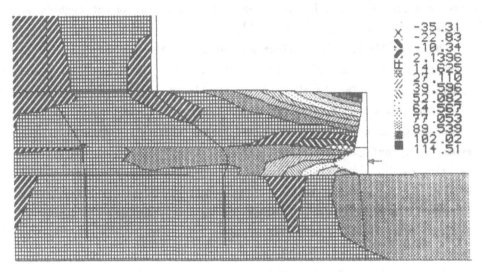

Figure 1. Compressive debonding force applied at
the composite layer.
Structure shown deformed x50 for clarity.
Maximum principal stress (MPa)

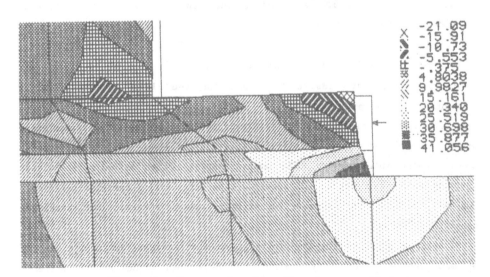

Figure 2. Compressive debonding force applied at
the metal bracket.
Structure shown deformed x50 for clarity.
Maximum principal stress (MPa).

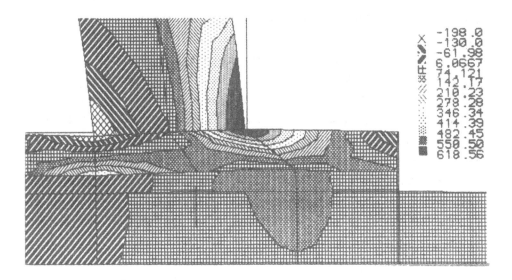

Figure 3.   Compression debonding with forces applied
            at the bracket wings.
            Structure shown deformed x10 for clarity.
            Maximum principal stress (MPa)

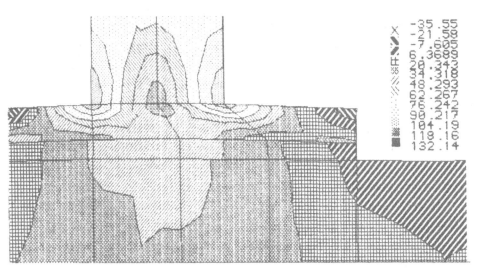

Figure 4.   Tensile debonding by applying forces at
            right angles to the enamel surface via the
            wings of the bracket.
            Structure shown deformed x 50 for clarity.
            Maximum principal stress (MPa).

to the enamel surface applied via the wings of the bracket (Method C, Figure 4).

The numerical programme provides results for a number of stress types over the described area, including local tensile and compressive in the x, y and z directions also local shear over the xy, yz and xz surfaces. However, these types of stress do not usually provide the most useful information in terms of the deformation and fracture at interfaces. In engineering terms it is the principal stresses that dictate plastic deformation and fracture sites in structures. (These are the maximum and minimum principal and effective stresses). Fracture of brittle materials is normally influenced by the maximum and minimum principals while shear deformation of ductile materials occurs at a specific effective stress. With the materials of interest in this examination, the composite and enamel are brittle solids at body temperature, whilst the metal (stainless steel) exhibits initial elastic behaviour, becoming a ductile deformable solid at higher stresses. Since we are interested in the fracture at interfaces the graphical output from the programme has concentrated on the maximum principal stresses.

The complex nature of the natural tissue and composite solids generally does not allow specific stresses to be quoted for deformation or fracture. In fact the fracture of brittle solids is generally examined using the fracture mechanics route; while deformation of ductile materials is normally analysed under a quite simple stress such as pure tension, or shear and with specimens of defined geometry. Nevertheless, the finite element method can provide indicative information and progress to a stage where fracture and deformation of complex materials and geometry are predicted.

In the present work we have used a plain strain pseudo three dimensional programme running on a P.C. with 512 Kb RAM and a 20 MB disc to analyse three debonding techniques. In addition, the experimental model has assumed :

1. Flat enamel surface

2. Flat bracket base (no undercut)

3. Uniform thickness of composite

4. For Method A a point source of load precisely applied to a flat surface parallel to the enamel surface.

Results were obtained using a 100N load for Method A and positioned at three different sites with two positions shown in Figure 1 and 2. Only the maximum principal ·stress is indicated in these figures. The original and deformed structures are also shown with the displacements magnified by x50 for clarity.

In Method B, a compression removal technique was used, where both wings of the bracket are forced together, with the force applied near the tips of the wing. Again, a 100N load was used which proved more than sufficient to bend and permanently deform the metal bracket (see Figure 3).

Finally, Method C used a 100N load at right angles to the enamel surface and positioned at the tip of the bracket in order to simulate tensile removal. The maximum principal stresses are shown in Figure 4 for this type of loading.

Although a good deal of graphical and numerical results are produced, the discussion essentially deals with the possible site of crack initiation.

An interesting point of note is that using debonding Method A for removal, both stresses move through a minimum as the point of loading moves away from the enamel, (Figures 1 and 2 illustrate the extremes).

If we assume the maximum bond strength of composite to enamel is 20 MPa then a sufficiently high maximum principle stress is generated with loading applied at the composite or metal to cause fracture of the enamel as illustrated in Figures 1 and 2. However, when the load is applied at the composite/metal interface the stresses are much more uniform and it is difficult to suggest a fracture route. Examination of the effective stress for this particular loading indicates a value of 510 MPa which is above the yield point and would plastically deform the stainless steel bracket at the composite/metal interface as indicated. It is therefore likely that debonding takes place initially along this interface not as a result of brittle fracture of the composite, but by gross distortion of the metal.

The compression removal route generates high deforming stresses in the metal (because of the large bending moments) and extremely large stress changes across the composite/metal surfaces, again suggesting failure at this interface, (see Figure 3).

Using the tensile debond technique, a general uniform stress pattern at the important interfaces (Figure 4) ensued.   It is only the modest change in stress from 6 - 20 MPa at the periphery of the bracket that is likely to initiate failure along the metal/composite interface.

Clearly, this static stress analysis can only suggest a possible site of initial fracture.   As the crack propogates under a changing load pattern, then the fracture path may well change from one interface to another.   Indeed this is known to happen in practice as indicated earlier[64].   It is therefore our intention to pursue a finite element fracture mechanics route to the debonding mechanism, in order to more clearly predict the debonding route.

ACKNOWLEDGEMENTS

We are grateful to Miss Karen Ball and Mrs. Lynette James for typing the manuscript.

REFERENCES

1.   Nordenvall, K.J., Brannstrom, M. and Malmgren, O., Etching of deciduous teeth and young and old permanent teeth. A comparison between 15 and 60 seconds of etching. Am. J. Orthod., 1980, **78**, 99-108.

2.   Oliver, R.G., The effects of differing etch times on the etch pattern on enamel of unerupted and erupted human teeth examined using the scanning electron microscope. Br. J. Orthod., 1987, **14**, 105-107.

3.   Oliver, R.G., The effects of differing acid concentrations, techniques and etch time on the etch pattern of enamel of erupted and unerupted human teeth examined using the scanning electron microscope. Br. J. Orthod., 1988, **15**, 45-49.

4.   Jacobs, G., Kuftinec, M.M., Showfety, K.J., Fraunhofer, J.A. von., Bonding characteristics of impacted versus erupted permanent teeth. Am. J. Orthod., 1986, **89**, 242-245.

5.   Oliver, R.G., Bond strength of orthodontic attachments to enamel from unerupted and erupted young permanent teeth. Eur. J. Orthod., 1986, **8**, 123-126.

6. Artun, J., Bergland, S., Clinical trials with crystal growth conditioning as an alternative to acid etch enamel pre-treatment. Am. J. Orthod., 1984, **85,**333-340.

7. Lehman, R., Davidson, C.L., Loss of surface enamel after acid etching procedures and its relation to fluoride content. Am. J. Orthod., 1981, **80,** 73-82.

8. Brannstrom, M., Nordenvall, K.J., Malmgren, O., The effects of various pre-treatment methods of the enamel in bonding procedures. Am. J. Orthod., 1978, **74,** 522-530.

9. Bryant, S., Retief, D.H., Bradley, E.L., Denys, F.R., The effect of topical fluoride treatment on enamel fluoride uptake and the tensile bond strength of an orthodontic bonding resin. Am. J. Orthod., 1985, **74,** 294-302.

10. Hirce, J.D., Sather, A.H., Chao, E.Y.S., The effect of topical fluorides, after acid etching of enamel, on the bond strength of directly bonded orthodontic brackets. Am. J. Orthod., 1980, **78,** 444-452.

11. Thornton, J.B., Retief, D.H., Bradley, E.L., Denys, F.R., The effect of fluoride in phosphoric acid on enamel fluoride uptake and the tensile bond strength of an orthodontic bonding resin. Am. J. Orthod. Dentofac. Orthop., 1986, **90,** 91-101.

12. Wright, F.A.C., Beck, D.J., Prevention of pit and fissure caries III. Fluoride and resin-enamel bonding. New Zealand Dent. J., 1973, **69,** 267-272.

13. Grajower, R., Glick, A., Gedalia, I., Kochavi, D., Tensile strength of the bond between resin to enamel etched with phosphoric acid containing fluoride. J. Oral Rehabil., 1979, **69,** 267-272.

14. Chumak, L., Galil, K.A., Way, D.C., Johnson, L., Hunter, W.S., An in vitro investigation of lingual bonding. Am. J. Orthod. Dentofac. Orthop., 1989, **95,** 20-28.

15. Denys, F.R., Retief, D.H., Variations in enamel etching patterns produced by different concentrations phosphoric acid. J. Dent. Assoc. South Africa, 1982, **37,** 185-189.

16. Soetopo, Beech, D.R., Hardwick, J.L., Mechanism of adhesion of polymers to acid-etched enamel. Effect of acid concentration and washing on bond strength. J. Oral Rehabil., 1978, **5,** 69-80.

17. Gottlieb, E.W., Retief, D.H., Jamison, H.C., An optimal concentration of phosphoric acid as an etching agent, Part 1 Tensile bond strength studies. J. Pros. Dent., 1982, **48,** 48-51.

18. Manson-Rahemtulla, B., Retief, D.H., Jamison, H.C., Effect of concentrations of phosphoric acid on enamel dissolution. J. Pros. Dent., 1984, **51,** 495-498.

19. Bryant, S., Retief, D.H., Russell, C.M., Denys, F.R., Tensile bond strengths of orthodontic bonding resins and attachments to etched enamel. Am. J. Orthod. Dentofac. Orthop., 1987, **92**, 225-231.

20. Brannstrom, M., Malmgren, O., Nordenvall, K.J., Etching of young permanent teeth with an acid gel. Am. J. Orthod., 1982, **82**, 379-383.

21. Gwinnett, A.J., The potential weak link in the bonding chain. New York State Dent. J., 1983, **49**, 392-395.

22. Sheykholeslam, Z., Brandt, S., Some factors affecting the bonding of orthodontic attachments to tooth surface. J. Clin. Orthod., 1977, **11**, 734-743.

23. Mardaga, W.J., Shannon, I.L., Decreasing the depth of etch for direct bonding in orthodontics. J. Clin. Orthod., 1982, **16**, 130-132.

24. Barkmeier, W.W., Gwinnett, A.J., Shaffer, S.E., Effects of enamel etching time on bond strength and morphology. J. Clin. Orthod., 1985, **19**, 36-38.

25. Barkmeier, W.W., Gwinnett, A.J., Shaffer, S.E., Effects of reduced acid concentration and etching time on bond strength and enamel morphology. J. Clin. Orthod., 1987, **21**, 395-398.

26. Carstensen, W., Clinical results after direct bonding of brackets using shorter etching times. Am. J. Orthod., 1986, **89**, 70-72.

27. Kinch, A.P., Taylor, H., Warltier, R., Oliver, R.G., Newcombe, R.G., A clinical trial comparing the failure rates of directly bonded brackets using etch times of 15 or 60 seconds. Am. J. Orthod. Dentofac. Orthop., 1988, **94**, 476-483.

28. Kinch, A.P., Taylor, H., Warltier, R., Oliver, R.G., Newcombe, R.G., A clinical study of amount of adhesive remaining on enamel after debonding, comparing etch times of 15 and 60 seconds. Am. J. Orthod. Dentofac. Orthop., 1989, **95**, 415-421.

29. Tyler, J.E., A scanning electron microscope study of factors influencing etch patterns of human enamel. Arch. Oral Biol., 1976, **21**, 765-769.

30. Hormati, A.A., Fuller, J.L., Denehy, G.E., Effects of contamination and mechanical distrubance on the quality of acid-etched enamel. J. Am. Dent. Assoc., 1980, **100**, 34-38.

31. Bates, D., Retief, D.H., Jamison, H.C., Denys, F.R., Effects of acid etch parameters on enamel topography and composite-resin-enamel bond strength. Paediatric Dentistry, 1982, **4**, 106-110.

32. Williams, B., Fraunhofer, J.A. von., The influence of the time of etching and washing on the bond strength of fissure sealants applied to enamel. J. Oral Rehabil., 1977, **4**, 139-143.

33. Gwinnett, A.J., Bonding of restorative resins to enamel. Inter-national Dent. J., 1988, 38, 91-96.

34. Thomson, J.L., Main, C., Gillespie, F.C., Stephen, K.W., The effect of salivary contamination on fissure sealant-enamel bond strength. J. Oral Rehabil., 1981, 8, 11-18.

35. Zachrissen, B.U., A post-treatment evaluation of direct bonding in orthodontics. Am. J. Orthod., 1977, 71, 173-189.

36. Retief, D.H., Woods, E., Is a low viscosity bonding resin necessary? J. Oral Rehabil., 1981, 8, 255-266.

37. Jassem, H.A., Retief, D.H., Jamison, H.C., Tensile and shear strengths of bonded and rebonded orthodontic attachments. Am. J. Orthod., 1981, 79, 661-668.

38. McCabe, J.F., Storer, R., Adaptation of resin restorative materials to etched enamel and the interfacial work of fracture. Br. Dent. J., 1980, 148, 155-159.

39. Asmussen, E., Clinical relevance of physical chemical and bonding properties of composite resins. Operative Dent., 1985, 10, 61-73.

40. Faust, J.B., Grego, G.N., Fan, P.L., Powers, J.M., Penetration co-efficient, tensile strength, and bond strength of thirteen direct bonding orthodontic cements. Am. J. Orthod., 1978, 73, 512-525.

41. Jergensen, K.D., Shimokobe, H., Adaptation of resinous restorative materials to acid etched enamel surfaces. Scand. J. Dent. Res., 1975, 83, 31-36.

42. Moin, K., Dogon, I.L., An evaluation of shear strength measurements of unfilled and filled resin combinations. Am. J. Orthod., 1978, 74, 531-536.

43. Delport, A., Grobler, S.R., A laboratory evaluation of the tensile bond strength of some orthodontic bonding resins to enamel. Am. J. Orthod. Dentofac. Orthop., 1988, 93, 133-137.

44. Pender, N., Dresner, D., Wilson, S., Vowles, R., Shear strength of orthodontic bonding agents. Eur. J. Orthod., 1988, 10, 374-379.

45. Wright, W.L., Powers, J.M., In vitro tensile bond strength of re-conditioned brackets. Am. J. Orthod., 1985, 87, 247-252.

46. Rezich, P.M., Panneton, M.J., Barkmeier, W.W., In vitro evaluation of fluoride and non-fluoride releasing orthodontic adhesives on bracket bond strength. J. Dent. Res., 1988, Abstract No. 1594, p. 312.

47. Eden, G.T., Craig, R.G., Peyton, F.A., Evaluation of a tensile test for direct filling resins. J. Dent. Res., 1970, 49, 428-434.

48. Schechter, G., Caputo, A.A., Chaconas, S.J., The effect of adhesive thickness on retention of direct bonded brackets. J. Dent. Res., 1980, **59**, 285, Abstract 72.

49. Evans, L.B., Powers, J.M., Factors affecting in vitro bond strength of no-mix orthodontic cements. Am. J. Orthod., 1985, **87**, 508-512.

50. Perry, A.C., Rebonding brackets. J. Clin. Orthod., 1980, **14**, 850-854.

51. Rosenstein, P., Binder, R.E., Bonding and rebonding peel testing of orthodontic brackets. Clin. Prev. Dent., 1980, **2**, 15-17.

52. Dickinson, P.T., Powers, J.M., Evaluation of fourteen direct bonding orthodontic bases. Am. J. Orthod., 1980, **78**, 630-639.

53. Thanos, C.E., Munholland, T., Caputo, A.A., Adhesion of mesh-base direct-bonding brackets. Am. J. Orthod., 1979, **75**, 421-430.

54. Lopez, J.I., Retentive shear strengths of various bonding attachment bases. Am. J. Orthod., 1980, **77**, 669-678.

55. Ferguson, J.W., Read, M.J.F., Watts, D.C., Bond strengths of an integral bracket-base combination : an in vitro study. Eur. J. Orthod., 1984, **6**, 267-276.

56. Maijer, R., Smith, D.C., Variables influencing the bond strength of metal orthodontic bracket bases. Am. J. Orthod., 1981, **79**, 20-34.

57. Hanson, G.H., Gibbon, W.M., Shimizu, H., Bonding bases coated with porous metal powder : a comparison with foil mesh. Am. J. Orthod., 1983, **83**, 1-4.

58. Siomka, L.V, Powers, J.M., In vitro bond strength of treated direct-bonding metal bases. Am. J. Orthod., 1985, **88**, 133-136.

59. Peutzfeldt, A., Asmussen, E., Silicoating : evaluation of a new method of bonding composite resin to metal. Scand. J. Dent. Res., 1988, **96**, 171-176.

60. Mascia, V.E., Chen, S-R., Shearing strengths of recycled direct-bonding brackets. Am. J. Orthod., 1982, **82**, 133-136.

61. Wheeler, J.J., Ackerman, R.J., Bond strength of thermally recycled metal brackets. Am. J. Orthod., 1983, **83**, 181-186.

62. Gwinnett, A.J., Gorelick, L., Microscopic evaluation of enamel after debonding : clinical application. Am. J. Orthod., 1977, **71**, 651-665.

63. Lovius, B.B.J., Pender, N., Hewage, S., O'Dowling, I., Tomkins, A., A clinical trial of a light activated bonding material over an 18 month period. Br. J. Orthod., 1987, **14**, 11-20.

64. Oliver, R.G., The effect of different methods of bracket removal on the amount of residual adhesive. Am. J. Orthod. Dentofac. Orthop., 1988, **93**, 196-200.

65. Buzzitta, V.AJ., Hallgren, S.E., Powers, J.M., Bond strength of orthodontic direct-bonding cement-bracket systems as studied in vitro. Am. J. Orthod., 1982, **81**, 87-92.

66. Sheridan, J.J., Brawlty, G., Hastings, J., Electro-thermal debracketing. Part 1. An in vitro study. Am. J. Orthod., 1986, **89**, 21-27.

67. Zidan, O., Asmussen, E., Jergensen, K.D., Micro-scopical analysis of fractured restorative resin/etched enamel bonds. Scand. J. Dent. Res., 1982, **90**, 286-291.

68. Bennett, C.G., Shen, C., Waldron, J.M., The effects of debonding on the enamel surface. J. Clin. Orthod., 1984, **18**, 330-334.

69. Mulville, D.R., Vaishnav, R.N., Interfacial crack propogation. J. Adhesion, 1975, **7**, 215-233.

70. Williams, K.R., Edmundson, J.T., Orthodontic tooth movement analysed by the finite element method. Biomaterials, 1984, **5**, 347-352.

71. Williams, K.R., Edmundson, J.T., Morgan, G., Jones, M.L., Richmond, S., Orthodontic movement of a canine into an adjoining extraction site. J. Biomed. Eng., 1986, **8**, 115-120.

72. Tanne, K., Sakuda, M., Burstone, C.J., Three dimensional finite element analysis for stress in the periodontal tissue by orthodontic forces. Am. J. Orthod. Dentofac. Orthop., 1987, **92**, 499-505.

73. Desai, C.S, Abel, J.F., An introduction to the finite element method. Van Nostrand-Reinhold, New York, 1972.

74. Hinton, E., Owen, D.R.J., An introduction to finite element computations. Pineridge Press Ltd., Swansea, U.K., 1979.

FORMATION OF EXTRACELLULAR DEPOSITS BY MACROPHAGES EXPOSED TO POORLY DEGRADABLE BIOMATERIALS

H.K. KOERTEN, C.A. VAN BLITTERSWIJK, S.C. HESSELING,
J.D. DE BRUIJN, J.J. GROTE, AND W.TH. DAEMS
Biomaterials Research Group, Laboratory for Electron
Microscopy, University of Leiden, Rijnsburgerweg 10, 2333 AA
Leiden, The Netherlands

ABSTRACT

Foreign-body granulomas formed in reaction to the intraperitoneal injection of suspensions containing granulated calcium phosphates or crocidolite asbestos fibers were studied with transmission electron microscopy and X-ray microanalysis. It was found that the structure of the granulomas was dependent on the nature of the injected material. Macrophages found in the granulomas contained iron-rich inclusion bodies. Asbestos bodies were plentiful in the crocidolite asbestos-stimulated animals. Structures similar to asbestos bodies were also detected in the tri-calcium phosphate stimulated animals and in smaller numbers in the hydroxyapatite-stimulated animals. Like the results of our studies on the mechanism of asbestos body formation [1,2] the present findings show that macrophages encountering relatively large, poorly digestible particles are triggered to exocytose their lysosomal contents into a micro-environment formed between cells and material.

INTRODUCTION

In reports on the biological performance of biomaterials, the macrophage/material interaction at the implantation site is generally considered to be of great importance [3-6]. However, analysis of the macrophage/material interaction is complicated by the wound reaction caused by surgical implantation of a medical device or an artificial organ made of such biomaterials. This wound reaction will initially exceed the tissue reaction [6] and may thus obscure the macrophage/material interaction. Such complications can be avoided experimentally by injecting the foreign substance into the peritoneal cavity, a technique commonly used in studies on the function and origin of macrophages [7] and on macrophage/material interactions [3,8-10].

In the present study, the effects of three intra-peritoneally injected materials were compared. For this purpose use was made of a suspension of crocidolite asbestos, known to be a strong inflammatory stimulus [8], and suspensions of granular hydroxyapatite and tri-calcium phosphate, both known as bio-active resorbable calcium phosphates [11-14] that give very low inflammatory reactions [15].

MATERIALS AND METHODS

Animals

SPF bred female Swiss mice, obtained from Harlan Central Animal Breeding Center (Zeist, The Netherlands), were 6 weeks old and weighed about 20 grams at the start of the experiments. All mice were allowed to adapt for at least three days before the experiment was started.

Materials

The calcium phosphates used were hydroxyapatite and tricalcium phosphate, both in a granulated form with a particle size smaller than 40 μm. The asbestos was UICC crocidolite asbestos with fiber lengths ranging from <0.1 μm to 1000 μm.

Administration of the materials

All three materials were suspended in Hanks' balanced salt solution (HBSS). Per animal, 1 ml of the suspension under study was injected into the peritoneal cavity on day 0. To obtain comparable numbers of particles, we used a suspension of crocidolite asbestos with a final concentration of 0.5 mg/ml, and hydroxyapatite and tri-calcium phosphate were suspended to a final concentration of 5 mg/ml. Control animals were given 1 ml HBSS alone.

Isolation and fixation

The procedures used for the isolation and fixation of granulomas have been described in detail [1]. In short, after decapitation of the animals the granulomas were dissected and fixed by immersion in 1.5% glutaraldehyde. For a second group fixation was performed by perfusion of the total body with 1.5% glutaraldehyde before removal of the granulomas. Tissue blocks were postfixed in a modified OsO_4 fixative according to de Bruijn et al. (1973) [16].

Transmission electron microscopy (TEM)

The fixed tissue blocks were dehydrated in a graded series of alcohol up to 100% and embedded in Epon. Ultrathin sections were cut on an LKB microtome, stained with lead hydroxide, and examined in a Philips EM 410.

X-ray microanalysis (XRMA)

X-ray microanalytical spot analyses were performed with a Tracor (TN) 2000 X-ray microanalysor attached to a Philips EM 400 scanning transmission electron microscope. Sections collected on copper grids were placed in a beryllium low-background holder which was tilted at an angle of

18 degrees to enhance the X-ray yield. X-ray point analyses were done with a spot diameter of 400 nm during 100 sec lifetime.

RESULTS

Peritoneal granulomas

Material/cell aggregates formed after the intra-peritoneal injection of all types of particulate material were mainly found in contact with the omentum, liver and diaphragm.

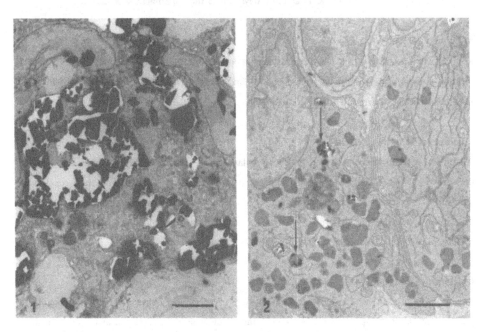

Figure 1. Transmission electron micrograph of a part of a giant cell in a tri-calcium phosphate granuloma at 2 months. Many endocytosed biomaterial particles are visible. Bar: 5 μm.
Figure 2. Part of a multinucleated giant cell in a crocidolite granuloma. Numerous iron-rich inclusion bodies, some of which containing asbestos fragments (arrows) are visible. Bar: 5 μm.

Transmission electron microscopy showed that aggregates of two or more weeks were encased in capsules formed by connective tissue. All granulomas showed basically the same organisation; a layer of mesothelial cells covering the connective tissue was observed. Besides fibroblasts and collagen, there were lymphocytes, plasma cells, mast cells, macrophages, and multinucleated giant cells. Large and small material particles, ingested or surrounded by the macrophages and multinucleated giant cells, were seen throughout the granulomas (Fig. 1). The majority of the macrophages contained iron-rich inclusion bodies (Fig. 2). The crocidolite-asbestos granulomas differed from the hydroxyapatite and tri-calcium phosphate granulomas by a higher density of the various cell types, especially in the center and relatively small intercellular spaces. Asbestos fibers were distributed throughout these granulomas. In the granulomas induced by hydroxyapatite and tri-calcium phosphate most of the biomaterial was encountered in the center, the intercellular spaces in these granulomas were larger. Further, the multi nucleated giant cells of the asbestos induced granulomas were larger (20 to 30 nuclei per sectioned syncytium) than those seen in the hydroxyapatite and tri-calcium phosphate granulomas (5 to 15 nuclei per sectioned syncytium). Particles too large to be phagocytosed were still in the extracellular space and were surrounded by macrophages, multinucleated giant cells, and collagen. Vascularization had occurred in all granulomas at one month and longer, the crocidolite asbestos granulomas were distinctly more vascularized than the others. No morphological differences between hydroxyapatite and tri-calcium phosphate granulomas were seen with TEM.

Extracellular depositions

As in an earlier study [1], four different types of asbestos body were found. The most frequently detected types were: type I with a homogeneous iron-rich coat (Fig. 3). Small asbestos particles were dispersed through the matrix of the coat of this type of asbestos body.

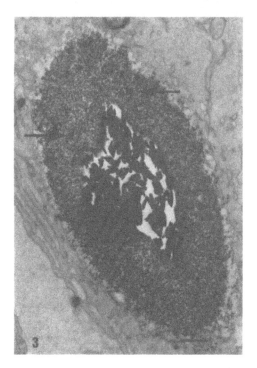

Another type (type II) had a lamellated coat in which concentric rings containing iron micelles with alternating densities were seen. This type too had small asbestos fragments but these were restricted to the In the granulomas containing crocidolite asbestos, asbestos bodies were frequently seen at the intervals of one month and longer. darker lamellae. Furthermore, small needle-like crystals resembling apatite crystals were sometimes seen on the edges of the lamellae and in the peripheral zones. XRMA of these needles gave peaks for calcium and phosphorus. Low concentrations of calcium were sometimes also detected by XRMA in the darker rings, but needle-like crystals like those just described were never found in these loci. XRMA of the rings showed that the darker rings had spectra with higher iron peaks than the lighter rings did. The third type, which was only seen occasionally, comprised asbestos bodies with an electron-lucid coat. Here the absence of electron density

Figure 3. Type I asbestos body. The central asbestos fiber is covered by a layer of iron-rich material. Small asbestos fragments are present in the coat (arrows). Bar: 1 μm.

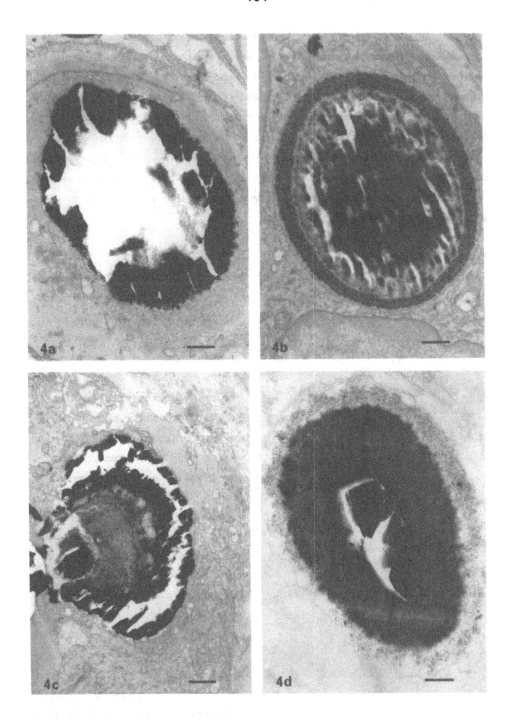

Figure 4, a-c. Extracellular deposits of crocidolite and tri-calcium phosphate granulomas.
a: Type IV asbestos body, b: combination of type IV and type I asbestos body. c: extracellular deposit type IV against tri-calcium phosphate granules. d: combnation of type IV and type I extracellular deposit against tri-calcium phosphate granules. Bar: 1 μm.

was due to the relatively low numbers of iron micelles in the various lamellae. Small asbestos fragments, as detected in asbestos bodies of types I and II, were never seen in the coat of type III asbestos bodies. The fourth type of asbestos body contained dark non-granular material. Remarkably, the coat of such asbestos bodies was always damaged by ultrathin sectioning. (Fig. 4a). XRMA of the non-granular coats showed the presence of calcium and phosphorus, iron was present in low concentrations in these areas. Finally, there were asbestos bodies with a morphology of type IV in the center and of type I in the periphery (Fig. 4b).

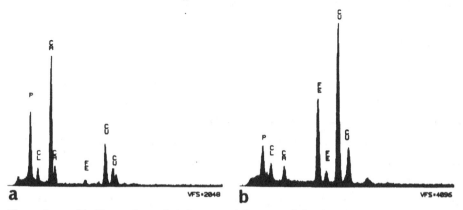

Figure 5, a and b. X-ray microanalytical spectr of the extracellular deposit in Fig. 4d.
a: Spectrum of the central part of the deposit. Note the distinct peaks for calcium and phosphorus. (Vertical full scale 2,048 counts). b: Spectrum of the granular ring. This spectrum shows that iron is present in a high concentration at this part. (Vertical full scale 4,096 counts).

Structures similar to asbestos bodies were also present in the granulomas containing one of the calcium phosphates. However, differences related to the type of calcium phosphate were observed. Granulomas of hydroxyapatite-stimulated animals seldom showed deposits resembling asbestos bodies, but tricalcium phosphate gave rise to the formation of numerous deposits with a morphological resemblance to asbestos bodies type IV (Fig. 4c). and to the combination of asbestos bodies of types I and IV (Fig. 4d). As in true asbestos-body formation, small fragments of tri-calcium phosphate were present in the granular rings of this type of deposit. Deposits resembling type I asbestos body were not seen in the calcium phosphate-induced granulomas.

XRMA of the matrices of the extracellular deposits showed that, like the type-IV asbestos body, they were composed of calcium phosphate-rich material containing iron in a low concentration (Fig. 5a). The deposits representing type I/IV gave similar XRMA patterns in a spot analysis of the center, whereas the granular rings showed high concentrations of iron (Fig. 5b).

DISCUSSION

The mouse peritoneal cavity has frequently been used to study the origin and function of macrophages [7,18,19] as well as the reaction of macrophages to the introduction of different stimuli

[1,7-10]. The demand for new biomaterials in reconstructive surgery has created a need for reliable tests to assess tissue tolerance [20,21]. Since macrophages generally come into direct contact with implanted biomaterials, this type of cell is assumed to be the most important of the cells functioning at the implant interface and reacting with the implant [3,4]. This means that extensive study of the macrophage/biomaterial interaction is required. The mouse peritoneal cavity has proven to be a suitable compartment for such studies [8-10,19].

The present investigation concerned processes occurring at the interface between three different materials and the macrophages present in peritoneal granulomas.

Iron accumulation

It is known that the iron concentration in phagocytes increases with age [22], but also as a reaction to the presence of poorly degradable materials [1,14]. The source of the iron is uncertain, but we assume that to a certain extent the iron pool in macrophages reflects the availability of iron in the extracellular environment. This assumption is supported by observations of other authors, who have shown that both a hemorrhage and injection of blood give rise to the formation of iron-rich inclusion bodies in macrophages [23-26].

An increased availability of iron in the present investigations could have been due to leakage of new blood vessels formed during the process of granuloma formation. The present study has shown that the amount of iron accumulated was dependent on the type of stimulus used. Crocidolite asbestos led to the formation of more iron-rich inclusion bodies than the two biomaterials did, the amount of iron present in macrophages in the direct vicinity of an implant must therefore to a certain extent be related to chemical or physical properties of the implant. Such correlation between the type of implant and the rate of iron accumulation could be valuable in studies on the biocompatibility of materials.

Extracelluar deposits

Extracellular deposition of materials normally present in the lysosomes of peritoneal macrophages has already been described in relation to the formation of asbestos bodies [1,2]. Those studies led us to conclude that where macrophages and/or multinucleated giant cells encounter particles too large to be completely ingested they seal off a part of that material to form a micro-environment between cell and material. Next, these cells exocytose their lysosomal content into this micro--environment in an attempt to digest the asbestos fibers extracellularly by a process similar to the resorption of bone by osteoclasts. Materials and iron present in the lysosomes by chance are also deposited in this micro-environment, and this leads to the formation of extracellular deposits. On the basis of morphological and chemical criteria, different types of extracellular deposits formed by this process can be distinguished. In one of our previous reports [1] we described four types of asbestos body, and in that study we saw combinations of types I and IV as well. In the present study, too asbestos bodies were formed after crocidolite asbestos stimulation and the four types were again recognized. Deposits resembling type-IV asbestos bodies and the combination of types I and IV were abundant in the tri-calcium phosphate-induced granulomas, but were seldom found in granulomas formed in reaction to hydroxyapatite.

The formation of deposits on non-asbestos fibrous materials has been reported [27,28], but such formation on non-fibrous materials has, to the best of our knowledge, never been described. Since the electron-dense deposits we saw in proximity with the calcium phosphates were similar to those seen in our previous studies and because small fragments of biomaterial were found in these deposits, we concluded that the structures seen in the present study are also related to a process of extracellular deposition.

Thus the present study has provided evidence supporting the hypothesis that the process leading to asbestos-body formation is not a specialized process occurring only when macrophages encounter asbestos fibers or a select group of other fibrous materials, but is a general form of behavior shown by macrophages and giant cells when they make contact with large particles of poorly digestible materials. The observations of the present study demonstrated that the various materials tested showed differences in the number of extracellular deposits formed. It therefore seems likely that the formation of extracellular depositions found in the present study is related to surface characteristics of the materials in question and thus provides direct information on the bioactivity of biomaterials.

REFERENCES

1. Koerten, H.K., Hazekamp, J., Kroon, M., Daems, W.Th., Asbestos body formation and iron accumulation in mouse peritoneal granulomas after the introduction of crocidolite asbestos fibers. Am. J. of Pathology 1990, 136(1):141-157

2. Koerten, H.K., de Bruijn, J.D., Daems, W.Th., The formation of asbestos bodies by mouse peritoneal macrophages: an in vitro study. Am. J. of Pathology 1990, 137(1):1-13

3. Harms, J., Mäusle, E., Tissue reaction to ceramic implant material. J. of Biomedical Materials Research 1979, 13:67-87

4. Chambers, T.J., The response of the macrophage to foreign material. In: Fundamental Aspects of Biocompatibility, Williams, D.F., Ed. CRC Press, Boca Raton, Fl, 1, 145, 1981

5. Anderson, J.M., Miller, K.M., Biomaterial biocompatibility and the macrophage. Biomaterials 1984, 5:5-10

6. Anderson, J.M., Inflammatory response to implant materials. ASAIO Trans (ASA) 1988, 34(2):101-107

7. Daems, W.Th. and Koerten, H.K., The effects of various stimuli on the cellular composition of peritoneal exudates in the mouse, Cell Tiss Res. 190, (1978) 47-60

8. Koerten, H.K., Brederoo, P., Ginsel, L.A., Daems, W.Th., The endocytosis of asbestos by mouse peritoneal macrophages and its long-term effect on iron accumulation and labyrinth formation. European Journal of Cell Biology 1986, 40:25-36

9. Pizzoferrato, A., Vespucci, A., Ciapetti, G., Stea, S., Tarabusi, C., The effect of injection of powdered biomaterials on mouse peritoneal cell populations. J. of Biomedical Materials Research 1987, 21:419-428

10. Koerten, H.K., Blitterswijk, C.A. van, Grote, J.J., Daems, W.Th., Accumulation of trace elements by macrophages during degradation of biomaterials. In: Implant Materials in Biofunction, C. de Putter, C.L. Lange, K. de Groot and A.J.C. Lee Eds. Advances in Biomaterials, 8, Elseviers Science Publishers, Amsterdam, 1988

11. Driessens, F.C.M., Formation and stability of calcium phosphates in relation to the phase composition of the mineral in calcified tissue. In: K. de Groot (ed.): Bioceramics of calcium phosphate, pp 1-26. CRC Press, Inc. (1983)

12. Blitterswijk, C.A. van, Grote, J.J., Kuijpers, W., Blok-van Hoek, C.J.G., Daems, W.Th., Bioreactions at the tissue/hydroxyapatite interface. Biomaterials 1985, 6:243-251

13. Klein, C.P.A.T., Driessen, A.A., Groot, K. de, Hooff, A. van den, Biodegradation behaviour of calcium phosphate materials in bone tissue. J. Biomedical Materials Research 1983, 17:769-784

14. Blitterswijk, C.A. van, Grote, J.J., Koerten, H.K., Kuijpers, W., The biological performance of calcium phosphate ceramics in an infected implantation site. III. Biological performance of ß-Whitlockite in the non-infected and infected rat middle ear. J. Biomedical Materials Research 1986, 20:1197-1218

15. Blitterswijk, C.A. van, Grote, J.J., Biological performance of ceramics during inflammation and infection. CRC critical reviews in biocompatibility 1989, 5(1):13-43

16. Bruijn, W.C. de, Glycogen, its chemistry and morphologic appearance in the electron microscope; I. A modified OsO4 fixative which selectively contrasts glycogen. J. Ultrastructure research 1973, 42:29-50

17. Bruijn, W.C. de, Koerten, H.K., Cleton-Soeteman, M.I., Blok-van Hoek, C.J.G., Image analysis and X-ray microanalysis in cytochemistry. Scanning Microscopy 1987, 1(4):1651-1667

18. Bakker, J.M. de, On the origin of peritoneal resident macrophages: An electron microscopical study on the heterogeneity of peritoneal macrophages. Thesis 1983, Leiden

19. Ginsel, L.A., Rijfkogel, L.P., Daems, W.Th., A dual origin of macrophages? Review and hypothesis. In: Macrophage biology (S. Reichard, M. Kojima, eds.), Alan R. Liss, Inc., New York 1985. p. 621-649

20. Hench, L.L., Ethridge, E.C., Biomaterials: An Interfacial Approach, Academic Press, New York, 1982

21. Grote, J.J., Biomaterials. In: Scott-Brown's Otolaryngology, Fifth Edition, A.G. Kerr (ed.), pp 612-618, Butterwords London (1987)

22. Leeuw, A.M. de, The ultrastructure of sinusoidal liver cells of aging rats in relation to function. Thesis 1985, Utrecht

23. Lalonde, J.M.A., Ghadially, F.N., Ultrastructure of experimentally produced subcutaneous haematomas in the rabbit. Virchows Arch B Cell Pathol 1977, 25:221-232

24. Lalonde, J.M.A., Ghadially, F.N., Ultrastructure of intramuscular haematomas and electron-probe X-ray analysis of extracellular and intracellular iron deposits. J Pathol 1978, 125:17-23

25. Ghadially, F.N., Schneider, R.J., Lalonde, J.M.A., Haemosiderin deposits in the human cornea. J Submicrosc Cytol 1981, 13(3):455-464

26. Ghadially, F.N., Haemorrhage and haemosiderin. J Submicrosc Cytol 1979, 11(2):271-291

27. Botham, S.K., Holt, P.F., The development of glass-fibre bodies in the lungs of guinea pigs. Journal of Pathology 1971, 103:149-156

28. Sebastien, P., Gaudichet, A., Bignon, J., Baris, Y.I., Zeolite bodies in human lungs from Turkey. Laboratory investigation 1981, 44(5):420-425

SUCCESSES AND FAILURES FOLLOWING 5 YEARS OF CLINICAL EXPERIENCE WITH THE SURGICRAFT ABC PROSTHETIC ANTERIOR CRUCIATE LIGAMENT

*T.K.O'Brien, PhD., +A. McLeod, Bsc., +W D Cooke, PhD., ++M Mowbray, MSc, FRCS., ++A Rees, FRCS., **B Shafighian FRCS., ***A E Strover, FRCS.

ABSTRACT

Successful ligament prosthetic surgery involves the rational design of a ligament "system". Such a "system" should have inputs from professionals in the fields of bioengineering, textile technology, and surgery. This paper reexamines the ABC scaffolding ligament design philosophy and surgical technique, from a multidisciplinary standpoint, in light of clinical results obtained from the follow-up of 198 patients over a 5 year period.

Retrosynovial positioning of the ligament has been shown to increase the likelihood of achieving tissue ingrowth into this scaffolding prosthesis. Such fibrous tissue penetrates through the entire substance of the intrarticular portion of the ligament. Tissue showing orientation in the longitudinal axis of the prosthesis. When achieved, full coverage of the ligament correlated to a reduction in synovial contamination with particulate debris, and a lower incidence of specific granulomatous reaction and consequent synovitis (0.5%).

Postoperative functional and arthrometer assessment, has shown that the ABC ligament restored stability to the majority of patients in this study. The mean Lysholm score increased from 66% preoperatively to 94% at follow-up, with 76% of patients having a score of >90%. Ligament fracture was the commonest complication (9%) and reason for return of anterior drawer to preoperative status. Electron microscopy studies of the morphology of textile damage, has identified the bone tunnel exit hole on the tibial plateau as a previously undescribed stress raiser; potentiating early fatigue failure of the ligament. Prosthesis failure due to inappropriate surgical technique eg notch-plasty has not proven to be the major source of breakage in this series.

The implications of these observations for both autologous and prosthetic reconstruction of the anterior cruciate are discussed and suggestions made to further improve the results obtained from such reconstructive surgery.

*Surgicraft Ltd, Redditch, Worcs
+Dept of Textiles, UMIST, Manchester
++Mayday Hospital, Croydon
**Ealing General Hospital, London
***Droitwich Private Hospital, Worcs

INTRODUCTION

Surgical repair of the ruptured anterior cruciate
ligament is one of the most challenging questions for the
orthopaedic surgeon. The majority of referred patients
will have had a long standing chronic injury, damage to
additional structures/constraints of the knee and be in
the process of developing early signs of osteoarthritis.
Such patient heterogeneity complicates the study of the
anterior cruciate deficient knee. Additionally, the
patient is usually young, sporting and demanding of a
repair to a severe injury, that will require minimal
rehabilitation and possibly need to last for more than 50
years. During that period the knee will flex and extend
under load at least 200 million times in the normal
individual and potentially more in a highly active adult.
Our understanding of the forces that the anterior
cruciate ligament is subjected to, are incomplete and
calculated from indirect measurements in cadaveric (1)
and forceplate analysis studies (2) on non pathological
knees.

An attractive approach to the repair of the cruciate
ligament is the use of a prosthetic/alloplastic material
(3,4,5), thereby avoiding the morbidity associated with
harvesting of autologous material(s) and lengthy post
operative rehabilitation. The development of this field
of prosthetic cruciate surgery will depend on the
parallel maturation of our knowledge of : biomaterials,
ligament and prosthetic biomechanics, preoperative and
postoperative assessment of pathology and surgery. The
authors of this study are professionals from the fields
of : biomechanics, textile technology and surgery and
form a multidisciplinary group, that have collaborated on
the study of the prosthetic reconstruction of the
anterior cruciate ligament with the ABC (Surgicraft Ltd)
ligament prosthesis. This collaboration has provided
objective results, that have suggested modifications to
both surgical technique and prosthetic design, that
should further enhance the success rate of both
autologous and prosthetic ligament surgery and reduce the
potential for ligament breakage.

MATERIALS AND METHODS

Patient Population
Subjects were all patients with chronic anterior cruciate ligament deficient knees and symptoms of instability (giving way and anterior drawer) incompatible with daily living, sporting expectations or work activity.

ABC Carbon Fibre and Polyester Ligament Prosthesis
The ABC (Surgicraft Ltd - Redditch, UK) anterior cruciate ligament prosthesis (6,7) is composed of a central core of carbon fibres partially over braided into a "Zig - Zag" configuration and reinforced at the fixation sites by a secondary over braiding of polyester. The ligament has been bioengineered to have defined mechanical characteristics that resemble those of the natural anterior cruciate ligament. Fixation is via a carbon fibre reinforced polysulphone bollard at the femoral end and second bollard or pure polysulphone button on the tibial bone.

Surgical Technique
The ABC ligament was implanted either under arthroscopic control or via an open arthrotomy technique. The ligament was passed into the joint through a medial tibial drill hole at an angle of 60 degrees to the tibial plateau. It then passes superiorly towards the posterior lateral femoral condyle and thence "over the top" and fixed on good cortical bone on the femur. In most cases an attempt was made to ensure the pathway of the ligament was positioned retrosynovially (8).

Patient Assessment
Patients were assessed with both objective (KSS knee arthrometer - Acufex, Portsmouth, UK) and subjective (Lysholm, Pivot Shift, Lachman Sign, Tegner Activity Score) methodology. Whenever possible patients were encouraged to undergo a "second look" postoperative arthroscopy, where the implant was inspected for ingrowth, impingement on the femoral condyle and synovial biopsy of the suprapatella pouch undertaken. Synovial samples were mounted, stained and examined under normal and polarised light for histology and the presence of carbon and polyester fibre particles.

Morphological Analysis of Textile Damage
Returned broken ligaments were photographed microscopically to record tissue penetration and position of damage. Ligaments were then soaked in a solution of collagenase, in a potassium based buffered solution, until free of tissue. The presence of tissue was determined by the application of the stain Rhodamine,

which clearly identified any residual tissue as a red colour against the background of the white prosthetic polyester material. Filament samples were then mounted and spluttered with a monomolecular layer of gold and examined under the scanning electron microscope.

RESULTS

Ligament Ingrowth at "Second Look" Arthroscopy
A group of 40 patients underwent "second look" arthroscopy at a mean post operative time of 5.5 +/- 3.4 months (range 3 - 19 months). At that time 27 patients were observed to have ligaments fully covered with autologous tissue, 6 patients had partial coverage of the ligament and in 7 patients the ligament was completely free from any tissue. Subgroup analysis of these results show that patients who had their ligaments approximated to, or covered by, autologous tissues at the time of operation had statistically greater chance (p<0.001, Chi 2) of the ligament being ingrown by host tissues at arthroscopy (85% fully covered), than those implanted without such a manoeuvre (38% fully covered).

Tissue Ingrowth Into Explanted Ligaments
Optical micrographs of partially digested/cleaned explanted ligaments shows clear evidence of ingrowth into the centre of the ligament (Figure 1) and the orientation of collagen fibres in the longitudinal axis of the ligament (Figure 2).

FIGURE 1
Partially Collagenase Treated Ligament - Showing the Interstand Penetration of Tissue

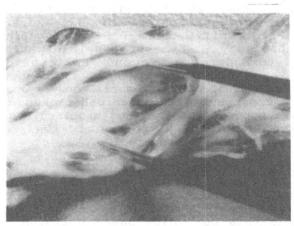

Notes: This ligament has been partially cleaned to expose the penetration of an organised tissue matrix.

FIGURE 2
SEM Evidence of Tissue Orientation

Notes: This fibre was obtained from an explanted
ligament that had been implanted for 32 months and shows
orientation of collagen fibres on a polyester filament of
approximately 20 um diameter.

Histological Analysis of Synovial Reaction to Particulate Debris

In a second study, 49 suprapatella synovial biopsy
specimens were obtained from 45 patients at a mean time
of 18 months (range 6 - 47 months) postoperatively. As
an negative control; synovial biopsies were also obtained
from 12 patients peroperatively during ligament
reconstruction.

Histological analysis (Table 1) shows that the synovial
appearance is influenced by the presence of instability
itself - 50% of the control biopsy specimens showing a
non-specific inflammation, with the presence of plasma
cells, macrophages and lymphocytes. The presence of
particulate debris from the implant appearing to be more
specifically, though not invariably so, associated with
the observation of a synovial chronic granulomatous
reaction and multinuclear giant cells.

TABLE 1

**Histological Appearance of the Synovium Peroperatively
and at Second Look Arthroscopy**

	Control Knees	Post Implantation Knees
Normal	50%	21%
Non Specific Reaction	50%	26%
Specific Reaction	0%	53%

Notes: Synovial samples were obtained from : 1. Control knees suffering from chronic instability immediately prior to implantation of the prosthesis and 2. Post ligament reconstruction knees at "second look" arthroscopy.

The presence of a specific granulomatous reaction to particulate debris is less likely in the presence of full coverage of the ligament at second look arthroscopy (p<0.01). In no patient was the presence of a synovial reaction associated with the death or focal necrosis of synovial cells.

Patient Follow-Up : - Subjective, Functional and Objective Results
The following results relate to the follow-up of 87 patients more than 2 years postoperatively. Mean follow-up was 35 months (range 2-5 years). There were 12 female and 75 males.

Subjective:- The mean preoperative Lysholm score was 66% (range 36-86%) and at the time of follow-up this had risen to 94% (range 60-100%). 76% of patients had a score of greater than 90% postoperatively.

When asked to rate the success of their reconstruction; 67% of patients rated it as excellent/good, 20% fair and 13% poor.

Functional:- Using the Tegner activity score 69 patients had a preinjury score of 7 and above, with a mean score of 8. The mean time from surgery to injury was 37 months and at the time of surgery no patient scored >7. Postoperatively 35 patients scored Tegner score 7 and over, with a mean score of 6. The mean decrease in score from the preinjury status was 1.46.

70 patients played competitive or recreational sports before injury, of these 49 returned to sport; 33 to their original level.

Objective: Using the KSS machine the mean preoperative side to side difference was 6.7 mm compared to 4 mm postoperatively at 200 N. Excluding patients with torn ligaments the postoperative side to side difference was 3mm.

Complications
These figures relate to the whole group of 198 patients. Overall there were 3 DVT's (1.5%), 1 synovitis (0.5%), 3 infections (1.5%), 18 torn ligaments (9%) and 20 ligaments required retightening (10%) due to inappropriate pretensioning of the ligament in the factory.

Studies on Textile Damage Morphology
To date the commonest site of ligament damage is adjacent to the tibial hole exit point. Four ligaments have been cleaned of tissue and examined under scanning electron microscopy for the mechanism of the break. 3 out of the 4 ligaments show clear signs of fatigue damage to the ligament (Figure 3). This is invariably localised to polyester fibres immediately adjacent to the tibial drill hole.

FIGURE 3
Multiple Splitting Fatigue of Polyester Fibre

Notes: This form of fatigue damage is associated with flex fatigue.

1 ligament has shown evidence of unambiguous crushing abrasion to the polyester fibres (Figure 4). This is the flattening and cracking of fibres due to their being forced against bone.

FIGURE 4
Crushing Abrasion of Polyester Fibres

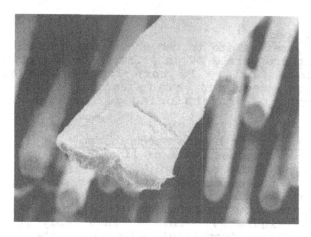

Notes: This ligament was described at reoperation as having failed at the notch and the textile morphology is consistent with failure through abrasion due to impingement on the lateral femoral condyle.

CONCLUSIONS

The repair of the anterior cruciate ligament, whether by autologous repair, augmentation or prosthetic reconstruction, is an extremely contentious and sometimes subjective issue. This appears to be especially true for the use of prosthetic materials in the knee. What has been lacking are clear objectives universally accepted criteria, for the evaluation of successes and failures in this field. This situation is, however, changing and with the development of knee rating systems, arthrometer devices, textile morphological analysis and non invasive techniques such as NMR, we are beginning to develop the tools that will allow further progress in this field. It is the belief of this group that there is a place for the use of prosthetics in the reconstruction of the anterior cruciate ligament. We have attempted to use all available methodology, from a variety of disciplines, to challenge some of our preconceptions and beliefs regarding the ABC ligament.

The ABC Ligament - A Biological Scaffold?

The results clearly show that the anatomical positioning of the ligament plays an essential role in the generation of a fibrotic/healing reaction.

It must be positioned retrosynovially and covered with the remnant of the old ligament and/or with a pedicle of fat pad. If the ligament passes through the joint in front of the synovial curtain then it will simply be covered with a layer of thin synovial tissue.

When approximated to vascular tissues the prosthesis progressively becomes infiltrated with a fibrous tissue, that penetrates through the intrarticular portion of the ligament. This tissue is not simply disorganised scar tissue, but shows organisation in the longitudinal axis of the ligament and is possibly load sharing with the prosthesis.

Ingrowth - What Use Is It?

The use of a scaffolding ligament is in our view crucial to long term success. The ingrowth achieved has many potential functions. Wear particles have been shown to induce the production of neutral proteinases that may be implicated in the degeneration of articular surfaces (9). This study shows that the incidence of a granulomatous reaction to the products of abrasion is markedly reduced if the ligament is fully covered. In no case was there observed focal necrosis of synovial cells as previously described by Ruston and Dandy (10) - when they reviewed the fate of the synovium following carbon fibre ligament reconstuction. Additionally, the abrasion potential of the notch on the ligament will be lessened by the intervention of autologous tissues. Peroperative notchplasty may only give a temporary solution to notch abrasion - as the presence of the cruciates maintains the presence of the notch and the interaction between condyle and ligament may be a normal/natural event in vivo.

The role of ingrowth to strengthen the prosthesis is not yet clear. However, the orientation of the collagenous tissue in the direction of applied force is encouraging.

Ligament Failure - Why Does This Occur?

The use of the ABC ligament has restored stability to the majority of patients in this series. The main reasons for return of instability were: stretching of the prosthesis due to inappropriate factory preconditioning in the early patients and ligament fracture. Ligament stretching necessitated repositioning of the femoral bollard in 20 patients. The ligament was subsequently factory preconditioned to remove this problem. The majority of ligaments that broke were involved in new traumatic incidents. However, it was difficult to see how a ligament, that was twice as strong as the natural ACL, could fail and usually at the same anatomical location. SEM work has clearly shown the potential of the tibial drill hole to act as a stress raiser and cause localised ligament fatigue at this point. Both fatigue damage and inter-filament abrasion have been demonstrated at the point of rupture. Analysis of polyester filament morphology at sites away from the rupture zone have not shown fatigue or abrasion damage.

Fatigue damage is most likely related to a number of factors: tibial spine hypertrophy in the chronic knee, the sudden transition from the ligament being totally constrained to unconstrained at this point and finally, the presence of the ligament overbraiding close to the exit hole. We are currently developing an in vitro model to reduplicate these events, in order to investigate potential improvements such as: increasing the radius of the exit hole and moving the prosthesis overbraiding away from the tibial plateau.

In conclusion, the ABC ligament has been shown to restore stability to a majority of operated patients. Attention to the important principles of retrosynovial position, close approximation to vascular tissues and the radiusing of the tibial drill hole, to avoid not only abrasion but also fatigue failure, are crucial for the long term effectiveness of this operation.

REFERENCES

1. Noyes F.R., Butler D.L., Grood E.S., et al. Biomechanical analysis of human grafts used in knee ligament repairs and reconstructions. J Bone Joint Surg. 66A: 344-352, 1984.

2. Morrison, J.B. Functions of the knee joint in various activities - Biomed Eng. 573 - 580, Dec 1969.

3. Jenkins D.H.R. and McKibbin B. The role of flexible carbon fibre implants as tendon and ligament substitutes in clinical practice. J Bone Joint Surg. 62B: 497-499, 1980.

4. Bolton C.W. and Bruchman W.C. The GORE-TEX expanded polytetrafluoroethylene ligament. An in vitro and in vivo evaluation. Clin. Orthop. 196: 202-213, 1985.

5. Amis A.A., Campbell J.R., Kempson S.A., et al. Comparison of the structure of neotendons induced by implantation of carbon or polyester fires. J. Bone Joint Surg. 66B 131-139, 1984.

6. Strover, A., Hughes F., O'Brien T., Minns R.J., Mechanical properties of the ABC carbon and polyester fibre anterior cruciate ligament. Eng. in Med. 97-101, June 1989.

7. O'Brien T., Hughes F., Strover A., Mowbray M., Shafighian B, Prosthetic anterior cruciate ligament reconstruction - Biomechanical and functional performance. Interfaces in Medicine and Mechanics, Pub. Dotesios, Ed. K. Williams and T Lesser, Trowbridge, Wilts, 1989.

8. Strover A.E., Ed Williams and Firer P. The use of carbon fibre implants in anterior cruciate ligament surgery. Clin. Orthop. 198: 43-49, 1985.

9. Olson E.J., James D., Kang M.D. The biochemical and histological effects of artificial ligament wear particles: In vitro and In vivo studies. AM. J. Sports Med. 16: 558-570, 1988.

10. Rushton N, Dandy D.J., Naylor C.P.E. The clinical arthroscopic and histological findings after replacement of the ACL ligament with carbon fibre - J Bone Joint Surg. 65B: 308, 1983.

BONE CEMENT: RETRIEVED WEAR PARTICLES VS FABRICATED WEAR PARTICLES

J. EMMANUAL, J.G. EMMANUAL, A. HEDLEY, B. SAUER
Harrington Arthritis Research Center
1800 East Van Buren, Phoenix, AZ 85006, U.S.A.

ABSTRACT

Wear particles of bone cement were retrieved by enzyme digestion of membranes from failed prosthesis during revision arthroplasty of cemented hip implants. Bone cement particles were generated in the laboratory using ASTM standards in preparing acrylic bone cement using the Howmedica Simplex P. Both the samples were prepared for viewing in the scanning electron microscope. There were differences in size and shape between the clinically retrieved bone cement wear particles and the laboratory fabricated bone cement particles. The retrieved wear particles were rounded, with pitting and cracks on the surface and had sizes ranging from 5 microns to 200 microns. The fabricated particles were always angular with jagged edges and ranged in size from 11 microns to 180 microns. Size, shape and surface characteristics of wear particles determine the biologic response and is critical when comparing in vitro versus in vivo reactions of fabricated particles.

INTRODUCTION

In orthopedic surgery, acrylic cement is widely used in many different parts of the world. Long term in vivo studies and in vitro studies with particulate bone cement have revealed many complications that can be produced as a result of bone cement wear.

A typical reaction to bone cement wear particles is a giant cell reaction, an indication of the instability of the tissue at the cellular level. Bone cement wear debris of a particular size has been known to interfere with DNA synthesis (1). Osteolysis has been commonly reported with the presence of interleukin-1 (IL-1) and prostaglandin E2 (PGE2) in the tissue adjacent to the interface. The presence of these minute bone cement wear debris ultimately lead to failure of the implant in the long term.

However many of the studies regarding wear debris of bone cement and its effect on the host have been done in vitro with particles fabricated in the laboratory (2). There is much bias in selection for in vitro assays and their results are not always satisfactory. These variations in results on the same biomaterial can be attributed to different methods of particle preparation and interpretation of results (3). The use of artificially produced wear particles and not the actual wear debris per se, raises questions of shape, size and alterations in surface characteristics as a result of changes in vivo. Results of future studies would be reproducible and favorable comparisons made, by using clinically retrieved wear particles for in vitro assays.

This study was undertaken to compare the size and shape of fabricated bone cement particles and actual bone cement wear particles retrieved from membranes during revision arthroplasty.

METHODS AND MATERIALS

Bone Cement

(Howmedica, Simplex P) was prepared according to instructions per insert. Monomer viscosity, mixing characteristics, doughing time and setting time and temperature were determined according to ASTM standards. The bone cement was made into 5mm thick cylindrical pellets. A number of pellets were placed in a platen press and compressed at about 20MPa and then further crushed in a hollow metal tube to generate varying sized particles. These particles were compared with clinically retrieved bone cement wear particles.

Actual Bone Cement

Actual bone cement wear particles were retrieved by enzyme digestion. Femoral and acetabular membranes were removed during revision arthroplasty. The method is outlined in Table 1. Five percent formic acid was used to decalcify and finally dissolve any bone spicules that may be present after enzyme digestion.

TABLE 1
ENZYME DIGESTION METHOD

Fix the membranes in 10% buffered neutral formalin.

Mince larger fragments of tissue into smaller pieces and centrifuge at 3000rpm for 10 minutes.

Discard the supernatant and digest the pellet with papain and sodium sulfite in a ration of 1:10 at a ph of 6.8. Incubate the sample for three days at 63°C.

Centrifuge the sample at 3000rpm for 10 minutes and discard the supernatant and treat the pellet with 5% formic acid for 16 hours.

Centrifuge the sample again at 3000rpm for 10 minutes.

TABLE 1
(Con't)

Wash the pellet three times with distilled
water.
Remove the supernatant and dry the pellet in a
weigh boat in an oven at 37°C.
The particles are now ready for analysis.

In the last stage there were polyethylene fragments sometimes
which were removed. The bone cement wear particles and the
fabricated cement particles were then prepared for analysis in the
scanning electron microscope to determine their size and shape. The
size distribution of larger particles of bone cement were determined
by a grid method using a light microscope at a total magnification of
630X.

RESULTS

Fabricated Wear Particles
Figure 1a shows that the fabricated bone cement particles were
angular with jagged edges. Tiny cracks were seen along the flat
planes. Sizes of particles ranged from between 11 microns to 180
microns with macro sizes of 3mm.

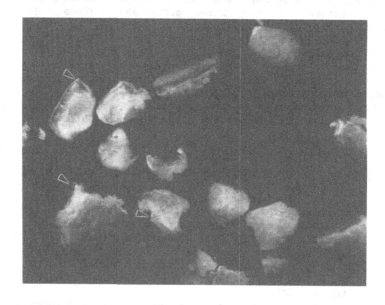

Figure 1a: Fabricated wear particles. (▲) Shows angular, jagged
              edges. Scanning electron micrograph 35X.

Actual Bone Cement Wear Particles

Figure 1b shows that the actual clinically retrieved wear particles were very different from the fabricated particles (see Table 2).

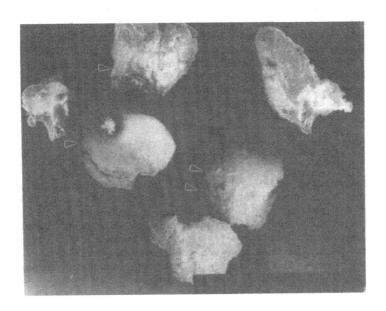

Figure 1b: Retrieved bone cement wear particles.   (▲) Shows rounded edges.   SEM 75X.

These particles were rounded with pits and cracks and some of them have the characteristic bead like appearance on the surface. The sizes ranged from 5 microns to 200 microns with larger particles being present, the largest being about 1cm in diameter.

TABLE 2

BONE CEMENT PARTICLE	COLOR	SIZE	SHAPE
Clinically Retrieved	Yellow	5um - 200um	Round
Fabricated	Grey/white	11um - 180um	Angular
			Jagged

DISCUSSION

Wear particle size and shape is a critical factor in determining the type of biological response to the particle, which could vary from an inflammatory response to a persistent immune response. There is an obvious difference in shape between the fabricated and actual bone cement wear particles. Fabricated particles were always angular and jagged. Sharp angular particles induce a different cellular response than spherical particles. In cemented prostheses the persistence of a macrophage response is due to the size of the bone cement particle (4). These particles induce enzyme degradation, inhibit DNA synthesis and cause necrosis of the cell. The difference in shape between fabricated bone cement particles and actual bone cement wear particles poses questions on the validity of past in vitro studies on the biologic effects of bone cement particles.

CONCLUSION

There is a marked difference between the size and shape of actual bone cement wear particles and those particles generated in the laboratory by various methods. These differences could have a profound effect on both in vitro and in vivo studies of the effects of wear debris on cells and tissues. The technique for retrieving actual wear particles is simple and the increasing number of revision arthroplasty makes it possible to retrieve adequate numbers of particles for in vitro and in vivo studies.

REFERENCES

1. L.C. Jones, D.S. Hungerford, Cement Disease. Clin. Orthop., 1987, 225, 192-206.

2. S.R. Goldring, In vitro cell culture system for characterizing biologic response to implant biomaterials. 34th Ann. Mtg., ORS, Georgia, 1988:54.

3. R.E. Sommerich, E.H. Chen, R.F. Jochen, Effects of wear particles on the joint: Literature review and a new surgical technique. J. Invest. Surg., 1989, 2, 3:341.

4. J. Charnley, Acrylic Cement In Orthopedic Surgery, Churchill/ Livingstone, Edinburgh, London, 1972.

ACKNOWLEDGMENT

This work is supported by the Del E. Webb Foundation.

CORTICAL BONE HYPERTROPHY AT THE TIP OF FEMORAL PROSTHESIS : BIOLOGICAL - MECHANICAL CORRELATION

M. PORTIGLIATTI BARBOS , C. VIGLINO

Dipartimento di Scienze Biologiche e Cliniche

Università di Torino ITALY

ABSTRACT

Bone adaptive remodelling around the tip of the hip prosthesis stem was studied from the histological point of view.

Diaphyseal bone thickening in 4 human prosthesized femurs is shown to be the result of new lamellar bone formation (periosteal bone) in which the degree of mineralization is lower than that of cortical bone.

All the phenomenon seems to be very similar to the normal way in which the long bones grow in section in young people (external circumferential system).

INTRODUCTION - MATERIALS AND METHOD

Cortical bone hypertrophy at the tip of femoral prosthesis or at the distal third of the stem is an item of frequent recurrence in orthopedic radiology (1, 3, 9, 14). Its occurrence may be correlated to the mechanical strains conveyed by the prosthesis to the surrounding diaphysis and interpreted as a functional adaptation to the redistribution of the stress and to the new mechanical conditions in which the bone tissue finds itself (2, 4, 7, 8, 10, 11, 12, 13, 15).

These consideration give rise to the interest in the histological study of diaphyseal bone hypertrophy at the tip of the femoral prosthesis.

Trasversal sections of four human specimens (fig. 1 and 2) were studied by means of polarized light to draw attention to the structure as well as by microradiography to determine the degree of mineralization. Diaphysis were transversally sectioned - without decalcification - at the point of typical fusiform thickening.

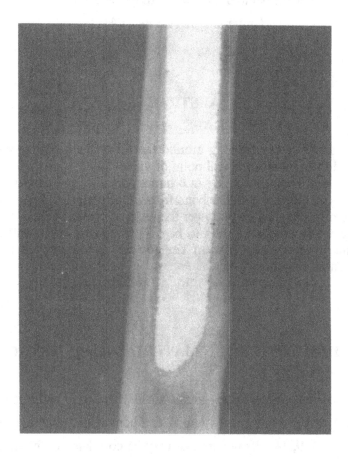

Figure 1. Cortical bone hypertrophy at the tip of femoral prosthesis : radiological finding

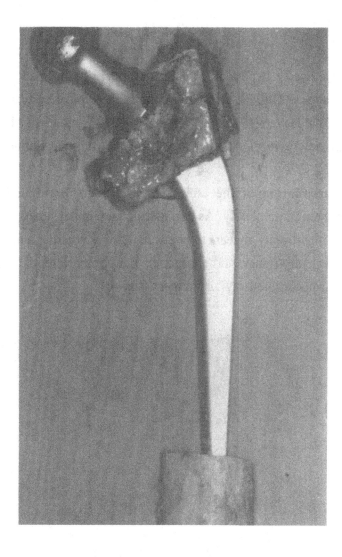

Figure 2. Human specimen of cortical remodelling around prosthesis stem

RESULTS

Histologically hypertrofic diaphysis appear to be made up of two distinct structures (fig. 3). Of the two one is the typical cortical diaphysis made up of the Haversian system; the other which is more peripheral, is the portion of newly formed bone, made up of a compact bone whithout osteons.

The orientation of the lamellae permits to distinguish the two structures even more clearly. As a matter of fact, under polarized light the diaphyseal hipertrophy appears to be made up of alternating lamellae which have a prevalently transversal orientation and circumrferential course; this make them appear intensely birifrangent (fig. 4).

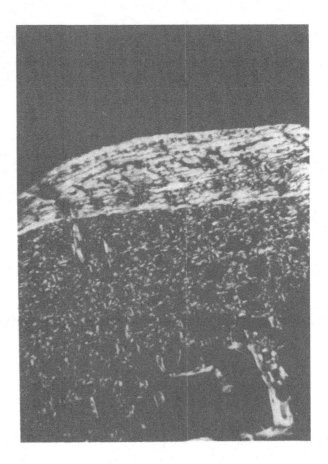

Figure 3.

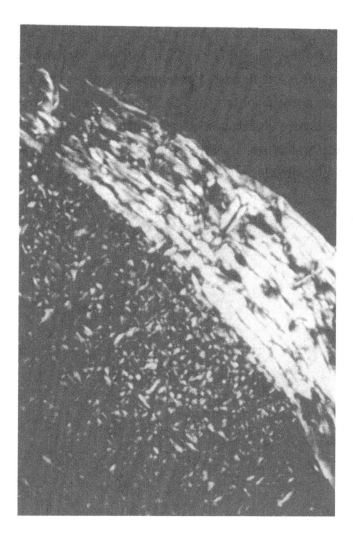

Figure 4

Figure 3 and 4. Femurs of two different patients. Non decalcified transversal sections, 80 micron thick: diaphyseal hypertrophy is histologically made up of circumferential lamellar systems (above). The intense birifrangence under polarized light makes it entirely distinguishable from the old cortical bone which is made up of Haversian system (below).

The phases of deposition of this newly formed bone are quite evident in microradiography. The process begin with the formation of a thin osseous layer with a low degree of mineralization that is deposited an the external surface of the diaphysis (fig . 5). This bone is of a periosteal type whose morphological characteristics are similar to those of the external circumferential system which is a residual trait in adults of physiological increase in width of the long bone.

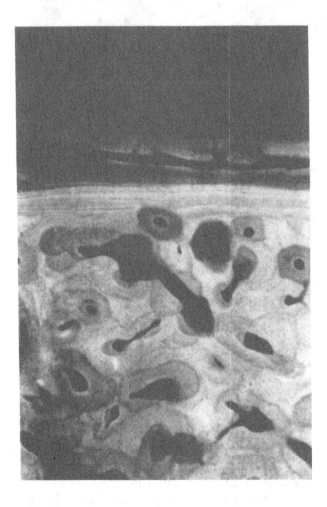

Figure 5. Femur. Microradiography of non decalcified transversal section, 100 micron thick: the layer of newly formed bone (above) presents a degree of mineralization inferior to theat of the diaphysis

This newly formed bone may thickens and, at the same time, the processes of the secondary mineralization begin and bring it to a degree of calcification which is similar to that of the compact diaphysis (fig. 6). Sometimes there is even a modest osteonic remodelling in the innermost portion of the hypertrophic zone (fig. 6). We have observed the same remodelling even in a case of clear resorption of the cortical bone around the prosthesis.

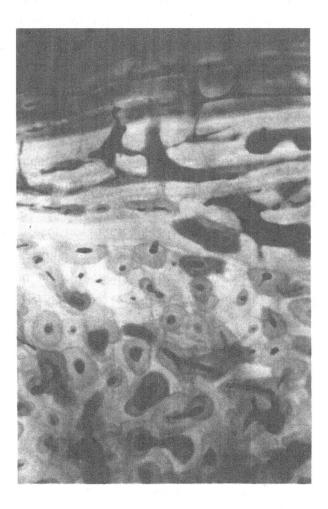

Figure 6. Femur. Microradiography of non decalcified transversal section, 100 micron thick: secondary mineralization of bone hypertrophy (above) in progress, with initial osteonal remodelling.

The circumferential lamellar system in this case are separated from the cortical bone by lacunar gaps which are allongated and parallel to the external surface of the diaphyisis.

CONCLUSIONS

The results of the histological study allow us to make several osservations: 1) Diaphyseal bone thickening is substantially made up of new lamellar bone formation (periosteal bone). All the phenomenon seems to be very similar to the normal way in which the long bones grow in section in young people (external circumferential system).

2) Clinical experience testifies for the presence of diaphyseal hypertrophy only in cases in which the transversal stresses at the tip of prosthesis are not too high for the cortical's resistance and therefore the bone remodelling can take place around it. These loads can be considered paraphysiological in the sense that the bone tissue is stimulated to structurally compensate for the ridistribution of stress.

3) The response of the bone to these paraphysiological loads is in itself paraphysiological and follows the same model of periosteal osteogenesis as bone in the stage of growth. In the whole process there is no signs of woven or coarse bone formation- which means emergency osteogenesis- while there is evidence of the slowness of deposition and the structural regularity of the secondary bone formation.

REFERENCES

1) Beckenbaug, R.D., Ilstrup, D. M., Total hip arthroplasty: a review of 333 cases with long follow up. J. Bone Joint Surg., 60 A, 306, 1978.

2) Berme, N., Paul, J.P., Load action transmitted by implants. J. Biomed. Engn., 1, 268, 1979.

3) Bertini, G., Calderale, M.P., Dettoni, A., Gallinaro, P., Lorenzi, G.L., Cortical bone changes after hip replacement: radiographic study. Acta Orthop. Belg., 46, 735, 1980.

4) Blaimont, P., Burny, F., Wagner, J., Application de l'extensiometrie à l'étude des réaction osseuses dans les arthroplasties de hanche. Acta Orthop. Belg., 2, 253, 1968.

5) Brach del Prever, E., Borroni, L., Portigliatti Barbos, M., Rottura dello stelo protesico. Rimodellamento osseo e problemi biomeccanici. Minerva Ortopedica, 38, 381-387, 1987.

6) Brach del Prever, E., Portigliatti Barbos, M., Protesi di Mittelmeier studio istologico e microradiografico a due anni dall'intervento. G.I.O.T., 10, 271-275, 1984.

7) Burny, F., Blaimont, P., Wagner, J., Précontrainte femorale liée à la prothèse d'Austin Moore. Acta Orthop. Belg., 34, 271, 1968.

8) Calderale, P.M., Gola, M.M., Gugliotta, A., New theoretical and experimental developments in the mechanical design of implant stems. Pawels Symp., Berlino, 1979.

9) Coventry, M.B., Beckenbaug, R.D., Nolan, D.R., 2012 total hip replacement: a study of postoperative course and early complications. J. Bone Joint Surg., 56, 273, 1984.

10) Gola, M.M., Gugliotta, A., A critical appraisal of theoretical and experimental analysis of the coupling of endoprosthesis and femour. I° China-Japan-USA Conf. on Biomechanics, Wuhan, 1983.

11) Jacob, H.A., Huggler, A.H., Spannungsanalysen an Kunstoffmodellen des menischen Beckens sowie des proximalen Femurendes mit und ohne Prothese. Pawels Symp., Berlino, 1979.

12) Jacob, H.A., Huggler, A.H., An investigation into biomechanical causes of prosthesis stem loosening within the proximal end of human femur. J. Biomechanics, 9, 1976.

13) Mizrahi, J., Livingstone, R.P., Rogan, I.M., An experimental analysis of the stresses at the surfaces of a Charnley hip prosthesis in different anatomical position. J. Biomechanics, 12, 491, 1979.

14) Oh, I., Harris, W.H., D'Errico, J, Effect of total hip replacement on the distribution of stress in the proximal femur. Harvard Med. School, pag.1, Boston, 1977.

15) Wolff, J., Das Gesetz der Transformation der Knochen. Hirshvald, Beerlin,1982

SUCCESS AND FAILURE OF EXPERIMENTAL ACTIVE GLASSES IN PERIODONTAL DISEASE REPAIR

*D. ZAFFE and °A.M. GATTI
*Institute of Human Anatomy, University of Modena
°Lab. of. Biomaterials, School of Dentistry, Univ. of Modena
Policlinico, Via del pozzo 71, 41100 MODENA, Italy

ABSTRACT

Four different ceramic materials, both inert and active, used in granule form for bone repair, were implanted in two sheep's jaws. After 3 months from the implant, the behaviour of the materials was studied by means of SEM and X-ray microprobe on thick sections of the embedded undecalcified jaws.

The unexpected success of the inert materials and the failure of the used bioactive glass opens the problem of the calibration of the degradation parameters that must be evaluated both on the massive material and also on the granules or powders.

INTRODUCTION

Nowadays active surface glasses are the most attractive materials in bone repair for their peculiarity to create a chemical bond with bone (1).

In periodontal disease repairs, some ceramic materials, like tricalcium phosphate (TCP) (2), hydroxyapatite (HA), are generally used. Powder or granules of these materials are forced into the cavity and encapsulated by new bone growing upon them thus creating a composite. The encapsulation occurs when the material is inert and does not degrade or solubilize. When a solubilization occurs after a certain time from the implant, as in the case of TCP, the bone cavity is reduced in size and it is impossible to distinguish the morphology of the granules.

When active glass particles are used (3, 4), only the external part of the granule interacts with the biological environment, whereas the core remains unchanged. In the outer part of the granule, a layer composed of hydroxylapatite binds the glass with the new-grown bone.

Aim of this work is to check the validity of these active materials in comparison with the inert ones.

MATERIALS AND METHODS

Four different materials were implanted in 2 Merinos sheep's jaws. The material used were:

- Tricalcium phosphate (TCP) in granule form (700-1400 microns) with a porosity of 60% developed by syntherization by Rob Mathis Company - Swiss (Ceros 82, art. n. 710.43 lot 2002).
- Bioglass as granules (250-425 microns), a bioactive glass developed by L.L. Henh (University of Florida) and manufactured by American Biomaterials Corporation (Princeton - NJ, USA) with the following composition: 45% SiO_2, 24.5% CaO, 24.5% Na_2O and 6% P_2O_5.
- GS 11 as granules (700-1500 microns), an inert glass developed by COVER (Italy) with the following composition: 66.9% SiO_2, 10.7% B_2O_3, 9,1% Na_2O, 6.2% Al_2O_3, 3.8% BaO, 1.4% K_2O, 0.9% ZnO, 0.6% CaO, 0.2% MgO and 0,06% Fe_2O_3.
- Zirconia as syntherized spheres (800 microns), developed by CERAM (USA)
The materials were sterilized in ethylene oxide gas before using.

With the sheep under general anaesthesia, the external left side of the jaw was transcutaneously exposed and five holes were drilled with a 4 mm diameter manual burr. One hole was left free as a reference and the other four were filled with the above-mentioned materials. The sequence of the materials was changed in the two sheep. At the end, the periosteum, muscles and skin were sutured according to the good medical practice. During the whole experiment time, radiographic controls of the zones were carried out monthly.

After 3 months from the implant, the sheep were sacrificed and the portions of the jaws containing the granules were fixed in 4% Paraformaldehyde for 4 hours, dehydrated and embedded in methylmethacrylate resin.

Cross sections (500 micron thick), taken from the levels of the implants by means of a diamond saw microtome (mod. 1600 - Leitz, Germany), were perfectly polished with emery paper and Alumina. A low resolution microradiograph of each section was made under X-ray generator (Italstructures, Italy) at 8 KV and 4 mA on EM Ilford Film. SEM observations (SEM 500 - Philips, The Netherlands) and X-ray Energy Dispersion System microanalyses (EDAX - Philips International) were carried out at 25 KV on these polished thick sections after sputtering their surface with a thin conductive carbon layer.

Sections of embedded non-implanted granules were prepared and analyzed like the implanted samples in order to compare the morphology and composition changes occurring in the implanted materials.

RESULTS

In almost all the cavities a certain amount of new bone growth was seen and in some cases this is similar to that observed in the reference hole.

The pocket containing the TCP granules appears to be closed already after 3 months from the implant (Fig. 1A). Few granules are still visible (in a smaller amount than the initial one) and the most part of them are completely surrounded by new formed bone. A close contact between TCP granules and the bone tissue is seen to occur without interposition of connective tissue (Fig. 1B). The new bone is not yet well structured, but it fits very well the rough surface of the granules. Some TCP granules disappeared due to biological degradation causing a release of Ca and P

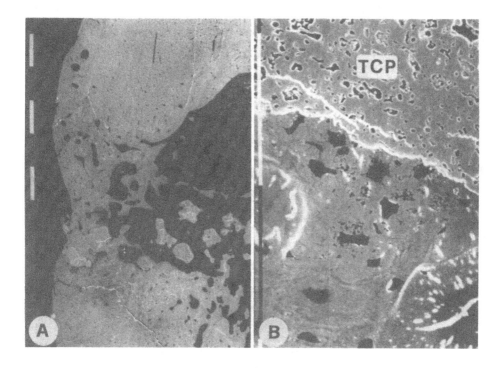

Figure 1. Microphotographs of the pocket containing the TCP granules. Note
in B several osteocyte lacunae in the bone in close contact
with the granule (TCP). (A, bar = 1 mm; B, bar = 0.1 mm).

ions.

A different result occurred in the Bioglass implants: only few
granules of the glass are still present in the hole and the diameter of
the cavity appears to be similar to the initial one (Fig. 2A). In the
center of the pocket few bone trabeculae are visible and some incorporate
glass fragments (Fig. 2B).

A change in the glass composition is apparent after 3 months from
the implant: Sodium and Silicon disappear from these degraded granules,
while the Calcium and the Phosphorus contents appear to be higher than
that of the non-implanted glass (Fig. 3). This evidence was confirmed by
the X-ray topographic analyses performed on cross sections of implanted
granules (Fig. 4). The microradiographic image of these sections shows

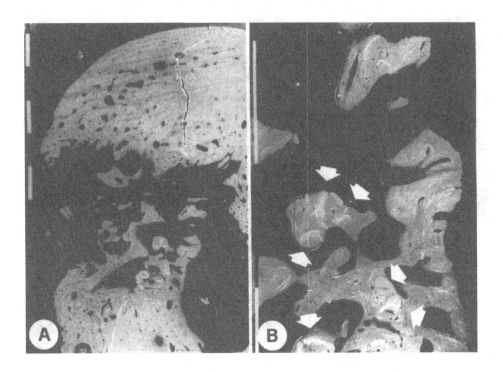

Figure 2. Microphotographs of the pocket containing the Bioglass granules. The arrows in B indicate some granules in the new-bone trabeculae. (A & B, bar = 1 mm).

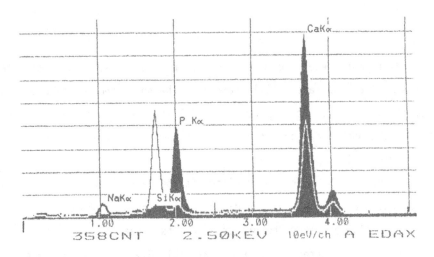

Figure 3. X-ray analyses of the Bioglass after 3 months from the implant (solid spectrum) and before the implant (superimposed blank spectrum plotted at the same scale).

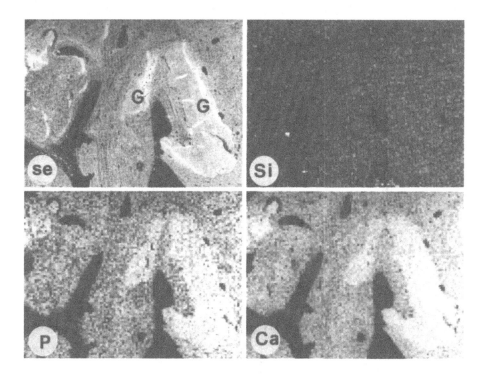

Figure 4. Digital scanning electron microphotograph (se) and digital X-ray dot maps for Silicon (Si), Phosphorus (P) and Calcium (Ca) of 3 month-implanted Bioglass granules (G). (x150.).

that the size of the granules corresponds to that of the SEM picture; i.e. the granule was sectioned across its center. The Calcium and Phosphorus appear to be uniformly distributed and Sodium and Silicon are not even present in the granule core. The Calcium content is now greater than that of the bone, whereas before the implant it was lower (5). A similar behaviour occurs for the Phosphorus that was usually detected with difficulty in the X-ray topographic analysis of the non-implanted glass.

The X-ray elemental analysis of the GS11 implanted glass shows an unchanged composition with respect to that of the non-implanted one (Fig. 5). This material has a formulation that corresponds to a glass not bonding the bone, but which undergoes a fibrous encapsulation by the connective tissue (region B in the glass compositional diagram (5) of

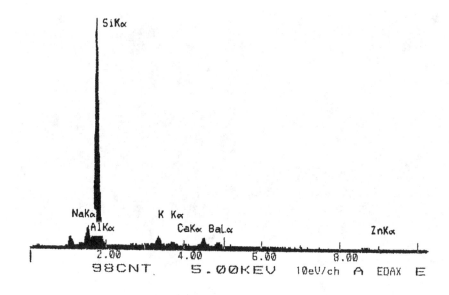

Figure 5. X-ray analysis of the GS11 glass after 3 months from implant.

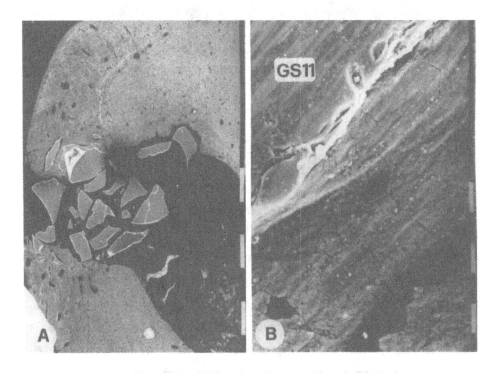

Figure 6. Microphotographs of the pocket containing the GS11 granules. In
B the bone appears to be in contact with the glass granule
(GS11). (A, bar = 1 mm; B, bar = 10 microns).

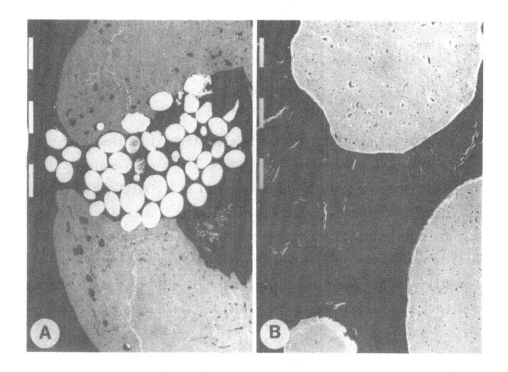

Figure 7. Microphotographs of the pocket containing the Zirconia spheres.
In B a fibrous layer separates the granules from the bone. (A,
bar = 1 mm; B, bar = 0.1 mm).

Hench). On the contrary, the SEM analyses of the implanted GS11 show a
bone growth only a little smaller than that of the TCP (Fig. 6A). Some
glass granules are entrapped by the new bone formation, but a thin
connective tissue layer can be seen in between. Portions of some granules
even appeared to be in contact with bone (Fig. 6B). The morphology of the
inner part of the glass granules is unchanged, thus confirming the absence
of glass degradation.

Zirconia spheres show a behaviour similar to that occurred in GS11
implants: no degradation of the ceramic material, good new bone growth and
unchanged internal   morphology (Fig.   7).   Probably due   to   their
comparatively large size, the spheres seem to be forced by the new-bone
growth into the mandibular canal or outside the hole. At   higher

magnification (Fig. 7B), a thick layer of connective tissue surrounding every sphere is seen.

CONCLUSIONS

This study underlines that, from the bone-growth point of view, also inert materials can work and sometimes do it better than bioactive glasses.

The tricalcium phosphate is the material that effectively increases the velocity of the bone growth. On the contrary, the velocity of the Bioglass degradation (interesting the whole granule and not only the outer part) is too high for a good bone growth. The ions released during the degradation change some local parameters, like the pH, that interact negatively with the bone deposition processes and lead to the failure of the bone repair.

The two inert materials, GS11 and Zirconia, different for chemical composition, morphology, size of the particles, show an identical result with respect to the bone growth. The spherical and smooth surface of Zirconia can be related to the presence of the thick fibrous layer, instead the rough surface of the GS11 glass can develop a direct bone contact.

The success and the failure of the active materials appear to be bonded to some parameters that must be calibrated accurately. In the case of inert materials a calibration does not seem necessary.

REFERENCES

1. Clark, A.E., Hench, L.L. and Paschall, H.A., The influence of surface chemistry on implant interface histology: a theoretical basis for implant materials selection. J. Biomed. Mater. Res., 1976, 10, 161-74.

2. de Groot, K., Degradable ceramics. In Biocompatibility of clinical implant, ed. D.F. Williams, Elsevier Sci. Publ., London, 199-222.

3. Klein, C.P.A.T., Abe, Y., Hosono, H. and de Groot K., comparison of calcium phosphate glass ceramics with apatite ceramics implanted in bone. An interface study II. Biomaterials, 1978, 8, 234-36.

4. Gross, U. and Strunz, V., The interface of various glasses and glass ceramics with a bony implantation bed. J. Biomed. Mater. Res., 1985, **19**, 251-71.

5. Hench, L.L. and Wilson. J., Surface-active biomaterials. Science, 1984, **226**, 630-35.

STRUCTURAL AND CELLULAR REACTIONS OF BONE TISSUE TO ORTHOPAEDIC IMPLANTS

Z.A.RALIS

Orthopaedic Research Laboratories
University of Wales College of Medicine
Cardiff Royal Infirmary,Cardiff, U.K.

ABSTRACT

Reaction of cortical bone and trabecular bone to implants were studied in sheep tibiae with experimental osteotomy plated by stainless steel and carbon - epoxy resin plates and in human femoral heads with inserted screws after femoral neck fracture. Monitoring of two new features - the osteocyte viability count and defective bone mineralisation allowed to detect not previously recognised, yet widespread degeneration and necrosis of bone cells accompanied by tinctorial and structural defects in the tissue. This generalised "distress" reaction generates the extensive resorption-reparative process which initiates the bone tissue rebuilding around the implants. Two simple methods for assessment of the Osteocyte viability count and for demonstration of the osteoid and Mineralisation defects in the ordinary decalcified paraffin sections are described. With these tools more new features, common in the remodelling bone tissue, were identified: bone tissue defect, repair bone malacia with splits,laminated bone and recurrent bone necrosis and repair.

INTRODUCTION

The reasons for aseptic (non-infective) loosening of orthopaedic prostheses and implants are not fully understood. The recognised microscopical changes at the interfaces are bone resorption and replacement by fibrous tissue or new bone remodelling but the rules which govern this process are not clear.

Bone response to implant is not dictated merely by its presence alone. It also involves a reaction to the bone damage for which the orthopaedic implant was inserted,usually a fracture or osteotomy, and to the injury caused by penetration or fixation of the implant.The response to the implant itself depends on its physical properties, shape, chemical composition and other aspects widely discussed in the literature, on the way in which the site is prepared and implant inserted, and how all this damages the bone blood supply.

In recent years during morphological studies of a large amount of surgical and experimental bone material, we included assessment of two important features: the differences in appearance of individual bone cells which reflect their functional state and viability, and the level of bone tissue mineralisation, including the defects in the mineral and matrix. Two simple histological techniques were developed which make it possible to distinguish, in ordinary decalcified paraffin sections, the non-mineralised osteoid tissue as well as defectively mineralised matrix against the normally mineralised bone [1,2].

For monitoring of the differences in morphology and viability of bone cells, whose fundamental importance for the metabolism, maintenance and reactions of the bone tissue had not received sufficient attention, an osteocyte viability count (OVC) was introduced [3].

By following the cellular morphology and matrix quality defects - features which had not been recorded before - we found that in bone specimens with various fixed orthopaedic implants, before any histological resorption or remodelling take place, the cells in a large part of the specimen, even at a distance from the implant react and show signs of damage, pyknosis and necrosis.This is followed by appearance of multiple differently stained and structurally altered areas in the tissue (Figure 2A,2B). This almost invisible but extensive reaction is limited to bone tissue elements and is not related to the local damage of blood supply by the surgical injury or implantation. Such generalised reaction of the bone tissue is in agreement with the very law of its existence.The attached implant interferes with the specific character, magnitude and direction of the physiological stresses normally imposed on the tissue, for which the bone is structured and without which it cannot thrive. The generalised early cellular and tissue quality changes due to the stress confusion were not previously reported. Yet they fundamentally change the overall structure of the bone and the way it will respond to the implant.

In the present work are described the following new cellular and structural reactions of the osteotomised cortical bone to AO plates and of the cancellous bone in human fractured femoral heads to the inserted screws: OVC and osteocyte degeneration, bone tissue defects (BTD) with cracks in the matrix, repair bone malacia (RBM) with tissue splits, laminated bone (LB) and recurrent bone necrosis and repair (RBNR). The known reactions to implants such as cortical necrosis and resorption under the plate, periosteal and endosteal callus formation, avascular bone necrosis in the femoral head, effect of inserted screws or nails on the blood supply or pathogenesis of the segmental collapse,etc, are not evaluated in this work.

MATERIAL AND METHODS

Experimental Material: Ten mature sheep tibiae,with transverse shaft osteotomy treated by AO technique with stainless steel (n=5) and carbon-epoxy (n=5) plates with transcortical screws for 8-26 weeks, were microscopically re-examined. They were part of the material from a previous study in which progress of the osteotomy healing after plating was evaluated radiographically and by mechanical testing [4,5]. After fixation in buffered formalin transverse and longitudinal blocks through the shaft were taken at intervals and from some of them slabs for microradiography and for undecalcified sections with Von Kossa staining were taken. All blocks were then decalcified and paraffin embedded. 8 and 14 micron sections were stained by Haematoxylin - eosin and Ralis' Tetrachrome 1 and 2 (see below).

Human Material: Five femoral heads from four females with femoral neck fractures (two aged 69 and 75 with subcapital fracture;two aged 78 and 80 sustained an intertrochanteric fracture [18]), and one male aged 82 with subcapital fracture, with previously inserted vitallium screw or nail, were fixed in buffered formalin and then decalcified. From coronal slabs numerous full-size sections were stained by Haematoxylin-eosin and Ralis Tetrachrome 1 and 2 methods [1],and see below.

METHODS

The Osteocyte Viability Count (OVC) It has recently been shown that bone tissue contains not just living osteocytes or their empty lacunae but that a spectrum of transient and degenerated forms can be seen in some bone diseases, e.g.in osteoporosis [3]. In well processed bone tissue and correctly stained nuclei the pleomorphism of osteocytes is not merely an artefact produced during the decalcification , embedding or staining of the tissue,but a biological reality. In disease also the bone cells,as do cells in other organs, go through a variety of abnormal stages .

Nuclear Staining and OVC Reading The bone specimens must be freshly fixed in buffered formalin, decalcified for the minimum time needed and then thoroughly washed in tap water - traces of acids in the tissue would otherwise impair the nuclear staining. From paraffin embedded blocks 6 -10 micron sections are stained by any good nuclear stain e.g. Weigert's, Harris' or Heidenhain's haematoxylin, toluidine blue or Feulgen reaction. We use Weigert's haematoxylin and the toluidine blue or haematoxylin on undecalcified sections as controls. The fine structure of the nuclear chromatin in normal osteocytes must be as clearly recognisable as the fine chromatin in osteoblasts, endothelial cells of capillaries and haemopoietic cells in the bone marrow. The perfect nuclear staining of the nearest bone marrow elements guarantees that also bone cells are stained correctly and their morphology is judged objectively.

The osteocytes are assessed in batches of hundreds, separately in the cortical Haversian and inter-Haversian bone and in the trabecular surface bone and the trabecular cores. The OVC is expressed as a percentage of osteocytes appearing in one of the six different categories: N = normal cells (with a large oval nucleus containing fine chromatin), P = pyknotic cells (with a small condensed round nucleus),F = fragments, S = shadows, E = empty lacunae, LE = large empty lacunae. In routine work we express the OVC in three figures which instantly disclose the total percentage of normal cells (N),abnormal cells (P) and dead (recently or some time ago) cells (F,S,E,LE). For example, an OVC recorded as N=30 %, P=10 %, F=4 %, S=16 %, E=40 %,LE=0 %, is expressed as " 30-10-60 ". This instantly informs us that in that particular field 30 % of cells were normal and 60 % were dead.

Methods for Demonstration of the Osteoid, Mineralisation and Bone Tissue Defects: Tetrachrome 1 and 2 method.

In the bone building or rebuilding process osteoid is the first tissue produced by the osteoblasts. Then it mineralises and changes into bone. Its excess or lack tells about the bone building and mineralisation capacity and about the speed or slowness of this process. If the mineralisation does not take place, or is very slow,osteoid appears as wide seams or forms a substantial part of the tissue, as in osteomalacia. On the other hand, in young or fast growing bone tissue, such as in foetuses or fracture healing, the wide seams of osteoid indicate that its production is fast and its mineralisation lags behind. In some pathological conditions the mineralisation takes place but is inadequate. Such badly mineralised (or demineralised) tissue is called "osteoid bone" [6].

How does one recognise osteoid and badly mineralised areas in ordinary paraffin sections? It is still the wide-spread belief, handed down in text books, that in order to achieve this the mineral in the bone sample must be preserved. Otherwise, once the tissue is decalcified,in most current stainings both osteoid and bone would look very much the same. This was true in the 1950ies and the need to recognise osteoid and to diagnose osteomalacia lead to invention of the undecalcified technique for bone: from a resin-embedded mineralised bone small sections are cut on a heavy duty microtome. After impregnation of the mineral by silver salts (black) or staining by dyes(green) the osteoid counterstained pink or red is clearly recognisable.Preservation of the mineral salts also allows to observe under fluorescent light the tetracycline label located in the mineralisation front. However, the undecalcified sections are small with frequent cracks and tears and most of the cellular and other bone tissue details are obscured by presence of the mineral. Polarised light for identification of the matrix collagen structure cannot be used. The method is also time consuming and, requiring special equipment, expensive and today, unnecessary (see below).

However, a long time ago it has been recognised in the biochemical and histochemical literature, that mineralisation of the osteoid tissue leaves in the matrix irreversible changes which can be detected even after the bone is decalcified (for details, see [1,2]). 15 years ago we confirmed and exploited this fact. After more than 200 staining tests (using undecalcified sections and microradiography as controls) we found that when the tissue is pre-treated with phosphotungstic acid (PTA) the osteoid will stain with some dyes,even in decalcified paraffin sections, differently from the mineralised bone. From the two published PTA methods [1,2] the simpler one,the Tetrachrome 1 which stains osteoid deep blue in good contrast to red mineralised bone, has been used on a large amount of experimental bone material and in over 1200 bone biopsies. Recently the method was tested in Japan against undecalcified methods and found superior [9].

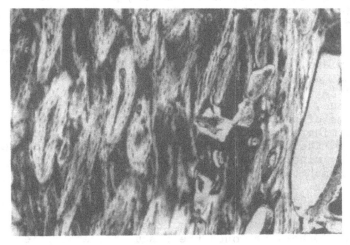

Figure 1. Bone tissue defect in the cortical bone. Cross section of the shaft of sheep tibia plated with stainless steel plate. Tinctorial and structural disintegration of the tissue can be seen in this section stained by Tetrachrome 2 and viewed in the polarised light. The BTD,seen here as dark (deep blue in the method) structureless patches, starts often in the inter-Haversian bone. The pale rings (bright red in the method) with the lamellar structure represent the living Haversian bone. Tetrachrome 2 in polarised light.

Later , with the Tetrachrome, also in combination with polarised light and in undecalcified sections , we first

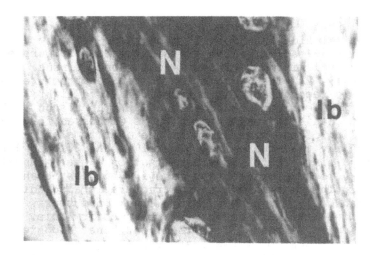

Figure 2A. Detail of the bone tissue defect from a similar field as in Figure 1. The dark (deep blue in the method) structure loosing necrotic interstitial bone (N) in the centre sharply contrasting with the preserved living Haversian bone on each side ("lb";bright red in the method).
Tetrachrome 2 in polarised light.

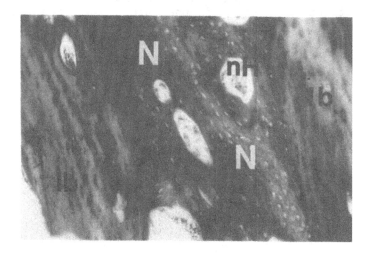

Figure 2B. Without the polarised light the Tetrachrome shows the cellular details: there is a patchy (blue-red) staining of the matrix in the dead area in the centre (N) with osteocytic lacunae empty or with nuclear shadows. The living Haversan bone (lb) on each side shows preserved bone cells nuclei. There is already a rebuliding new Haversian system (nH) tunelling into the necrotic bone (N) area.
Tetrachrome 2.

detected that in specimens from old patients with osteoporosis inside the "normal" bone there are many areas of patchy mineralisation which stain dirty blue or purple and show textural irregularities - the Bone tissue defect (BTD),(Fig.1, 2A).The degradation of tissue structure in the BTD areas is recognisable also under polarised light, in undecalcified sections and under the SEM. The tissue in these areas is mechanically weakened which is witnessed,for example by crumbling of the mineral under the microtome knife in a different way than the normal mineral around. These defects are not a simple patchy demineralisation - the mineral is there, as the microradiography shows, but has changed in composition.The matrix collagen structure is also disturbed and mechanically weakened. Many cracks and tears in the demineralised matrix in defective areas confirm this.

BTDs were already described in bone biopsies from patients with osteoporotic fractures [7,8] and recognised elsewhere [9]. But as a part of the bone tissue reaction to osteotomy and implants they have not been previously reported.

Since in the Tetrachrome method even imperfectly mineralised bone is stained red (see Figure 4), the areas of the BTD sometimes come out in less contrast to the normal bone - unlike the non-mineralised osteoid. In order to stain the tissue defects more intensively we have, after further staining experiments, prolonged the timing and increased the concentrations in some of the steps of the original

Figure 3. Tetrachrome 2 stains the osteoid (small arrows) and defective mineralisation (curved arrows) in deep blue color (dark on this photograph) against the red (pale) normally mineralised tissue. A necrotic trabecula (N) in the femoral head had been repaired by accretion of a malatic repair bone. Tetrachrome 2.

(Tetrachrome 1) method. This new Tetrachrome 2 stains only the normal and mature mineralised bone orange-red. Any incompletely or less maturely mineralised tissue, such as the primary bone mineral during endo-chondral ossification,calcification front,immature mineral of woven bone,undermineralised or defectively mineralised BTDs, are easily recognisable, all being stained blue (Fig.3). Because the youngest mineral of the bone - the mineralisation front- is also stained blue,the "osteoid" seams in the Tetrachrome 2 are wider than in the Tetrachrome 1, representing, in fact, the osteoid + calcification front together (see Fig.4).

Figure 4.

STAINABILITY OF BONE MINERAL BY TETRACHROME METHODS (T₁ AND T₂)

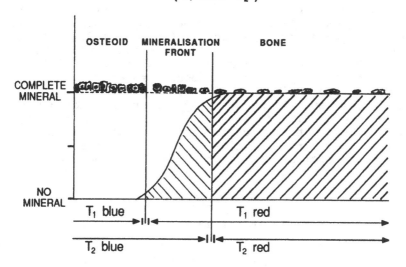

Between the non-mineralised osteoid and a mature lalellar bone not only the quantity of the mineral has changed (as shown in this diagram) but also its quality. Tetrachrome 1 stains blue only the non-mineralised osteoid, most of all other mineralised tissues will show in red. Tetrachrome 2 shows only fully mineralised mature bone in red, all other non-mineralised, defective and undermineralised tissues will stain in deep blue.

Tetrachrome 2 has been used routinely on biopsies for diagnosis of osteomalacia,of the bone tissue defects and in a number of pathological and experimental studies, e.g.of the defective mineralisation of denervated bones [10], callus maturation in fracture healing [11], repair bone malacia [12] and BTDs [8]. Tetrachrome 1 and 2 can be used in any routine hospital laboratory, on ordinary decalcified paraffin sections of "unlimited" size (e.g.the whole femoral head),in which cellular and tissue details are preserved. It can also be combined with polarised light for identification of the lamellar, woven or abnormal collagen patterns in the matrix.

TETRACHROME 1 - Method for Osteoid in Paraffin Sections.

*1. Fix fresh bone samples in 10% formalin or 10% formol-saline; *2. Decalcify in formic acid - sodium citrate, nitric acid or E.D.T.A.,shortest possible time. *3. Wash well 24-48 hrs in tap water. *4. Cut paraffin sections at 14 microns (important). *5. Take sections to water. *6.Stain in Weigert"s hematoxylin - 10 mins. *7. Wash and blue in tap water. *8.Differentiate in 1% acid alcohol. *9.Wash and blue in tap water - 10 mins. *10. Immerse in alcohol - 2 mins. *11. 1 % phosphotungstic acid - 5 mins. *12. Rinse in lukewarm tap water - 20 sec. *13. Stain in 0.1% Soluble blue solution in 1% acetic acid at 23 dgr C - 8 mins. *14. Wash in lukewarm water - 10 sec. *15. Stain in Picro-orange(1 part 1% aqueous orange G : 9 parts saturated aqueous picric acid) - 30 secs. *16. Wash in lukewarm tap water - 10 secs. *17. Stain in Ponceau mixture (dissolve separately 4 parts 2% Ponceau 2R and 1 part 2% Crystal Ponceau in 1% acetic acid, filter and then mix together), at 23 dgr C - 5 mins. *18. Wash briefly in water, dehydrate,clear and mount. Results: see Table 1.

TABLE 1

Staining results of Tetrachrome 1 and 2 methods

tissue stained:	TETRACHROME 1 for osteoid	TETRACHROME 2 for osteoid and tissue mineralisation defects
mature lamellar bone	red	orange-red
osteoid	deep blue	deep blue
mineralisation front	pale blue or pink	deep blue
immature woven bone	blue	deep blue
mmature woven bone	red	blue
osteoid bone underminer.	blue patchy	deep blue
osteoid bone de-mineral.	pale blue	blue
bone tissue defects	dirty red-blue	deep blue
connective tissue	blue	deep blue

TETRACHROME 2 - Method for Osteoid and Mineralisation
Defects in Paraffin Sections

Steps   *1-*10,   *12,*14   *16 and *18 are the same as   in
Tetrachrome 1. The following steps are different:
    *11.Stain in 1% PTA - 15 mins.   *13. Stain   in   0.5%
Soluble blue in 1% acetic acid at 23 dgr C - 8 mins.   *17.
Stain   in   2%   Ponceau 2R in 1% acetic acid at 23 dgr   C - 8
mins.   Results: see the Table 1.

MONITORED TISSUE REACTIONS

Areas Selected. From the transverse histological sections of
the   sheep   tibiae   and   the   whole   femoral   heads,
histophotographs   or enlarged Xerograms [13.14]   were   taken.
Based on these a chart of the specimen was drawn, with marked
topographical   relationships of the main features, i.e.   the
implant,   bone   tissue resorbed and newly   built,   areas   of
necrosis,concentrations   or   lack   of   stresses,   etc.
Recording of the cellular and structural reactions was   done
in   several   typical   areas which   differed   in   histological
character and likely distribution of loads (see Figure 5).

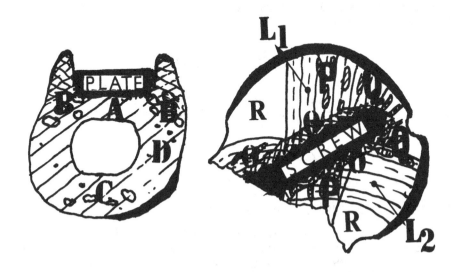

Figure 5.   Different areas in plated tibial shaft and femoral
head with   screw in which the bone   tissue   reactions   were
evaluated. L1 and L2 = the original trabecular   trajectorial
systems in the femoral head. Areas A-D and O-R: see the Text.

In the tibial cortical shaft: "A": a stress-shielded area
with cortical thinning, advanced rebuilding and some
remaining absorption cavities. "B": areas with high stress
concentrations at the edges of the plate and the base of the
periosteal callus; there is a continuing rebuilding with a
number of large resorption cavities. "C": - similar to area
"B" but less "intensive" in rebuilding activity and with some
resorption cavities; "D": an area showing no gross micro-
anatomical yet widespread cellular changes; presumably a low
stress or stress-confusion area .
In the femoral heads: the topography of changes was less
uniform, much dependent on the direction and area into which
the screw was inserted and on other factors such as the post-
implant fracture stability and healing,and the state of the
blood supply. In a typical case tissue reactions were
recorded in areas "O" - encasement of the implant by a
thickened trabecular or cancellised cortical bone, with or
without fibrous layer at the inerface.Loads are transmitted
from the articular surface on to the implant through this
bony capsule which shows extensive remodelling activity;
Areas "P" in which the dead remnants of trabeculae of the two
original trajectorial systems in the head are used,where the
loads persisted, as scaffolding for building of a thick,
trabecular bone; Area "Q" in which the loads from the
articular surface are concentrated on to the proximal tip of
the screw. Usually there are thick trabeculae with mixed
areas of bone necrosis and active rebuiding; Areas "R"-a
quiet thin cancellous bone being even before the implantation
outside the main load trajectories; surfaces are covered by a
thin repair bone.

FEATURES MONITORED

In ten different fields of each area listed above, after
the OVC percentage was counted, the frequency, size and
intensity of the following features were assessed semiquantitatively (as " -, +, ++ and +++ ").

Osteocyte Viability Count (OVC ,see p.3.)

Bone Tissue Defect (BTD,see p.5.)

Repair Bone Malacia (RBM) and splits

With the Tetrachrome and PTAIH methods and
microradiography as control we noticed in the past that
whenever a new repairing bone is built on scaffold of a
necrotic bone the interface between them is incompletely
mineralised. In H & E staining this appears as "prominent
cement lines" or "marginal basophilia" but the
undermineralisation often goes into depth of the repairing
tissue (Fig.1,6,7).Polarised light shows that in this Repair
bone malacia (RBM) the margin of the tissue facing the dead
fragments to a depth of several rows of bone cells is a woven

bone. Normal osteocytes live and work in a syncytium and ,probably, the new coming repairing osteoblasts need, in order to maintain their normal function, i.e. osteoid production and mineralisation, to meet in the old bone living osteocytes, as they do during normal appositional bone growth. The RBM could be a nutritional phenomenon but a different explanation is more likely: the repair bone will treat the dead tissue as being without loads - either because the dead fragment no longer carries the loads or, even if it would, the dead cells in the fragment cannot transfer this information to the repairing tissue. And so, while normal appositional growth of a living, loads carrying lamellar bone is materialised again by the lamellar tissue,on the dead bone the repair would start as woven bone.This primary, "starting" bone tissue is normally converted to lamellar bone when subjected to orientated loads. In 1978 the existence of RBM ("local osteomalacia") was proved experimentally [12] and it was found to be a widespread phenomenon in the bone remodelling process in a variety of clinincal and experimental conditions.

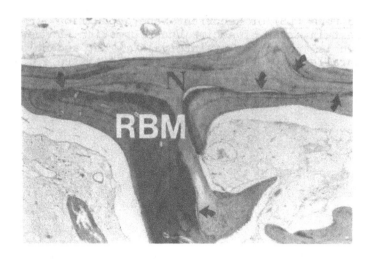

Figure 6. Repair bone malacia: necrotic trabecula (N) in a highly loaded area in the femoral head is repaired by several layers of malatic repair bone (RBM). This shows mineralisation defect at the repair interfaces (dark; blue in the method) against the normally mineralised dead bone (pale; red in the method), and common splits (arrows). Tetrachrome 2

In the RBM the edges of the repair bone are not only undermineralised but the living and dead bone are not properly "cemented" together. Results are the common splits along the accretion lines, which accompany this condition (Figure 6). These were evaluated too.

A question remains open if the numerous cracks seen in BTD areas and splits along the accretion lines in RBM are an intra-vital phenomenon or if they are artifacts incurred during the histological processing, embedding or cutting of the tissue. Against the possibility that they occur in vivo speaks the fact that there is no obvious repairing tissue around them as is,e.g., the woven bone microcallus around fractured bone trabecula. For their in vivo origin speak: a/ Frost's staining experiments [15] proving that cracks in inter-Haversian bone could exist before the histological processing; b/ our findings that the cracks and splits are more frequent in areas with high stresses. That there is no tissue around repairing them as it is in trabecular microfractures can be explained by the fact that since both cracks and splits occur in a devitalised bone tissue there are no living osteocyte messengers to pass the S.O.S. call for repair. It seems, therefore, more likely that the splits and cracks do exist in vivo. But regardless when they occured, in vivo or in vitro during the cutting process, they indicate a mechanical inferiority of the tissue in the areas which, undoubtly, contributes to the repairing bone fragility.

Laminated Bone (LB) or Recurrent bone apposition (Fig.7).
In certain areas of active bone formation the new bone is laid down in succession at intervals. The accumulation of close parallel accretion lines gives the tissue a laminated, onion-like appearance. If the bone tissue in the previous layers was dead or dying, then along the accretion lines the bone would be malatic and could contain splits. The splits

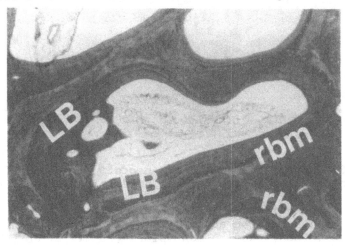

Figure 7. A thick trabecular network in highly concentrated loads area in the femoral head, near the tip of the screw. Onion-like layers of the laminated bone (LB) show also dark (blue in the method) mineralisation defects along the interfaces- a repair bone malacia (rbm). Tetrachrome 2.

are most common in the thickened repairing cancellous bone
but could be seen also in the cortical bone at the outer
circumference of the Haversian bone along the reversal line.
We saw many such "dislocated" Haversian systems in the
structurally disintegrating tibial shaft. The likely reason
for this osteogenesis at intervals and laminated bone
formation are unstable or interrupted high loads or
mechanical unsoundness of the previous layers requiring
further re-inforcing layers of bone.

Recurrent Bone Necrosis Repair (RBNR), (Figure 8).
This condition is caused by repeated interruptions of the
blood supply and was first described in pathological specimen
in Perthes' disease [16]. The appearance is that of a
necrotic area in which both bone tissue and bone marrow are
dead but the bone fragments show signs of a previous repair -
either signs of resorption (with recognisable Howship's
lacunae, less often osteoclasts) or accretion of a new
bone.The latter is easily recognisable when the repair bone
was malatic, with first layers of woven basophilic tissue or
thick accretion lines with possible splits.

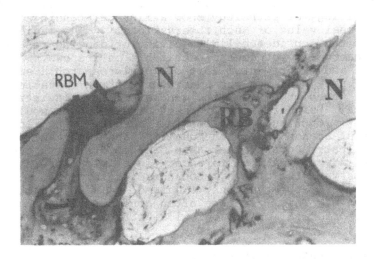

Figure 8. Recurrent bone necrosis repair (RBNR). A necrotic
area in the region of loads concentration in the femoral head
The dead bone trabeculae (N) had been repaired in the past by
the repair bone (RB) which also showed a malacia (RBM)
with undermineralised (dark) tissue on some inerfaces
(arrows). Both bone components and the bone marrow then died
again. Tetrachrome 2.

Sometimes there are repetitions of the RBNR . The cause is
undoubtedly episodes of interrupted blood supply to the area

- the bone marrow suffers equally. In Perthes'disease, though cause of the interrupted blood supply remains unknown, the RBNR could explain the prolonged recovery and slow re-building of the femoral head. In femoral heads with implanted screws the RBNR heralds not only the unreliability of the blood supply in this mode of fracture treatment but also the uncertainty of the outcome of the bone - implant adaptation.

RESULTS

[Note The following are biological and semi-quantitative, not mathematically exact observations. Their interpretation and definition of a typical or an "average" picture inevitably suffers from subjective judgement by differences between the individual implant receivers, the mode of the implantation and by the histological sampling problems. No attempt was made to attach any statistical significance to the results.

When quantitatively expressing complex biological reactions of non-identical subjects it is acceptable, as practised, e.g.in the routine of diagnostic pathology, to evaluate the features on a semi-quantitative scale, e.g. " none-mild-moderate-severe ". Demands to express more convincingly in "exact" figures events in inexact sciences such as biology or pathology show a misunderstanding of where lies the real value of such observations. The magic figure of 95% of the confidence limits,giving or denying results the "significance" is ,of course, debatable on its own. In surgical and experimental morphological research a small "n" number of say, subjects who underwent certain operation or of dissected specimens, is the general rule. Insistence on statistical "significance" in such circumstances would mean that many important experimental results with logically significant trends would not "qualify" and important messages would be lost. On the other hand many papers were published in which a very small difference - in clinical or laboratory terms - between two groups, because of a large "n" number of subjects or observations, gained a high statistical significance. This is then often confused with the real, often low practical or biological significance, or value, of the work.]

SHEEP TIBIAL SHAFTS WITH PLATES .

An example of intensity of reactions in different areas in sheep tibiae plated by carbon-epoxy plates is shown in Table 2. Average reactions on both types of plates - see Table 3.

The bed of the plate.
In nine out of ten samples in the bed around or under the plate there was a mild-moderately thick fibrous capsule and on both sides of the plate a woven bone callus of differing amount.

The cortical shaft

The OVC. In the original bone tissue, i.e. Haversian and inter-Haversian bone, in all investigated areas the bone cells are dying. The OVC death rate is highest in the high stressed area - 90% - 95%; in other areas it is around 70% with 25% - 30% of cells remaining pyknotic. The new Haversian bone replacing the dead bone is viable in the high and low stressed areas 60% - 70 %, but again a relatively high proportion of cells are pyknotic. The earliest shifts of the OVC to the right (towards the degenerated and dead cells) occur in the inter-Haversian bone (Figure 2B).

The BTD In all areas with increased, diminished or confused stresses the BTD defect starts often as a "lamellar BTD," i.e. alternating lammellae will stain blue in Tetrachrome 2 and later will show structural deterioration (" lamellar dissociation"). Parallel with this a more intensive BTD develops in the inter-Haversian bone. Such cortical bone will appear in Tetrachrome 2 and in polarised light as blue, dull areas of structureless inter-Haversian bone with shining rings of red Haversian bone with accentuated lamelllar structure (Figure 1,2A). In the low stressed areas the inter-Haversian BTD seems to be less intensive. There is a positive correlation between the prevailing OVC and the intensity of the BTD. This is supporting evidence that BTD defects are a consequence of the osteocyte ill-health and death (Figure 2B). Cracks within the BTD and peri-Haversian splits at the reverse lines were recorded mainly in areas with high and confused stresses. Repair starts by appearance of resorption cavities and building units throughout tissue to replace the dead bone (Figure 2B). Open resorption cavities (i.e. bone reconstruction in progress) are predominantly in the high stressed areas but the overall replacement of the dead by the new Haversian bone took place within the timing of the experiment in most of the regions.

Comparison between plates

The operation technique being the same, the plates differ in the material and in their stability in bone. Stainless steel is more rigid than the carbon-epoxy plates [4,17] and evidently produced more intensive stress protection and stress concentrations. This is reflected in the observed more widespread cellular death and more intensive BTD. In stainless steel-plated shafts the open resorption cavities were more common and the new Haversian repair bone lining the cavities was thinner - indications of a slower rebuilding recovery. The more intensive cancellisation, thinning and post-implant weakness of the cortical shafts after stainless steel-plates against carbon-epoxy plates were already recorded [5,17].

As to the stability of the plate fixation - in any of the specimens we have bio-mechanical evidence of this. But we

TABLE 2

Reactions in different areas of the sheep tibial shaft
plated by the carbon-epoxy plates

area	OVC			BTD			peri -Hav. splits	rebuild. cavities
	orig.Havers. systems	inter-Hav. bone	new Hav. systems	lamell.	interst.	cracks		
A	?	0-30-70	70-30-0	++	++	-	-	++
B	?	10-30-60	60-40-0	++	+++	++	++	++
C	?	10-10-80	30-70-0	++	+++	+	++	++
D	10-70-20	50-0-50	-	-	+	-+	-	-
con-trol	18-41-41	3-15-82	-	-	-	-	-	-

TABLE 3

Average cortical bone reactions in sheep tibial shaft
plated by stainless steel and carbon-epoxy plates

area	OVC			BTD			peri -Hav splits	rebuild. cavities
	orig.Hav. systems	interH. bone	new Hav. systems	lamell.	interst.	cracks		
stress increase	05-05-90	0-05-95	60-40-0	++	++	+	++	+++
stress shielded	05-30-65	0-30-70	70-30-0	++	-+	-	-	+
stress confusion	05-25-70	0-30-70	70-30-0	++	++	+/++	++	-+

compared the cortical bone reaction in two different sites at the
interface with a transcortical fixation screw: in one with no fibrous
capsule and almost a perfect osseointegration of the screw the situation
being presumably "stable". In the other site there was a moderately
thick lining connective tissue capsule around the screw and the bone did
not reach the implant. This was a less stable situation. In the stable
area the mineralisation of the woven and lamellar bone seemed to be

faster and bone remodelling quicker. Near the less stable implant the bone repair looked slow, woven bone and mineral remain immature and there were layers of unmineralised osteoid on the surfaces.

FEMORAL HEADS WITH IMPLANTED SCREWS

Because of the differences in the clinical aspects, type of fracture, and area and direction in which the screw was inserted, there was a greater variability in the cancellous bone reaction (Table 4).

Most of the original trabeculae suffered massive cellular death especially in areas of the post-implant high loads concentration, including that of the cushioned loads. The OVC dead cell values exceed those counted in the iliac crest biopsies from patients with a femoral neck fracture of similar age. The bone cell necrosis was accompanied by the BTD, mainly in the trabecular cores. The intensity

TABLE 4

Average reactions of cancellous bone in the femoral heads
with inserted screw

area	ORIGINAL TRABECULAE			REMODELLING BONE						
	OVC	BTD	cracks	thickness	OVC	BTD	RBM	splits	LB	RBNR
concentrated high loads	0-04-96	++/+++	+++	++/+++	15-45-40	++	+++	++/+++	++	+++
high cushioned loads	0-05-95	++	++	++/+++	30-50-20	+/++	+++	++	++	++
low loads	5-25-70	++	+	+	50-15-35	-	+/+	+	+	-

OVC=osteocyte viability count; BTD=bone tissue defect; RBM=repair bone malacia; LB=laminated bone; RBNR=recurrent bone necrosis repair

of the defect and frequency of the cracks rose with the concentration of loads. The new repairing bone shows the largest volume and activity in the most loaded areas. But even this recently built bone suffers from some loss of osteocytes and retains a relatively high percentage of pyknotic nuclei. It is more viable in cushioned and especially low-loaded areas. It also contains frequent high loads-related BTD.

The mixture of living and necrotic areas in the repair bone leads to malacia on the repairing interfaces and frequent splits. The high uncushioned and probably changing loads also produce a large amount of LB and this, when combined with malacia, might also show splits along the accretion lines.

The most serious setbacks to this, in overall, somewhat confused and not very satisfactorily progressing cancellous bone - implant accommodation are undoubtedly the episodes of recurrent bone necrosis (RBNR), appearing predominantly in areas of high and concentrated loads. Signs of a past repair and then necrosis again are not good omens and indicate an instability and unwelcome readiness to interrupt the blood supply. Thus the RBNR further jeopardise and interrupts the effort of the strong yet structurally very delicate and sensitive bone tissue to accomodate a rough intruder into its finely balanced enviroment.

SUMMARY

1/ The reaction of cortical and trabecular bone to implants was studied in 10 sheep tibial shafts, plated after osteotomy by stainless steel and carbon-epoxy plates and five human femoral heads with inserted screws following a femoral neck fracture.

2/ Early, previously unrecognised, yet fundamental changes occur in the tissue: a widespread response and degeneration of bone cells accompanied by the tinctorial and structural defects in the mineral and matrix.

3/ Two simple methods for assessment of the cellular and bone matrix "pre-histological" changes in ordinary paraffin sections are described.

4/ With these tools the following, not before reported features were demonstrated in the bone tissue reacting to an implant: lowered Osteocyte viability count (OVC); Bone tissue defect (BTD): Repair bone malacia (RBM) and splits: Laminated bone (LB) and Recurrent bone necrosis and repair (RBNR).

5/ Frequency and distribution of these features were semiquantitatively evaluated, separately in areas of high stress conentration, of low stresses and in areas of stress confusion.

6/ Summary of the general bone tissue reaction:

a/ After fracture or osteotomy and fixation of the implant to the bone, the very structure of the tissue is shattered: an extensive bone cells pyknosis,degeneration and death occurs along the whole specimen, accompanied by widespread mineralisation and structural defects, first confined to interstitial and inter-Haversian bone. These changes are due to the gross interfence with the physiological stresses for which the bone is built. They are most pronounced in areas with concentrated loads but can be seen everywhere. It is resorption of this dead bone which starts the peri-implant bone rebuilding. In the cortical shafts the dead bone is gradually replaced by new Haversian remodelling. In the femoral heads the necrotic original trabeculae are dealt with by layers of the new repair bone whose further activity is directed by the magnitude and direction of the new post-implant loads and stresses.

b/ The remodelling bone tissue contains many of the the newly described irregular features listed above. The very low osteocyte viability, bone tissue defect, RBM with splits and RBNR are more pronounced in the high loads areas. These mineralisation and structural problems undoubtedly add to the mechanical instability of the peri-implant bone.

c/ Probably the most serious complications for the repairing tissue are the episodes of RBNR occuring frequently in previously adequately vascularised regions. They further interrupt the elaborate efforts of the bone tissue to accomodate the implant.

7/ In most of the specimens studied the implants were presumably in less than stable position as judged by presence of fibrous capsule of variable thickness around or under the implant, by the amount of external callus along the plates and by numerous areas of LB indicating the presence of intermittent or changing loads.

8/ The more rigid stainless steel implants caused a more widespread cellular death and tissue defects and a slower bone repairing process. Near the less stable implants the bone remodelling is slow and mineralisation of the new woven and lamellar bone is immature.

ACKOWLEDGEMENT

The author thanks his colleagues,the orthopaedic surgeons at Cardiff Royal Infirmary for letting him to study the bone samples from their patients and Mrs Heather M.Ralis for preparation of the manuscript. The work was partially supported by the Action Research for the Crippled Child, Grant No.A/8/1550.

REFERENCES

1. Ralis,Z.A. and Ralis,Heather M., A simple method for demonstration of osteoid in paraffin sections. Med. Lab. Technol. 1975,32,3, 203-213.

2. Ralis,Z.A. and Ralis, Heather M.,Phosphotungstic acid - iron haematoxylin (PTAIH) method for osteoid, boundary bone and bone components in paraffin sections. Microscopica Acta(Basel), 1976,78 5,407-425;

3. Ralis,Z.A., Low osteocyte viability rate in biopsies from patients with femoral neck fractures.In: Ring EFJ, Ed.:Current Research in Osteoporosis and Bone Mineral Measurement. British Institute of Radiology, London 1990,1-3.

4. Bradley,J.S.,Hastings G.W. and Johnson-Nurse,C.,Carbon fibre reinforced epoxy as a high strength, low modulus material for internal fixation plates. Biomaterials, 1980,1, 38-40.

5. Johnson-Nurse,C.,Ralis,Z.A.,Tayton,K.,Bradley,J.S., Evaluation of carbon fibre epoxy interrnal fixation plates on healing of osteotomised sheep tibiae. (Unpublished observations, 1980).

6. Ralis,H.M., Ralis,Z.A., Poorly mineralised osteoid bone and its staining in paraffin sections - (Prize for Histopathology 1977 of the Institute of Medical Laboratory Sciences winning paper). Med Lab Sciences,1978, 35,3,293-303.

7. Ralis,Z.A., Bone weakness in the elderly: new developments. Hosp. Update 1983,9,12,1381-91.

8. Ralis,Z.A., Jane Lane, Watkins G., Ralis, Heather M., Diagnostic features in bone biopsies from patients with FN fracture.In: Ring EFJ, Ed.:Current Research in Osteoporosis and Bone Mineral Measurement. British Institute of Radiology, London 1990, 66-67.

9. Iwasaki,K., Hirano,T. Yamamoto, N,: Ralis' Tetrachrome method for demonstration of osteoid in decalcified sections. Seikei Geka - Orthopaedic Surgery, Tokyo, 1986,37,7, 917-923.

10.Ralis,Z.A., Ralis, Heather M., Randall Margaret, Watkins, G., Blake, P.D., Changes in shape, ossification and quality of long bones in children with spina bifida. A clinico-pathological and experimental study. Develop. Med. Child. Neurol.,1976, 18, Suppl 37, 29-41;

11.Phillips,G., Ralis,Z.A.,Healing of experimental osteotomy of rat metatarsals. Unpublished observations (1982).

12.Ralis, Z.A., Experimental production of local osteomalacia. Experientia ,1978,34,1203-04.

13.Ralis,Z.A. and Blake,P.D., The use of histophotography in bone histology and morphometry. Microscopica Acta (Basel)1976, 79, 5, 443.

14.Ralis,Z.A., Recording of hard tissue morphology by electrostatic (Xerox) copying. Med. Lab. Sciences , 1979,36, 391-392.

15.Frost,H.L., Presence of microscopic cracks in vivo in bone. Henry Ford Hosp.Bull.,1960,8,25-35.

16.McKibbin,B., Ralis,Z.A., Pathological changes in a case of Perthes disease. <u>Journal of Bone and Joint Surgery</u>, 56-B, 3, 438-447.

17.Claes,L., The mechanical and morphological properties of bone beneath internal fixation plates of differing rigidity. <u>J.Orthop.Research</u>,1989,7,170-7.

18.Howard,C., Ralis,Z.A., Unpublished observations (1986).

NUMERICAL AND CLINICAL ANALYSIS OF ORTHODONTIC TOOTH MOVEMENT

KEITH R. WILLIAMS
Senior Lecturer
Department of Basic Dental Science
Heath Park
Cardiff   CF4 4XY

ABSTRACT

The orthodontic movement of teeth has been calculated using both 2-D and 3-D finite element analysis and compared with recent clinical measurements. The wide variation in the observed tooth behaviour suggests that centres of rotation are difficult to define clinically. Several factors markedly influence this behaviour, the most important being the angle of load application and the exact geometry of the alveolar process.

Furthermore, the orthodontic load magnitude appears not to influence the rate of tooth retraction above some threshold limit. It is suggested that only a compressive or hydrostatic component of stress transmitted through the periodontal membrane to the lamina dura is responsible for bone resorption, resulting in a gradually enlarging tooth socket in both mesial and distal directions. Simultaneous bone resorption and deposition leading to tooth movement is therefore highly unlikely during orthodontic treatment.

The deposition phase of bone remodelling must necessarily follow the completion of treatment when bone is gradually laid down on the mesial aspect of the alveolar process where stretched periodontal fibres induce osteoblastic activity.

INTRODUCTION

The finite element method (fem), although only recently applied to orthodontic tooth movement (1,2,3,4,5,6) has successfully indicated the position of the instantaneous centre of rotation (ICR) of a variety of teeth for a number of loading situations. The method has also allowed the analysis of the complex stress and displacement patterns in and around the tooth environment which are ultimately responsible for the direction and rate of tooth movement (7).

It is generally assumed by the Orthodontist that a point loaded, single rooted tooth will tip about points lying within the apical third of the root about an ICR. Displacements of the teeth taken directly after loading are generally in agreement with the mathematical and numerically calculated ICR (6).

However, tooth movement continues after the initial elastic displacement through a process of bone remodelling and under such circumstances it is the orthodontic centre of rotation (OCR) which is of fundamental importance to the Clinician. It has been proposed (1) that a correlation exists between the ICR and the subsequent remodelling of alveolar bone, while Stephens (8) from clinical measurements, suggests that the ICR and OCR are usually one and the same. In his work, Stephens (8) suggested a relationship between the axis of rotation (ℓ) and the rate of retraction (d) of the form -

$$\ell = 1.06 - 0.13 \log d \qquad (1)$$

By combining these data with the numerical data of Williams and Edmundson (2), it can be shown that the rate of retraction is inversely proportional to the height at which the load acts on the tooth crown, the tooth tending to retract faster the more apical the position of the ICR. Whether or not this relationship is realised during tooth movement requires more detailed clinical measurement with time. However, analysis of the local stress and displacement patterns observed in the 2-D calculation suggest that when the ICR is at the apex or below, the whole of the periodontal membrane on the distal aspect of the tooth is in compression. At intermediate positions of the ICR, a more complex stress pattern exists with both compressive and tension stress components existing on the distal surface of the tooth and alveolar process. It has been suggested that one type of stress leads to bone resorption and the other to bone deposition. Thus an ICR at or below the apex would inevitably lead to more uniform bone remodelling and hence a faster tooth retraction.

The final position of the tooth however depends on several other additional factors, including the load type and magnitude, load position and angle, the initial tooth position in the alveolar process as well as time dependant physiological changes. These will not be simple relationships and will be further complicated by the variability of biological factors.

In an effort towards a better understanding of these instantaneous and time dependent variables, the initial 2-D analysis (2,3,4) has been extended to a full 3-D analysis which includes a more realistic modelling of the alveolar process, tooth and surrounding structures. The earlier work has been further analysed in order to incorporate and explain the new results towards an understanding of the time dependent orthodontic tooth movement.

METHOD

The three cases to be analysed have been fully explained earlier and include -
(i)   central incisor rotation   Williams and Edmundson (2)
(ii) canine rotation and translation Williams et al (3)
(iii) canine movement by 3-D analysis Wilson (7)

At this point we simply wish to reiterate some of the main features of the numerical programme for clarity. In each example, main frame computing was employed with the additional use of a PC in order to assess local fluctuations in stress and displacement as an aid towards an understanding of the time dependent analysis.

The methodology follows the normal fem approach in which the structure to be analysed is discretised into a series of elements connected at certain nodal points (see for example Figure 1). It is the displacement at the nodes during loading which are regarded as the basic unknowns in the problem and the stress, strains and strain rates in the body can be related to these.

The problem of elastic tooth movement involves the solution of the system displacement, δ, and the resulting stresses, σ, under a known set of applied forces. In order to solve this problem, the following conditions must be satisfied:-

$$\sigma = D\epsilon$$

(a) relating stress to strain (ϵ) in terms of the material elastic modulus, E, and poissons ratio, ν

(b) continuity and differentiability of δ throughout the domain together with satisfaction of boundary displacement conditions and the strain-displacement law

$$\epsilon = L\delta$$

where L is a linear operator

(c) and equilibrium where

$$L^T\sigma + F = 0$$

where F are the body forces per unit volume.

The solution to the deformation problem can be formulated through a virtual work equation. If the stress vector is in equilibrium with body forces [b] and surface tractions [q] then

$$\int_v \left([\epsilon]^e\right)^T [\sigma]^e \, dv - \int_v \left([u]^e\right)^T [b] \, dv - \int_s \left([u]^e\right)^T [q] \, dS = 0 \qquad (2)$$

where $[u]^e$ is the matrix of displacements within an individual element.

Since displacements are related to strain through the linear operator, L, and the generalised displacements within the element to nodal displacements through a shape function matrix, N, then equation 2 becomes;

$$\int [B]^T [\sigma] \, dv - [F] = 0$$

$$[B]^e = [L]^e [N]^e \qquad (3)$$

i.e. the B matrix relates strain at any point within the continuum to the discretised variables, δ. Equation 3 gives rise to a set of simultaneous equations.

$$[K] [\delta] = [F] \qquad (4)$$

where $[K]$ is the assembled stiffness matrix for the structure. Equation (4) can be solved directly to provide the set of displacements $[\delta]$, which through the application of the trial or shape functions can be interpolated throughout the discretised continuum. The element nature of the calculation means that attention can be focused entirely on small regions of the physiological structures one at a time. Thus the varying properties of the materials making up the alveolar process present no difficulties in the analysis. Furthermore, arbitrary loading conditions can be applied which is an important factor if orthodontic tooth movement is to be accurately predicted. In the case of the 2-D work, a quadrilateral isoparametric element was chosen, allowing the discretisation of complex geometry including curved boundaries. The element chosen for the 3-D analysis was the well tried eight-noded quadrilateral with tri-linear shape functions.

RESULTS

The material property values of the tissues making up the structure have been given previously (2,3,4,7). The effect of changes in material properties on ICR can be conveniently summarised in Figure 2. The ICR values are in good agreement with those measured (6,8) and calculated (1,2,3) in earlier work. It can be seen that poissons ratio effects are relatively unimportant except near values of 0.5 where incompressibility effects may become important.

A further important finding concerns the change in ICR with position of loading on the tooth crown, and this is indicated in Figure 3 which suggests a gradual increase of the centre of rotation with increasing height of loading above the cervical margin. However, the most marked change in ICR occurs with load angulation. These data are shown in Figures 4 and 5 where the change in tooth position with loading angle at the cervical margin and ICR with angle are plotted respectively. In Figure 5, two singularities appear when the load is along the tooth axis in the positive and negative y direction.

Loads applied at positions other than the cervical margin will tend to displace the points of singularity in the positive or negative x direction.

The introduction of a third dimension in the 3-D analysis, allows the examination of surface and interface stresses and displacements. In this work we have concentrated on the loadings within and around the alveolar process, since it is in this area that bone remodelling allows tooth movement. The new coordinate system is shown in Figure 6, together with the sampling points for stress and strain. Thus tooth displacements in the direction of load application (distal) are identified with the z axis, while displacements along the tooth axis remain along the y axis as in the 2-D work.

The post-processing facilities used for the 3-D analysis allows the extraction of the loads and displacements (stress and strains) over selected surfaces, volumes and tissue structures. Two loading conditions were examined, i.e. a 1N horizontal point load at the cervical margin and tooth tip respectively. The principal stresses in the periodontal ligament at position 1 (cf Figure 6) are shown in Figures 7 and 8. Figures 9a, b show the principal stresses in the lamina dura for positions 1 and 3 under a tipping load at the tooth tip, while Figures 10a, b reflect the displacements in the same tissue for these same stresses.

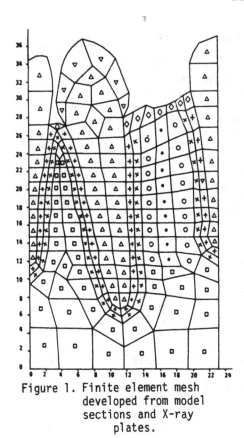

Figure 1. Finite element mesh
developed from model
sections and X-ray
plates.

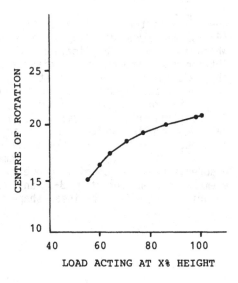

Figure 2. Effect of changes in
material properties
on ICR.

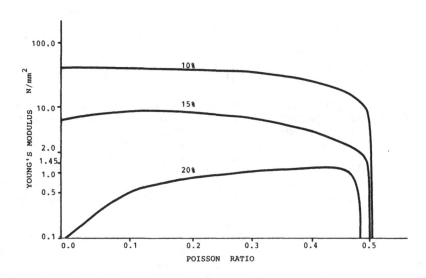

Figure 3. Relationship between ICR and the height at which the load
acts above the apex.

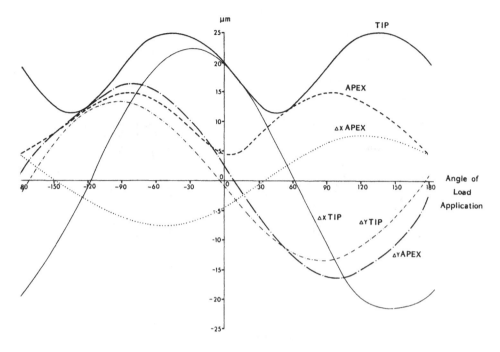

Figure 4. Distribution of apex and tip movements with angle of load application.

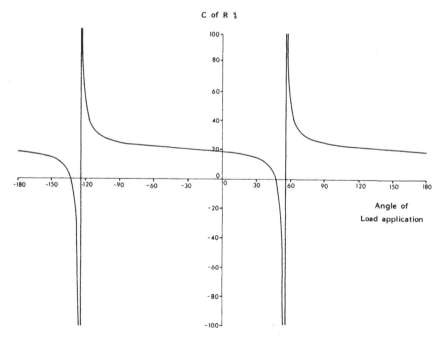

Figure 5. Translation of the displacements shown in Figure 4 into centres of rotation.

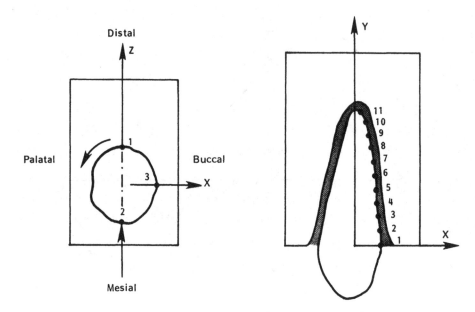

Figure 6. Schematic representation of a canine showing sampling points for the finite element results.

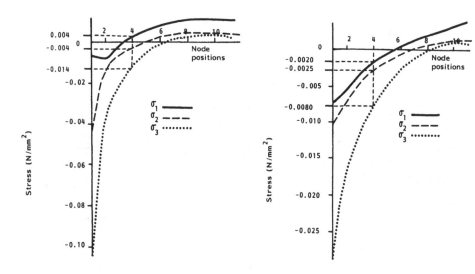

Figure 7. Principle stresses in the periodontal ligament at position 1 (Fig 6) with a point load at the tooth crown.

Figure 8. Principle stresses in the periodontal ligament at position 1 (Fig 6) with a point load at the cervical margin.

233

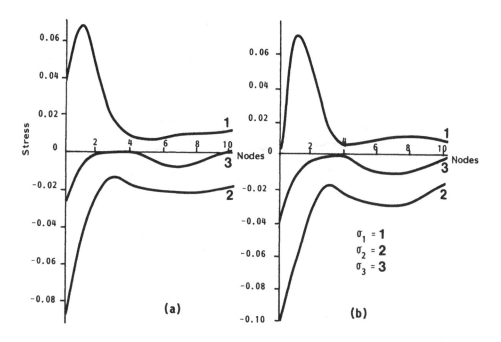

Figure 9a, b. Principle stresses in the lamina dura for positions 1 and 3,
point load at tooth crown.

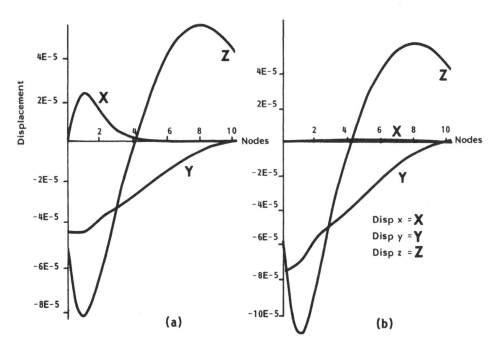

Figure 10a, b. Corresponding displacements produced at positions 1 and 3
for stresses shown in Figure 9.

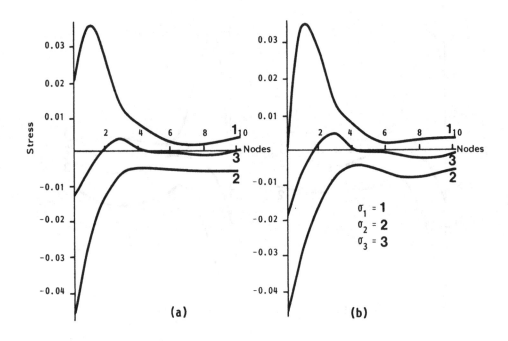

Figure 11a, b. Principle stresses in the lamina dura for positions 1 and 3, point load at cervical margin.

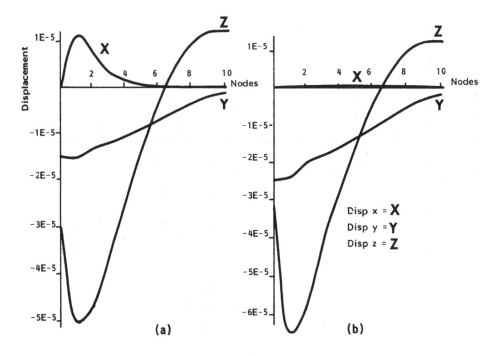

Figure 12a, b. Corresponding displacements produced at positions 1 and 3 for stresses shown in Figure 11.

Figures 11a, b and Figures 12a, b illustrate stresses and displacements at the same positions as Figures 9 and 10 with the orthodontic load applied at the cervical margin.

DISCUSSION

The additional information provided by 3-D analysis enables the conclusions gained from earlier 2-D work (2,3,4) to be re-examined as follows.

It is clear from this and the previous numerical analysis, that the instantaneous tooth position can be calculated for a variety of geometries, loading conditions and tissue properties. In clinical practice, large discrepancies are observed outside the classical orthodontic tooth rotation, from rotations in the apical third of the root to gross lateral translations under point load (4).

Furthermore, it is known that teeth move at different rates suggesting some relationship between load magnitude and rate of retraction. This is the basis of orthodontic tooth movement and the Clinician, by experience, has to judge the correct load and direction in order to bring about a satisfactory treatment. The reliance on the ICR is not however always successful as indicated earlier (3,4) (cf Figures 4 and 5).

If we rearrange equation (1) as follows;-

$$d = \alpha e^{-\beta l}$$

(5)

where α, β are geometrical constants

Integrating this expression with respect to time we have -

$$(a - a_0) = \frac{\alpha t}{e^{\beta l}}$$

(6)

where a_0 is the initial tooth position.

The rate of tooth movement is therefore independent of load magnitude and depends only on load angulation (Figure 4). Numerically this has been shown by Yettram (1) and Williams and Edmundson (2).

The applied orthodontic force initially produces an instantaneous elastic stress pattern around the tooth root and alveolar process. This stress pattern is dependent on the position of the load (Figures 7 and 8) and in turn will result in an axis of rotation proportional to the height of loading above the cervical margin (Figure 3). It can be seen that the principal stress at the cervical margin is approximately x4 greater with the load applied at the tooth tip (Figure 7) compared with the load at the cervical margin (Figure 8) and yet according to Figure 3 and the Stevens (8) data, (equations 5 and 6), the retraction is faster for the lower local stress. Furthermore, as the orthodontic point load is increased, the ICR is invariant (1,2).

This is a further contradiction to the hypothesised behaviour indicating that the rate of bone resorption is independent of the load magnitude above some unspecified threshold. Thus the rate of retraction is dependent only on the local stress distribution pattern around the alveolar process.

Clearly, the angulation of load at a point can promote wide variations in local stress pattern (3) resulting in wide variations in the type of tooth movement and rate (Figures 4 and 5).

Generally, stresses are compressive in the periodontal membrane to the point of rotation (1) (Figures 7 and 8) and these compressive forces can be readily transferred to the lamina dura by the viscous membrane. However, tensile forces are less easily transferred because of the low E value of the periodontal ligament. The resulting stress patterns in the x and y directions (cf Figure 6) are therefore largely different on the mesial and distal aspects of the lamina dura and the only stress producing active bone response is the compressive component or hydrostatic component.

Movement of the tissues in the y direction is generally in the negative direction. Thus the alveolar process moves downwards (see Figures 10a,b and 12a,b) under the influence of the general compressive components and seen in Figures 9a, b and 11a, b. The uniform elastic downward displacement of the lamina dura suggests a gradual time dependent enlargement of the alveolar socket regardless of load position above the cervical margin. Thus it is expected that partial remodelling of bone by resorption occurs on the mesial aspect where compressive stresses are also generated in response to the applied load, together with the normal gradual distal enlargement of the socket. The tooth is thus moved distally and partially downward.

Bone remodelling in response to compressive or hydrostatic stress can provide a physiological explanation for the necessity for substantial tooth support following orthodontic completion in order to prevent relaxation of tooth in a mesial direction. Failure to provide such support allows stretched fibres on the mesial aspect to relax, hence pulling the tooth back in the enlarged mesial socket area. Presumably during this period of tooth support, osteoblastic activity can commence on the mesial aspect of the lamina dura until stresses within the membrane are relaxed and the normal unstressed resting state of the newly displaced tooth is restored.

The general pattern of bone resorption under a compressive (hydrostatic) stress is probably far more localised on a microscale. Figure 13 indicates a typical bone surface which is of an undulating nature, and varying in its degree of mineralisation from point to point. Applying a compressive stress to such a surface results in again a complex changing stress pattern. Here we have chosen the   component of the principal stress and Figure 13 shows nuclei of high local compressive stress are evident. Resorption may thus commence from centres away from the bone surface, eventually spreading and changing the local morphology. As this morphological and hence material property change takes place, the local stress pattern will change in sympathy. The rate of bone resorption is therefore an extremely complex function of stress, geometry, material properties and associated physiological factors.

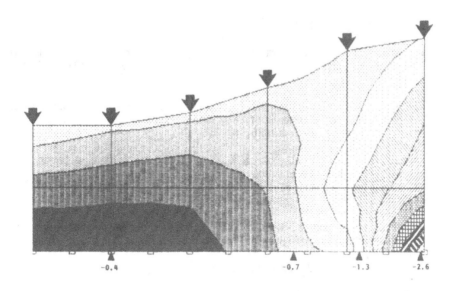

Figure 13. 2-D representation of the local stress pattern at the bone surface under compressive loading.

CONCLUSIONS

The present analysis of our fem work allows an accurate description of the local stress patterns surrounding teeth for a number of geometries and loading conditions. However, more relevantly the complex stress and strain patterns in the alveolar process leading to bone remodelling and tooth movement can be calculated for a real clinical situation.

Combining these data with previous findings, the fem suggested that:-

(i)   rate of tooth movement is independent of load magnitude.
(ii)   the rate is controlled by the local compressive stress pattern at the bone surface.
(iii)   the bone resorption is inhomogeneous and delicately balanced with respect to local changing stress patterns with time and space.

These findings do not rule out the possibility of the prediction of tooth position with time, load and geometry but only serve to enhance the need for further reliable bone resorption data and materials property data of hard and soft tissue.

REFERENCES

1. Yettram, A.L., Wright, K.W.J. and Houston, W.J.B., Centre of rotation of a maxillary central incisor under orthodontic loading. Brit. J.

Ortho., 1977, **4,** 23-27.

2. Williams, K.R. and Edmundson, J.T., Orthodontic tooth movement analysed by the finite element method. Biomaterials, 1984, **5,** 347-351.

3. Williams, K.R., Edmundson, J.T., Morgan, G., Jones, M.L. and Richmond, S., Orthodontic movement of a canine into an adjoining extraction site. J. Biomed. Eng., 1986, **8,** 115-120.

4. Jones, M.L. and Williams, K.R., The use of finite element stress analysis in the assessment of tooth movement. In Interfaces in Medicine and Mechanics, ed. K.R. Williams and T.H.J Lesser, Elsevier Applied Science Publishers, London, 1989, pp. 270-281.

5. Tanne, K. and Sakuda, M., Initial stress induced in the periodontal tissue at the time of the application of various types of orthodontic force: Three dimensional analysis by means of the finite element method. J. Osaka Univ. Dent. Sch., 1983, **23,** 143-171.

6. Tanne, K., Sakuda, M. and Burstone, C.J., Three dimensional finite element analysis for stress in the periodontal tissue by orthodontic forces. Am. J. Orthod. Dentofac. Orthop., 1987, **92,** 499-505.

7. Wilson, A.N., Finite element analysis of orthodontic tooth movement. University of Wales, M.Sc. Thesis, Department of Civil Engineering, Swansea, November 1988.

8. Stephens, C.D., The orthodontic centre of rotation of the maxillary central incisor. Am. J. Orthod., 1979, **76,** 209-217.

CARBON FIBER REINFORCED PLASTIC (CFRP) KNEE-ANKLE-FOOT-ORTHOSIS (KAFO) PROTOTYPE FOR MYOPATHIC PATIENTS.

C.Granata, A.De Lollis*, G.Campo, L.Piancastelli*, A.Ballestrazzi, L.Merlini
The Muscle Clinic of the Istituto Ortopedico Rizzoli.
*The Research Centre of the Officine Rizzoli, the Department of Mechanical Engineering,
University of Bologna.
Bologna, Italy.

ABSTRACT

A traditional Knee-Ankle-Foot-Orthosis (KAFO) for myopathic patients has been studied for the assessment of loads and fatigue resistance.

Starting from this basis a thermoplastic matrix carbon fiber reinforced composite (CFRP) knee-ankle-foot-orthosis has been developed in order to reduce the weight. A finite element simulation program for deformation analysis was used to compare the behaviour of conventional and CFRP orthosis. At the test stresses there were no breakages either of the prototype or of its parts. The CFRP orthosis allows a weight reduction of more than 40%.

INTRODUCTION

Knee-ankle-foot-orthoses (KAFO) are used in several neuromuscular diseases. In Duchenne Muscular Dystrophy they are applied on loss of walking to permit prolongation of function which otherwise would be lost because of the progressive muscular weakness. In Spinal Muscular Atrophy, intermediate form, in which unaided walking is not achieved, they are used to promote upright position and walking.

KAFO orthoses support the patient fixing the knee joint thus allowing walking by limiting the muscular load. They are at present made of traditional materials, such as polyethylene and aluminium with high strength steel for the critical zones (Fig. 1).

The use of advanced composites allows reducing the weight and increasing the stiffness (endurance) of these orthoses, on which the patient's sense of security depends. Because of the use of advanced composites the KAFO orthosis had to be redesigned, not least to ensure that the appropriate technologies would be used in the making.

LOAD ANALYSIS

A complete load analysis was necessary as the equilibrium of the orthosis is assured by the forces applied onto it, deriving from two different interactions: ground to orthosis and patient to orthosis. Once the loads had been analyzed a structure was designed that would be lighter in weight and as rigid, or more so, than a conventional orthosis.

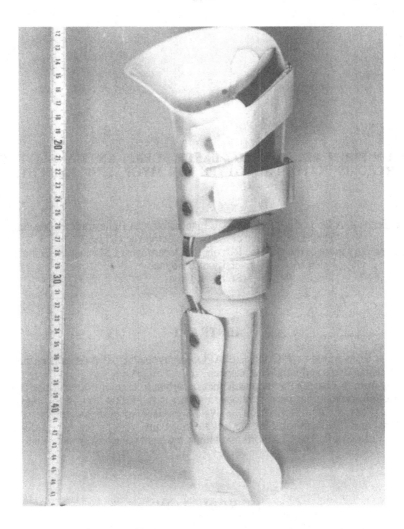

Figure 1. KAFO orthosis made of traditional materials (polypropilene shells with aluminium rods) for myopathic children.

CHOOSING THE MATERIAL

Since strength and light weight are paramount requirements, a thermoplastic matrix carbon fibre reinforced composite (CFRP) was chosen in order to ensure the workability necessary for the manufacture and final adjustment of the orthosis. A comparative analysis of the various resins available and the need for fairly low forming temperatures led to the choice of an acrylic resin, a polymethyilmethacrylate with a monomer-polymer ratio of 80/20.

Two technologies for polymerizing the composite were devised: hot press moulding and room temperature vacuum bag method using a catalyst. Tests were done for resistance to solar radiation and to immersion in a saline solution simulating human sweat. Specimens were exposed to these conditions for over 3 months without significant variations in the mechanical properties of the material.

DESIGNING ORTHOSIS IN COMPOSITE

The magnitude of the stresses arising in the weak zones of the orthosis was studied at the design stage by simulating the membrane and tensile stress behaviour of a part by means of a finite element programme. This programme was used initially for simulating the behaviour of an orthosis made of conventional material to see whether the problem had been correctly stated, defining the finite element mesh and the pattern of loads and constraints. The material was then replaced by the chosen composite, with adjustments to the shape and thickness of the various parts to optimize weight reduction. The orthosis prototype thus designed (Fig. 2) reduced the weight by some 40%.

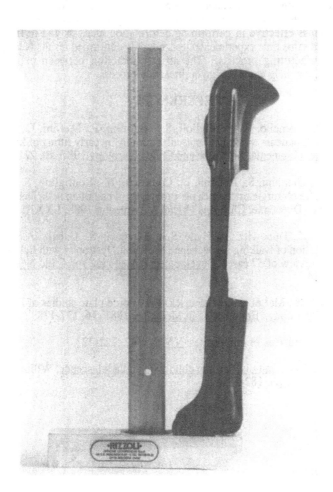

Figure 2. Prototype of CFRP orthosis for myopathic children.

TESTING

The object of testing was to compare a conventional orthosis with the newly designed one. First to be compared were ultimate static strength and ultimate fatigue strength of the aluminium rods and the composite rods, these being the fundamental structural elements of the orthosis. Then, the prototype was subjected to a fatigue test on a dummy lower limb. The results were fairly close (±10%) to those obtained with finite element simulation.

RESULTS AND DISCUSSION

At the test stresses there were no breakages either of the prototype or of its parts, showing that the procedure adopted was broadly correct. However, the experimental tests demonstrated that despite a weight reduction of more than 40% a certain degree of overdimensioning still exists, since all the specimens and the prototypes tested were not damaged before reaching 1.5 times the design load even under the most severe conditions. So there is scope for further development in optimising weight reduction.

The results obtained indicate that the design method points to the right direction: finite element simulation is effective in performing deformation analysis and in finding critical failure points; at the same time experimental tests not only confirmed results but also helped in refining the manufacturing process. The strict interaction between experiments and simulation was of great help to the design optimisation process.

REFERENCES

1.   Granata, C., Cornelio, F., Bonfiglioli, S., Mattutini, P., Merlini, L., Promotion of ambulation of patients with spinal muscular atrophy by early fitting of knee-ankle-foot orthoses. Developmental Medicine and Child Neurology,1987, 29, 221-224.

2.   Granata, C., Giannini, S., Rubbini, L., Corbascio, M., Bonfiglioli, S., Sabattini, L., Merlini, L., La chirurgia ortopedica per prolungare il cammino nella Distrofia Muscolare di Duchenne. Chirurgia Organi Movimento, 1988, LXXIII, 237-248.

3.   Heckmatt, J.Z., Dubowitz, V., Hyde, S.A., Florence, S., Gabain, A.S., Thompson, N., Prolongation of walking in Duchenne Muscular Dystrophy with lightweight orthoses: a review of 57 cases. Developmental Medicine and Child Neurology, 1985, 27, 149-154.

4.   Khodadadeh, S., McLelland, M., Patrick, J.H., Force plate studies of Duchenne Muscular Dystrophy. Engineering in Medicine, 1987, 16, 177-178.

5.   Lubin, G., Handbook of composites. VNR, 1982, 722-723.

6.   Sutherland, D.H., Gait disorders in childhood and adolescence, Williams & Wilkins, Baltimore, 1983, pp. 182-183.

IMPROVEMENT OF FRACTURE HEALING BY APPLIED AXIAL MICROMOVEMENT : A CLINICAL STUDY

S H WHITE, J L CUNNINGHAM, J B RICHARDSON, M EVANS and
J KENWRIGHT (OXFORD)*, and M A ADAMS, A E GOODSHIP,
E SMITH and J H NEWMAN (BRISTOL)+
*Oxford Orthopaedic Engineering Centre, University of Oxford, Nuffield
Orthopaedic Centre, Headington, Oxford OX3 7LD, UK.
+Bristol Royal Infirmary, Marlborough Street, Bristol, BS2 8HW

ABSTRACT

A multicentre, randomised, controlled trial of the application of axial micromovement via external skeletal fixation to open tibial fractures is described. The clinical results showed a shorter time to bony union, and the mechanical results showed a steeper gradient in the rise of fracture stiffness measurements than with rigid fixation. This specific regime of short daily periods of imposed axial cyclical strain is thus shown to be osteogenic to tibial fractures in man.

INTRODUCTION

Open tibial fractures are still a major challenge and most surgeons opt for external fixation to stabilise the fracture and yet permit access for management of the soft tissues. There is however concern that the degree of rigidity imposed by external fixation is detrimental to healing for one often sees an inhibited callus response. We have designed an external fixator which allows the surgeon to change the axial rigidity of fixation, by application of axial micromovement through the pins to the fracture interface, in a precise and predictable manner. In a series of experiments in ovine osteotomies, we have defined an axial strain magnitude and rate of application that is osteogenic [1]. This study tests the hypothesis that the same stimulus of axial micromovement applied early after injury, and continued throughout healing, will enhance the healing of tibial fractures in patients.

MATERIALS AND METHODS

Patients at two centres (Bristol and Oxford) with open tibial fractures were randomly selected for treatment using either a rigid external fixator, or the same fixator but with a micromovement module attached via sliding clamps. Children were excluded from the study as were patients with Grade 3:C fractures. Of the 82 patients entering the trial, one patient died and another emigrated, leaving 80 patients who were followed through to radiological consolidation of the fracture. The allocation of patients took account of injury severity using a stratified randomisation, based upon the soft tissue injury grading of Gustilo and Anderson [2], combined with a bone injury classification of Johner and Wruhs [3]:

TABLE 1

Classification of Injury Severity

Group	Soft Tissue Injury	Bone Injury
A	Grade 1	Not comminuted
B	Grade 2/3	Not comminuted
C	Grade 1	Comminuted
D	Grade 2/3	Comminuted

All patients were treated with a unilateral frame (Dynabrace, Richards Medical (UK) Ltd), applied to the medial surface of the tibia. The two pins on one side of the fracture were held by the micromovement module which allowed the option for rigid treatment using locking nuts, or micromovement treatment using sliding clamps. In the micromovement group a pneumatic pump was attached to the sliding clamps and small amounts of inter-fragmentary cyclical axial displacement were applied (Figure 1). A daily regime of applied axial micromovement was started as soon as the leg was comfortable and always within seven days of frame attachment. This daily mechanical stimulus had the following characteristics:

1. Maximal initial longitudinal axial displacement of 1mm.

2. Constant low force of 400N, so that the fatigue stress limit of the bone screw interface was not exceeded.

3. A frequency loading of 0.5Hz, approximately that of physiological walking.

4. 500 cycles per day (i.e. continuously over 17 minutes)

This externally imposed micromovement was discontinued when the patient was able to reproduce this axial excursion through his or her own activities. The spring in the micromovement module was pretensioned to 10kgs to allow movement when the patient started at least 10kgs weight bearing (Figure 2). In contrast, the patients treated with rigid fixation had all their clamps locked against the beam.

The external fixator was removed when clinical and radiological fracture union was diagnosed. Most surgeons applied a light-weight functional cast after frame removal until it was considered safe to allow unsupported weight bearing. The time to unsupported weight bearing without fixator or cast was used as one measure of fracture healing. The stiffness of the fracture was monitored throughout healing, using a strain gauge transducer which we attached to the fixator column at two weekly intervals (Figure 3). Fracture stiffness typically increases exponentially with time, and from a logarithmic plot of stiffness values the time taken to reach a bending stiffness of 15NM/deg was calculated.

Results were analysed using non-orthogonal three-way analysis of variance.

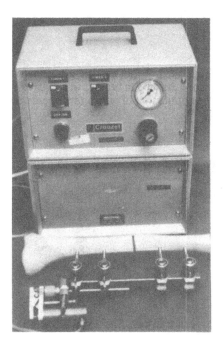

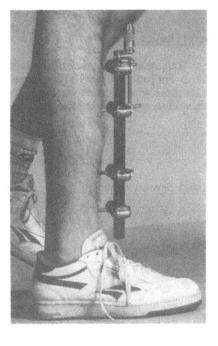

Figure 1. A pneumatic pump was used to activate the micromovement module whilst the patient was in hospital. The clamps were able to move along the beam by means of sliding plates. For patients treated in rigid mode, the clamps were held firmly fixed to the beam with locking nuts.

Figure 2. The two uppermost pins are connected to the micromovement unit, and their clamps have sliding plates to allow micromovement of the pins along the beam on weight bearing.

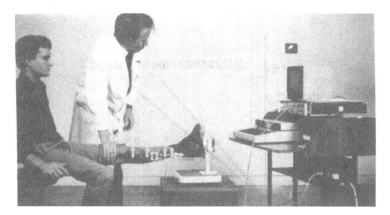

Figure 3. At two weekly intervals the bending stiffness of the fracture was measured indirectly by using a strain gauge that was clamped onto the beam during loading tests.

RESULTS

<u>Clinical Healing Time</u>

The mean healing times from injury to unsupported weight bearing are shown in Table 2. For patients treated with a frame in fixed mode the time to healing was 29 weeks, and for those with micromovement, 23 weeks; the application of micromovement was independently related to the faster healing time ($p<0.05$).

Table 2

Mean time to unsupported weight bearing in weeks

<u>Injury Severity Group</u>	<u>Micromovement</u>	<u>Rigid</u>
A	16.15	22.3
B	25.5	48.6
C	20.6	22.2
D	26.8	28.9
ALL	**23**	**29 (P<0.05)**

The grading of soft tissue injury independently influenced healing; the mean bone healing time for grade 1 soft tissue injuries was 20 weeks and for more serious open wounds was 30 weeks ($p<0.01$).

No significant difference in healing times was observed when the degree of cominution was analysed as an independent variable.

<u>Mechanical Healing Time</u>

The results are shown in Figure 4. The mean time to reach a bending stiffness of

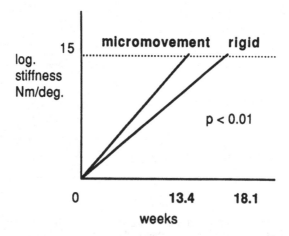

Figure 4. Results of the stiffness measurements of tibial fractures treated with micromovement or with rigid fixation.

15NM/deg in the fixed mode group was 18 weeks, and in the micromovement group was significantly shorter at 13 weeks (p<0.01).

Complications

There was no significant difference in the incidence of pin tract infections between treatment groups. Refracture occurred in three patients in the fixed mode treatment group. The need for secondary surgery in a total of five patients was not significantly different in either treatment group.

DISCUSSION

In patients with long bone fractures, the bone is, in the early days after injury, deprived of its normal cyclical loading, as a result of the enforced immobilisation of conventional methods of treatment. It is possible that these first few days or weeks are a crucial period in which to reintroduce the normal mechanical stimulus to osteogenesis. The results of this study show that such an effect may be achieved by the early active application of an appropriate mechanical stimulus.

There is now a trend in treatment methods towards the use of either more flexible frames, or the adjustment of frames to allow "dynamisation" of the fracture so as to increase the stress and strain acting at the fracture site [4, 5, 6]. Dynamisation in usually prescribed several or many weeks after a fracture when bridging has occurred, though Burny has advocated the use of flexible frames throughout treatment. Sliding frames have also been devised to stabilise a fracture yet allow axial stress to be applied through weight bearing [7]. In the present study the stimulus was applied early whilst at the same time providing sufficient stability to retain fracture reduction.

The optimal mechanical conditions required for the different stages of healing for different patterns and sites of fracture have yet to be defined. The time of application of a mechanical stimulus has been shown to influence bone healing in experimental conditions [1, 8]. The amount of strain is also important, for in experimental ovine osteotomies we have increased the initial displacement within a 3mm transverse osteotomy gap from 1mm as used in the present regime to 2mm (66% initial strain). Displacement of 2mm applied at 0.5Hz for 500 cycles was statistically significantly detrimental to the rate of healing [9]. Clearly, the differentiating callus following a fracture is extremely sensitive to the mechanical environment. It may prove important to adjust the prescribed regime according to the phase reached in healing for an individual fracture and patient to achieve a faster bony union. The use of such a micromovement module on an external fixator provides the surgeon with a new therapeutic tool to achieve this goal.

REFERENCES

1.   Goodship, A.E. and Kenwright J. The influence of induced micromovement upon the healing of experimental tibial fractures. J. Bone Joint Surg. (Br.), 1985, 67-B, 650-655.

2.   Gustilo, R.B. and Anderson, J.T. Prevention of infection in the treatment of one thousand and twenty-five open fractures of long bones. J. Bone Joint Surg. (Am.), 1976, 58-A, 453-458.

3.   Johner, R. and Wruhs, O. Classification of tibial shaft fractures and correlation with results after rigid internal fixation. Clin. Orthop., 1983, **178**, 7-25.

4.   Burny, F.L. Strain gauge measurement of fracture healing. In External Fixation: the Current State of the Art, ed. A.F. Brooker, C.C. Edwards, William and Wilkins, Baltimore, 1979, pp. 371-82.

5.   Behrens, F. and Searls, K. External fixation of the tibia. Basic concepts and prospective evaluation. J. Bone Joint Surg. (Br.), 1986, **68-B**, 246-54.

6.   De Bastiani, G., Aldegheri, R. and Brivio, L.R. The treatment of fractures with a dynamic axial fixator. J. Bone Joint Surg. (Br.), 1984, **66-B**, 538-45.

7.   Lazo-Zbibowski, J., Aguilar, F., Mozo, F., Gonzalez-Buendia, R. and Lazo, J.M. Biocompression External Fixation, Clin. Orthop., 1986, **206**, 169-184.

8.   White, S.H. and Kenwright, J. The timing of distraction of an osteotomy. J. Bone Joint Surg. (Br.), 1990, **72-B**, 356-361.

9.   Kenwright, J., Goodship, A.E., Kelly, D.J., Rigby, H.S. and Watkins, P.E. The effect of different regimes of axial micromovement on the healing of tibial fractures; an experimental study. Abstract: International Conference on Hoffmann External Fixation, Garmisch-Partenkirchen, Murnau, Bavaria, West Germany, 1986, pp.50.

IMPROVEMENT IN THE DESIGN OF BONE SCREWS FOR EXTERNAL FIXATION

M EVANS, S H WHITE and J L CUNNINGHAM*
Oxford Orthopaedic Engineering Centre, University of Oxford,
Nuffield Orthopaedic Centre, Headington, Oxford OX3 7LD, UK
*School of Engineering, Science Laboratories,
University of Durham, South Road, Durham DH1 3LE, UK

ABSTRACT

Bone screw loosening presents a major clinical problem with external fixation. In this study, design factors influencing the mechanics of the bone-screw interface were analysed and various experimental screws designed with the intention of optimising the holding strength and stiffness of the inserted screw. Push-in, pull-out and bending tests were carried out on three experimental screws, and on two commercially available screws in a synthetic material and cadaveric bone. The results of these tests indicate that the screw threadform and cutting head have a significant effect on the holding strength of the screw.

INTRODUCTION

Bone screw loosening is a major clinical factor with external fixation [1], and the decision to remove the fixator is often governed by screw loosening and/or screw tract infection rather than fracture healing. Axial loading of a unilateral external fixator results in bending loads being carried by the screws in addition to tensile and compressive loads. Stresses at the sites of screw insertion may exceed the fatigue stress limit of the bone and so lead to loosening. Cyclic loading of the screws results in an increased incidence of screw loosening over unloaded or statically loaded screws [2]. This may result from micromotion of the screw relative to the surrounding bone causing local bone resorption [3, 4].

In this study, the factors influencing the mechanical conditions at the bone screw interface were assessed. Experimental screws, were designed with the intention of reducing micromovement and stress at the screw-bone interface and so maximising the holding strength of the screw and minimising screw loosening. They were then tested, together with two commercial screws, in both a bone substitute and in cadaveric bone.

DESIGN CONSIDERATIONS

The analysis of a unilateral external fixator when stabilising a limb carrying axial load (Figure 1) shows that the bone screws are subjected to both longitudinal forces Z, and bending moments M, the values of which can be determined from the applied loading and the fixator geometry [5].

The axial screw force Z produces a shearing action within the bone and is responsible for the pull out failure of bone screws. The bending moment M generates compressive stresses in the bone which may overload the bone locally. The failure of screw fixation usually results from bone resorption around the screw in the proximal cortex and this will result from many factors including the bone/screw interface, fixator geometry and patient activity. Of these factors, the bone/screw interface can be significantly influenced by screw design and in this study a number of aspects of bone screw design and screw insertion were studied with the object of providing the optimum mechanical conditions at the bone screw interface. Various features influence the overall performance of the bone screw - these are inter-related in their effect, and are therefore difficult to study in isolation. The main design features considered are shown in Figure 2:

(i) **The Screw Diameter**: In general terms, the larger the screw diameter the stiffer is the bone screw/fixator system and the greater the load bearing areas for both shear and compression.
(ii) **The Screw Core Diameter**: This also influences the screw stiffness and the load bearing areas for shear and compression.
(iii) **The Thread Form**: This is particularly important in providing the optimal load distribution especially with regard to pull out strength. The thread form will also affect the insertion torque required for the screw.

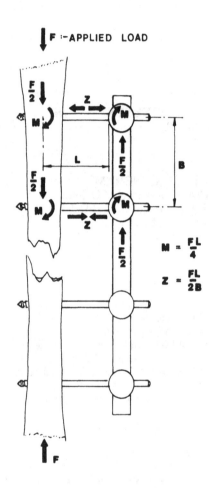

Figure 1. Diagram of external fixator stabilising fracture showing forces applied to bone screws

(iv) **Pre-Drilling**: For a self tapping bone screw this is important to ensure a correct fit of the screw thread. Drilling oversize reduces the load bearing capacity of the screw, particularly when considering compressive loads applied by the screws to the bone. Alternatively, undersize drilling increases the insertion torque and will also increase the risk of thermal necrosis of the bone.
(v) **Thread Cutting**: For a self tapping bone screw, this is of great importance in providing the correct fit of the screw in the tapped hole. With hand insertion, the design of the thread cutting portion of the screw influences the alignment of the screw when passing through one pre-drilled cortex of the bone into the other. The thread cutting portion of the bone screw will also influence the insertion torque of the screw.

Screws of 6mm O.D. were designed with three features intended to improve their performance:- these were; a) by increasing the number of cutting heads on the screw to minimise bone damage on screw insertion, b) by utilising a fine threadform to maximise the screw threads in contact with cortical bone and c) by increasing the core diameter and hence bending stiffness of the screw. The influence of each design feature was assessed

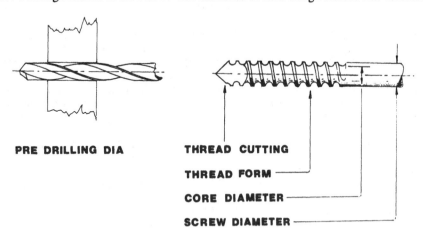

PRE DRILLING DIA   THREAD CUTTING

THREAD FORM

CORE DIAMETER

SCREW DIAMETER

Figure 2. Design features of bone screw.

separately by using different experimental screws, each with a successive design feature added. Hence the first experimental screw (termed OOEC 1) utilises a coarse threadform similar to the Schanz screw (Richards Medical (U.K.) Ltd., 6 The Technopark, Newmarket Road, Cambridge CB5 8PB, U.K.), with the number of cutting heads increased from two to three. The second experimental screw (OOEC 2) has three cutting heads and uses a fine threadform similar to that of the Aesculap screw (Aesculap A.G., Postfach 40, D-7200 Tuttlingen, West Germany). The third experimental screw (OOEC 3) has three cutting heads, a fine threadform and an increased core diameter of 4.5mm compared to the 4.1mm core diameter of the OOEC 1 and OOEC 2 screws. Each experimental screw had a rounded tip to minimise damage to the soft tissues beyond the bone. A summary of the main features of the experimental and Schanz and Aesculap screws is given in Table 1.

TABLE 1
Comparison of Features of Screws Tested

SCREW TYPE	MAJOR DIAMETER (mm)	MINOR DIAMETER (mm)	THREAD PITCH (mm)	NO. OF THREAD CUTTERS	LENGTH OF CUTTING END OF SCREW (mm)
SCHANZ	6.00	4.10	3.00	2	12.00
AESCULAP	6.00	4.10	2.00	2	7.00
OOEC1	6.00	4.10	3.00	3	9.00
OOEC2	6.00	4.10	2.00	3	10.00
OOEC3	6.00	4.50	2.00	3	10.00

The experimental screw designs (OOEC 1, 2 and 3) and 6mm Schanz and Aesculap screws were tested in 'Tufnol' tube and in cadaveric bone. Variables measured were insertion torque and push-in and pull-out failure loads.

MATERIALS AND METHODS

Because of the wide variety of the mechanical properties of cadaveric bone [6, 7], the screws were initially tested in phenolic resin coated paper laminate tube (Tufnol Ltd., Perry Barr, Birmingham B42 2TB, U.K.) of 29mm outside diameter and 4mm wall thickness since this material has consistent mechanical properties reasonably similar to that of cortical bone. Once consistent differences between screws inserted in Tufnol were established , it was hoped to be able to confirm these differences in the limited quantities of available cadaveric bone (Table 2).

The cadaveric bone was from eight femurs obtained postmortem in the age range 31 - 82 years at death and which were free from pathological bone disorders. The femurs were frozen at $-20^{\circ}C$ until required and were thawed at room temperature for at least 4 hours. Each femur was divided into three and the middle third sub-divided into four sections each

TABLE 2
Comparison of mechanical properties of Tufnol and cortical bone.

	Tufnol	Cortical Bone [12, 13]
Compressive Stress (MPa)	350	108 – 200
Shear Stress (MPa)	100	53 – 82
Tensile Stess (MPa)	140 –190	54 – 172

of approximately 4cm in length. The soft tissue was removed although the bone marrow was left in-situ. The sections were numbered from the proximal end of the bone, Section 1 being the most proximal and Section 4 the most distal.

Each screw was inserted into either the Tufnol tube or bone using a specially designed jig which bolts on to the stand of an upright drill. The jig was strain gauged to measure the torque required to insert the screw at a constant insertion speed of 45 revs/min. Part of this jig containing the inserted bone screw was then mounted on an Instron 1122 materials testing machine (Instron Ltd., Coronation Road, High Wycombe, Bucks, U.K.) and the bone screw loaded axially to failure at a rate of 1mm/min, in either tension or in compression.

Ten tests were carried out on each type of screw in Tufnol, and eight tests were carried out for each screw in cadaveric bone. The mean and standard deviation of the insertion torque and failure load were calculated for each test and the results for each screw type were compared using an unpaired Student's t-test.

RESULTS

Initial push-in and pull-out tests of the Schanz screw in Tufnol indicated similar failure loads (all the threadforms tested were symmetrical); subsequent tests only used push-in loading. The results for push-in tests are summarised for the various screws tested in Tufnol in Figure 3. The results for the OOEC1 screw were only marginally better than the standard Schanz screw and considerably less than the other two experimental screws, and so this screw was dropped from subsequent tests.

The results of these push-in tests indicate that both the cutting head and the threadform of the screw have a marked effect on the holding strength. For example, two screws with significantly different threadforms (Schanz cf. Aesculap) exhibited a marked difference in push-in strength. If the number of thread cutting flukes on the screw tip is increased from two to three for a similar threadform (Schanz cf. OOEC 1, Aesculap cf.

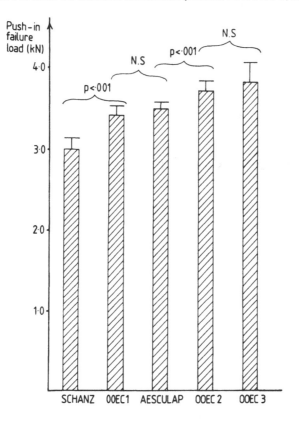

Figure 3. Push-in test results for the various screws tested in Tufnol.

OOEC 2) then the push-in failure strength can also be increased.

All of the above tests were carried out using a pilot hole diameter of 4.5mm. further push-in tests on one screw (OOEC 2) in which the pilot hole diameter was varied, showed no discernible effect on the failure strength up to a pilot hole diameter ratio of 0.93 (Figure 4). Hence a small increase in screw core diameter, which results in a marked increase in the bending stiffness of a screw, gives no significant reduction in the holding strength. Reducing the pilot hole diameter from 4.5mm to 3.8mm for the Aesculap screw,

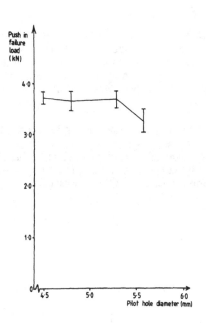

Figure 4. Variation of push-in strength with pilot hole diameter of OOEC 2 screw tested in Tufnol.

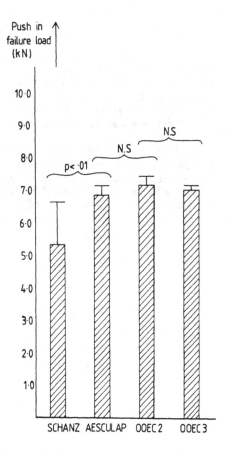

Figure 5. Push-in test results for the various screws tested in cadaveric bone.

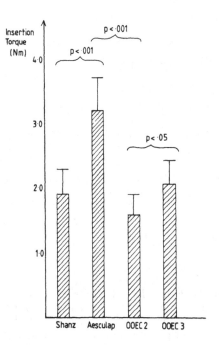

Figure 6. Maximum insertion torques for the various screws tested in cadaveric bone.

as recommended by the manufacturer, had no significant effect on the push-in failure load.

The results of push-in tests using cadaveric bone are given in Figure 5, and the torques measured on insertion of these screws given in Figure 6. The magnitude of the failure loads is greater in the cadaveric bone due to increased cortical or wall thickness of the bone compared to the Tufnol tube, despite the reasonably similar mechanical properties of the two materials. Some of the significant differences found in failure strength between the screws tested in Tufnol are also seen in these tests. However, some of the differences found in the Tufnol tests are not seen in the tests in bone, due to the expected variation in properties between the bones tested which can be inferred from the increased standard deviations seen with these tests. Higher maximum insertion torques are seen for the fine threaded Aesculap screw and for the OOEC 3 screw with its increased core diameter. It is interesting to note that the use of three thread cutting heads instead of two reduces the maximum insertion torque for a similar threadform (OOEC 2 cf. Aesculap).

DISCUSSION

The geometry of an external fixator results in large bending loads being carried by the screws in addition to tensile and compressive loads. The cyclical bending of the screws during normal weightbearing can result in micro-movement between the screw and the bone. Schatzker et al. [3], Perren [9] and Uhthoff [4] have all independently shown for screws holding internal fixation plates that relative movement between the screw and the bone causes the migrating cells which fill the microscopic spaces between the screw threads and the bone to differentiate into osteoclasts, fibroblasts and chondroblasts. This results in the resorption of existing or newly laid down bone and its replacement by fibrous and cartilaginous tissue, thus causing the screw to loosen. Conversely, in the absence of movement between the screw and bone the migrating cells differentiate into osteogenic cells and produce a solid callus which firmly anchors the screw. Schatzker et al. [8] and Perren [9] have also shown that cortical bone subjected to static compression by screw threads retains it integrity and is not resorbed. Pettine et al. [2] have shown that in external fixation screw loosening also occurs in the presence of a moderate value of cyclically applied load, loosening being minimal under high static loading. Uhthoff et al. [10] have shown this process of loosening to be reversible, i.e. if the relative movement of screw and bone is reduced or eliminated the osteoclastic activity ceases and osteoblastic activity appears, resulting in new bone formation.

In external fixation, the potential for relative movement between screw and bone is high due to the high bending moments carried by the screw, by the fact that the screw is not tensioned as in internal fixation and by the instability of fracture fixation resulting in cyclic loading of the screws on normal weightbearing [11]. All of these factors will increase the incidence of bone screw loosening in external over internal fixation. Hence any improvement in the mechanical interface of the bone and screw which reduces stress and relative movement will decrease the incidence of bone screw loosening.

The results presented above indicate that large increases in core diameter are possible without compromising the holding strength of the screw in the bone (bending stiffness varies as D^4). The in-vivo effect of increasing the screw stiffness has yet to be assessed, but it is probable that for similar cyclical loading conditions, the stiffer screw will survive for longer without loosening.

ACKNOWLEDGEMENTS

The Authors would like to thank Aesculap A.G. for supplying some of the screws used in these tests.

REFERENCES

1.   Seligson D., Donald G.D., Stanwyck T.S. and Pope M.H., Consideration of pin diameter and insertion technique for external fixation in diaphyseal bone. Acta Orthop. Belgica, 1984, **50**, 441-450.

2.   Pettine K.A., Kelly P.J., Chao E.Y.S. and Huiskes R., Histologic and biomechanical analysis of external fixator pin-bone interface. Orthop. Trans., 1986, **10**, 337.

3.   Schatzker J., Horne J.G. and Sumner-Smith G., The effect of movement on the holding power of screws in bone. Clin. Orthop. & Rel. Res., 1975, **111**, 257-262.

4.   Unthoff H.K., Mechanical factors influencing the holding power of screws in compact bone. J. Bone & Jt. Surg., 1973, **55B**, 633-639.

5.   Evans M. and Kenwright J., The Oxford external skeletal fixation system. In The Severely Injured Limb, Eds. Ackroyd C.E., O'Connor B.T. and de Bruyn P.F., Churchill Livingstone, 1983.

6.   Carter D.R. and Hayes W.C., The compressive behaviour of bone as a two-phase porough structure. J. Bone & Jt. Surg., 1977, **59A**, 954-962.

7.   Goldstein S.A., Wilson D.L., Sonstegard D.A. and Matthews L.S., The mechanical Properties of human tibial trabecular bone as a function of metaphyseal location. J. Biomech., 1983, **16**, 965-969.

8.   Schatzker J., Horne J.G. and Sumner-Smith G., The reaction of cortical bone to compression by screw threads. Clin. Orthop. & Rel. Res., 1975, **111**, 263-265.

9.   Perren S.M., Physical and biological aspects of fracture healing with special reference to internal fixation. Clin. Orthop. & Rel. Res., 1979, **138**, 175-196.

10.   Uhthoff H.K. and Germain J-P., The reversal of tissue differentiation around screws. Clin. Orthop. & Rel. Res., 1977, **123**, 248-252.

11.   Cunningham J.L., Evans M. and Kenwright J., Measurement of fracture site movement in patients treated with external fixation. J. Biomed. Eng., 1989, **11**, 118-122.

12.   Reilly D.T. and Burstein A.H., Mechanical properties of cortical bone. J. Bone Joint Surg., 1974, **56A**, 1001-1022.

13.   Currey J.D., The mechanical properties of bone. Clin. Orthop. & Rel. Res., 1970, **73**, 210-231.

CLINICAL APPLICATION OF TOTAL ARTIFICIAL HEART (TAH) AND BLOOD RHEOLOGY

TING-CHENG HUNG, HARVEY S. BOROVETZ, ROBERT L. KORMOS,
BARTLEY P. GRIFFITH, AND ROBERT L. HARDESTY
School of Medicine
University of Pittsburgh
Pittsburgh, PA 15261, USA

ABSTRACT

Hemorheology was studied in 18 patients ranging in age from 27-57, who received the Jarvik-7 total artificial heart (TAH) implant from 1-48 days as a bridge to cardiac transplantation. Thrombus and embolus formation leading to transient ischemic attacks has been a major problem in these patients, and formation of thrombosis on the artificial heart valve housing is also frequently seen. Our studies in these patients were aimed at the rheologic impact of TAH and the role of rheology in thrombus formation. Blood viscosity, red cell rigidity and plasma fibrinogen were evaluated during TAH implantation. Observed abnormalities in blood rheology may be crucial factors in thrombus formation on artificial heart valves as well. Our results show that the therapeutic management of rheologic parameters should prove to be a unique and important advancement in the clinical course of treatment of these patients.

INTRODUCTION

Blood alterations induced by high shear stress and exposure to foreign surface are known to be major hemorheological complications involved in the use of artificial organs, artificial heart valve and small prosthetic vessels. In the first clinical use of the TAH by Cooley et al [1] in 1969, infection and multi-organ failure as the result of blood trauma were the primary problems associated with TAH implantation. Over the past 30 years, however, a great deal of work has been done on resolution of these problems [2-11]. Progressive improvements in TAH design, fabrication, and biocompatibility have dramatically increased the feasibility of its clinical application in recent years. Between 1982 and the present time, long-term TAH support was documented in five patients [12]. Support length has even been reported at 112 days in one patient at the University of Utah, and at 197 days in another patient at the Karolinska Institute in Sweden.

Nevertheless, persistent complications, such as stroke, limited the TAH to short-term use as a bridge to cardiac transplantation. The success of the TAH as an interim device was reported in animal experiments by several centers [5,8,10,12], leading to widespread acceptance of the use of the TAH as an interim device. In August 1985, the clinical "bridge-to-transplantation" program using the Jarvik-7 TAH began at the University of Arizona. In October of that same year, the University of Pittsburgh along with six other centers [13,14] joined the program.

The increased use of cardiac transplantation and prolonged waiting period even for patients who are most critically ill has established the need for an interim means of support for these patients. The implementation of the Jarvik artificial heart as a bridge to cardiac transplantation at the University of Pittsburgh Health Center is based on the fact that there is a group of mortally ill patients who might survive if their circulation could be temporarily and safely supported by a mechanical system.

Since the beginning of the University of Pittsburgh program, the Jarvik-7 has been implanted into 19 critically ill, moribund cardiac transplant candidates for a range of 1 to 48 days prior to transplantation. Sixteen of nineteen patients survived to receive cardiac transplants. Subsequent to transplantation, one patient died of donor organ failure, two died of acute rejection and five others died as a consequence of infection. Our involvement in rheological monitoring in this program did not include the first transplant patient. This paper presents our study of the blood rheology of 18 of these 19 patients during their support on the Jarvik-7 TAH in an effort to gain insight into the pathology of the hemorheological alterations induced by the Jarvik-7.

MATERIALS AND METHOD

Overall blood rheology was evaluated as a function of blood viscosity, red cell (erythrocyte) rigidity, and plasma fibrinogen. All patients were anticoagulated within 24 h of implantation using continuous intravenous heparin and oral dipyridamole (75 mg three times a day). The partial thromboplastin time was maintained at 1.5 to 2 times control. Peripheral blood samples were drawn for rheologic studies on all but the first patient during the periods of Jarvik-7 TAH support.

Blood Viscosity Measurements

Blood viscosity was determined with a Wells-Brookfield cone-plate viscometer with a cone angle of 0.8 degree and cone radius of 2.5 cm [14-16]. An aliquot (0.5 ml) of blood, anticoagulated with sodium EDTA, was introduced into the viscometer and a uniform shear flow established by rotation of the viscometer cone. Viscosity measurements for each sample were made at 37C. The apparent blood viscosity (μ) was calculated from the ratio of shear stress (τ) to shear rate (du/dr) and can be expressed in terms of blood yield shear stress (τ_y), ultimate Newtonian viscosity (η) and shear rate (du/dr) by the non-Newtonian Casson fluid model [15,16] as

$$\mu = [\ \tau_y^{1/2}\eta'\ (du/dr)^{1/2} + \eta^{1/2}]^{1/2} \qquad (1)$$

Values of $\tau_y^{1/2}$ and $\eta^{1/2}$ can be obtained from the intercept and slope respectively of the plot of measured $\tau^{1/2}$ vs. $(du/dr)^{1/2}$. The blood viscosity at any given shear rate for TAH patients is then normalized to that of hematocrit-matched controls.

Erythrocyte Rigidity

Erythrocyte rigidity was measured following a modification of the filtration method of Chien et al [17], Lessin et al [18] and Patwa et al [19]. Heparinized whole blood was first centrifuged at 3000 rpm for 12 minutes and the buffy coat removed. The packed red blood cells were resuspended in autologous plasma and passed by gravitation through a 1.0 gm nylon column packed in a 1 cc tuberculin syringe to further remove leukocyte and platelet contamination. The filtered suspension was recentrifuged, the plasma removed and the packed red cells resuspended in PBS buffer with 0.35% bovine albumin and 0.1% dextrose. A dilute red cell suspension with a cell count of 250,000 ± 50,000 cells/μl was prepared and filtered through 3 μm Nuclepore filters at a constant flow rate of 3.4 ml/min [20]. The filtration pressure was monitored by means of a Statham pressure transducer and recorded on a Grass polygraph. Red cell rigidity index was defined as the initial passage pressure of the red cell suspension, extrapolated from the slope of the pressure-time curve to time zero [17-20].

Plasma Fibrinogen Assay

Plasma fibrinogen concentration was assayed according to the method of Clauss [21] by measuring the rate of fibrinogen to fibrin conversion in the presence of excess thrombin. A calibration curve prepared from a fibrinogen reference is obtained and used to determine the fibrinogen concentration in the test sample.

RESULTS AND DISCUSSION

Overall blood rheology for Jarvik-7 patients was complicated due to a large number of rheologic parameters, such as hematocrit, plasma and whole blood viscosity, plasma protein constituents, red cell rigidity, white blood cell and/or platelet activities. Hematocrit is a major factor controlling the magnitude of blood viscosity. Therefore, for all the patients in our series, hematocrit was maintained at 30.8% $\pm$ 3.6% during the entire period of TAH support. Because of this low level, absolute whole blood viscosity was consequently low for these patients.

To evaluate the changes in blood rheology of recipients' response to implanted TAH, the blood viscosity measured from all patients was normalized to that of hematocrit-matched controls. Therefore, any blood viscosity alterations could be evaluated independent of hematocrit. For all Jarvik patients, elevated blood viscosity was seen at low shear rates. In Figure 1, we illustrate the data for daily average values of normalized blood viscosity for the first three week period of support at a shear rate of 0.1 sec^{-1}. Data are presented as the mean $\pm$ the standard error of the mean. The total number of patient data points on each day of support is also indicated above the error bar.

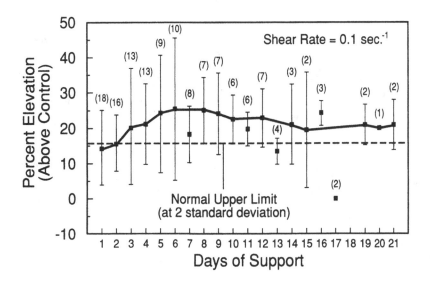

Figure 1. Variation of relative blood viscosity for patients with Jarvik-7 TAH.

As illustrated in Figure 1, on the first two days post-implant, a slight increase (14% above normal) in blood viscosity at a shear rate of 0.1 sec^{-1} was noted. However, blood viscosity increased sharply from 20% above normal on day 3 to a peak of 25% above normal on days 5-9. It remained at 23%-25% above normal until day 12. A slight decrease, to 20% above normal, was observed after day 12; viscosity remained at that level for the third-week period following implant. We found that increases in blood viscosity were primarily attributed to increased τ_y since η was normal for all patients except two who had elevated plasma viscosity. Specifically, this elevated plasma viscosity occurred during days and/or surrounding days of neurological deficit in one (HL) and of multi-organ (i.e. both liver and kidney) failure in the other (GB). Increases in τ_y were caused in part by both elevated plasma fibrinogen concentrations (Figure 2) and red cell rigidity (Figure 3).

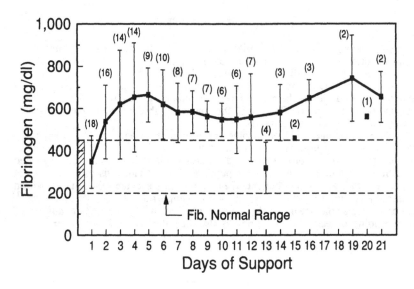

Figure 2. Fibrinogen levels for patients during the first three-week period of Jarvik-7 TAH support

We found that, for all Jarvik patients, plasma fibrinogen was always elevated throughout the entire Jarvik transplantation support. The mean values of plasma fibrinogen concentration during the first three weeks of support are shown in Figure 2. On day 1 post-implant, there was a fairly normal plasma fibrinogen concentration with a mean value of approximately 350 mg/dl. However, by the second day post-implant, plasma fibrinogen (530 mg/dl) had sharply increased beyond the normal range (200-450 mg/dl). It continued to increase to markedly elevated concentrations over 600 mg/dl on days 3,4,5 and 6 of the implant period. The daily trend in this parameter often demonstrated a marked increase, reaching levels above 600 mg/dl in seven patients. In particular, patient HL's fibrinogen level was elevated to 800 mg/dl on the day before his transplant dysphasia, while patient GB had fibrinogen levels of 900 md/dl during the first week of TAH support. This plasma protein is a major factor responsible for elevated blood viscosity and τ_y seen in these patients, because high plasma fibrinogen concentration strongly increases cellular interaction and aggregation at low shear flow conditions.

In addition, elevated plasma fibrinogen is believed to be a major component contributing to thrombus formation on heart valves of Jarvik-7 devices. In fact, most of the explained Jarvik-7 devices demonstrated deposition of small thrombi (1 to 3 um diameter) in the inner and outer parts of the device between the valve rings and their housing, and occasionally in the groove formed by the graft valve housing [22]. Valve thrombus formation in devices and loosely adherent large red thrombi found on two of the Jarvik-7 TAH valves were associated with high plasma fibrinogen concentrations during the implant period.

A plot of mean values of red cell rigidity versus days of support for TAH implant patients is presented in Figure 3. A normal upper limit of red cell rigidity obtained from 19 normal subjects is also plotted as a dash line in Figure 1. For Jarvik patients, mean red cell rigidity on each day of TAH support is slightly high and remains at a constant value (27 mm.Hg) of filtration pressure throughout the first three-week period of support.

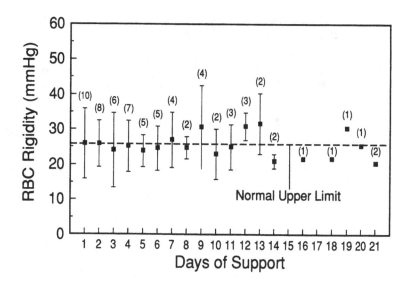

Figure 3. Red blood cell rigidity in Jarvik-7 TAH support

As illustrated in Figures 1, 2, and 3, increased whole blood viscosity along with the contributing factors of red cell rigidity and plasma fibrinogen strongly influenced these Jarvik patients' blood rheology. Both red cell rigidity and plasma fibrinogen play major roles in the microvasculature where individual cell flow is critical. Therefore, the increased flow resistance, increased cellular interaction as well as aggregation and decreased flow all produced simultaneously, resulted in transient neurological deficit for patient HL [14] and in multi-organ failure for patient GB [13, 15].

CONCLUSION

In the present study, there were many significantly abnormal rheologic parameters in the eighteen patients implanted with a Jarvik-7 TAH as a bridge to cardiac transplantation. Increased blood viscosity, red cell rigidity and plasma fibrinogen concentration were all observed. Overall blood rheology results were found to be unique to each patient both in terms of rheological parameters affected and the magnitude of alterations. In addition, the measured rheological alterations correlated well with the individual patient's clinical condition. Our results suggest that the monitoring and control of blood rheology in these patients may provide considerable assistance in managing and perhaps even preventing any complications in these patients.

ACKNOWLEDGMENT

The authors would like to thank Joan N. Nilson for her editorial assistance in the preparation of this manuscript.

REFERENCES

1.   Cooley, D.A.,Liotta, D., Hallman, G.L., Bloodwell, R.D., Leachman, R.D. and Milam, J.D., Orthotopic cardiac prosthesis for two-staged cardiac replacement. Am. J. Cardiol., 1969, 24:723-730.

2. Akutsu, T. and Kolff, W.J., Permanent substitute for valves and hearts. Trans. Am. Soc. Artif. Intern. Organs., 1958, 4:230-235.

3. Jarvik, R.K., Smith, L.M., Lawson, J.H., Sandquist, G.M., Fukumasu, H., Olsen, D.B., Iwaya, K. and Kolff, W.J., Comparison of pneumatic and electrically powered total artificial hearts in vivo. Trans. Am. Soc. Artif. Intern. Organs, 1978, 24:593-599.

4. Landis, D.L., Rosenberg, G., Donachy, J.H. and Pierce, W.S., Automatic control for the artificial heart. In IEEE 1980 Frontiers of Engineering in Health Care, 1980, pp. 305-310.

5. Atsumi, K., Fujimasa, I., Imachi, K., Miyake, H., Takido, N., Nakajima, M., Kouno, A., Ono, T., Yuasa, S., Mori, Y., Nagaoka, S., Kawase, S. and Kibuchi, T., Three goats survived for 288 days, 243 days, and 232 days with hybrid total artificial heart (HTAH). Trans. Am. Soc. Artif. Intern. Organs, 1981, 27:77-83.

6. Olsen, D.B., Kessler, T.R., Pons, A.B., Razzeca, K., Lawson, J.H. and Kolff, W.J., Fabrication, implantation, and pathophysiology of the total artificial heart in calves in six months. In US-USSR Joint Symposium on Circulatory Assistance and the Artificial Heart, ed. W.S. Pierce, Tbilisi,USSR. Bethesda, Maryland, U.S. Government Printing Office,(NIH publication No. 80-2032), 1981, p. 155.

7. DeVries, W.C., Anderson, J.L., Joyce, L.D., Anderson, F.L., Hammond, E.H., Jarvik, R.K. and Kolff, W.J., Clinical use of the total artificial heart. N. Engl. J. Med., 1984, 310:273-278.

8. Bucherl, E.S., Hennig, E., Baer, P., Frank, J., Lemm, W. and Zartnack, F., Status of the artificial heart program in Berlin. World J. Surg., 1985, 9:103-115.

9. Dew, P.A., Holfert, J.W., Burns, G.L., Taenaka, Y. and Olsen, D.W.B., Reduced thrombosis, improved durability and fit:A new artificial heart design. Am. Soc. of Artif. Intern. Organs., 1986, 15:11 (abstract).

10. Olsen, D.B. and Taenaka, Y., State-of-the-art and clinical applied pneumatic artificial hearts. In Critical Care Clinics, ed. D. Bregman, WB Saunders, Philadelphia, 1986, pp. 195-207.

11. Relman, A.S., Artificial hearts-Permanent and temporary. N. Engl. J. Med., 1986, 314:644-645.

12. Pierce, W.S., The artificial heart-1986:Partial fulfillment of a promise. Trans. Am. Soc. Artif. Intern. Organs, 1986, 32:5-10.

13. Borovetz, H.S., Kormos, R.L., Griffith, B.P. and Hung, T.C., Clinical utilization of the artificial heart. Critical Rew. in Biomed. Engr., 1989, 17:179-201.

14. Kormos, R.L., Borovetz, H.S., Griffith, B.P. and Hung, T.C., Rheologic abnormalities in patients with Jarvik-7 total artificial heart. Trans. Am. Soc. Artif. Intern. Organs,1987, 33:413-417.

15. Hung, T.C., Kormos, R.L., Borovetz, H.S., Griffith, B.P. and Hardesty, R.L., Blood rheology in total artificial heart implantation. In Biomedical Engineering: An International Symposium., ed. W.J. Yang and C.J. Lee, Hemisphere Publishing Corp., New York, 1989, pp. 247-257.

16. Hung, T.C., Butter, D.B., Yie, C.L., Yonas, H. and Sekhar, L.N., The role of hemorheology in cerebral vasospasm following sub-arachnoid hemorrhage (SAH). in Biofluid Mechanics and Biorheology, ed. Dieter Liepsch, 1990 (in press).

17. Chien, S., Usami, S. and Bertles, J.F., Abnormal rheology of oxygenated blood in sickle cell anemia. J. Clin. Invest., 1970, 49:623-624.

18. Lessin, L.S., Kurantsin-Mills, J. and Weems, H.B., Deformability of normal and sickle erythrocytes in a pressure-flow filtration system. Blood Cells, 1977, 3:241-262.

19. Patwa, D.C., Abraham, D.J. and Hung, T.C., Design, synthesis, and testing of potential antisickling agents. 6. Rheologic studies with active phenoxy and benzyloxy acids. Blood Cells, 1987, 12:589-601.

20. Hung, T.C., Pham, S., Steed, D.L., Webster, M.W. and Butter, D.B., Alterations in erythrocyte rheology in patients with severe pheripheral vascular disease. 1. Cell volume dependence of erythrocyte rigidity. Angiology, 1990 (in press).

21. Clauss A, Gerinnungs physiologische schnell methods zur bestimmung des fibrinogen. Acta Haematol., 1957,17:237.

22. Griffith, B.P., Interim use of the Jarvik-7 artificial heart: Lessons learned at Presbyterian-University Hospital of Pittsburgh. Ann. Thorac. Surg., 1989, 47:158-166.

ANTERIOR CRUCIATE LIGAMENT PROSTHESIS: "PROGRAMMED LOOSENING"
AUGMENTATION AS A SOLUTION TO THE PROBLEM.

M.MARCACCI, R.BUDA, S.ZAFFAGNINI, A.VISANI, D.DA VALLE.
Medical Doctors / 2nd Orthopaedic Department
Istituto Ortopedico Rizzoli
Via G.Pupilli, 1, Bologna,40136,Italy

ABSTRACT

The experience gained in recent years has shown that true prosthetic replacement of the A.C.L; must be considered a temporary solution of the
problem.
The mechanical stresses that develop during walking or running inevitably
cause more or less rapid rupture of the synthetic material so that the
reconstruction gesture fails.
However, the prosthetic ligament concept is not to be thought of as exhau-
sted if one considers the possibility of using this prosthesis associated
to the implantation of autologous tendons in accordance with the augmentation concept.
In practice the synthetic ligament guarantees stability of the implant
during the early re-education period while the autologous ligament assumes
the lasting mechanical characteristics of over the years.
It is precisely the yielding of the artificial prosthesis which, paradoxi-
cally favours progressive transmission in time of the articular mechanical
stresses to the biological implant, thus encouraging maturation of a proper neoligament.
In practice, the association of an artificial ligament to a biological
graft makes it possible to rapidly achieve re-education times similar to
those of plastics with prosthetic ligaments, but possessing the unlimited
duration typical of biological reconstructions.

INTRODUCTION

The need to seek techniques to replace the now consolidated and reliable

A.C.L. biological reconstruction, is linked to the long re-education period this technique requires. These lengthy periods are bound to the long biological times required for anchoring and maturation of the implanted neo-ligament and can in no manner be shortened without risking failure of the implant. (1,5)

From the data in literature it comes out that the patellar tendon graft is the most experimented. (3)

In studying the behaviour of patellar tendon grafts, implanted pro-A.C.L., in the monkey Clancy evaluated the mechanical resistance of the grafts at varying times after implant, comparing it with the healthy controlateral A.C.L. resistance. (2)

It was , in practice, possible to observe that the resistance of the implanted patellar tendon progressively decreases up to the third month after the operation, till it reaches a resistance equal to 30% of the healthy cruciate ligament.

After this period there is slow, progressive maturation of the implant until it reaches its maximum after about 12 months.

Referring these data back to man it seems reasonable to think of a return to sport without risk by respecting the 12 months of biological maturation. Due to this, however, there are long rehabilitation times and a period of absence from participation in sport which is often unaccetable, especially for professional sports man.

In order to appreciably reduce the recovery period of patients operated for A.C.L. reconstruction, the concept of augmentation or, better, temporary protection of the biological implant has been introduced.

The aim of this technique is to guarantee the resistance of the neo-implant up to 100% of A.C.L. during the whole post-operative course in order to ensure safe, rapid re-education and resumption of the sport.

In practice, at the time of implantation as well as during the first phases of re-education the prosthesis bears all the stresses, gradually loosing its mechanical efficiency so that the stresses pass to the autologous ligament as it matures.

On the basis of this implant philosophy we have, for over 4 years now, done pro-A.C.L. reconstructions using the central third of the patellar tendon, backed up by prosthetic augmentation.

Not having available ideal programmed loosening prostheses for this clinical study, we used LAD and SEM prostheses which, in our opinion, have characteristics compatible with the aims of the research.

On the basis of the clinical data obtained, and presented in this paper, a study was undertaken for the construction and utilization of "programmed loosening" prostheses. Three types have been discovered which provide for progressive loss of 80% of their mechanical resistance 2, 6 and 12 months after implantation .

MATERIAL AND METHOD

From January 1986 to December 1987 two random groups of 30 sports people suffering for chronic instability of the knee due to lesion of the A.C.L., were operated at the 2nd Orthopaedic Clinic of the Rizzoli Orthopaedic Institute in Bologna.

One group of patients was treated with biological technique, using the central third of the patellar tendon with the association of traditional re-education and return to sports activity 1" months after the operation. The other group was treated using the augmentation concept.

In these cases a prosthesis, trade name LAD or SEM, having a resistance equal to that of the cruciate ligament, supports the tendon-removed and prepared in accordance with the traditional method. The loss of mechanical resistance after the operation is known.

The tendon and the prosthesis are mechanically independent and work in parallel.

The prosthesis is rigidly fixed to both extremities to ensure that, immediately after the implant, the load is exclusively borne by the prosthesis, thus relieving the graft of stresses.

The second group of patients followed a more aggressive re-education protocol which provided for a resumption of sports activity 6 months after the operation.

They were evaluated from the clinical standpoint using the 5é point I.O.R. card, while the Genucom computerized system was used for the lassitometric evaluation.

RESULTS

All the patients were recontrolled from the clinical and lassitometric viewpoint with a mean follow-up of 34 months (minimum 24, maximum 48).

Analysis of the clinical results according to the 50-point I.O.R. card (4) showed a mean value of 38,7 points for the first group, treated with the traditional method, and 38,2 points for the second group.

In particular, 13 excellent results, 13 good, 2 fair and 2 bad were obtained from the first group while, for the second, the results were: 12 excellent, 15 good, 2 fair and 1 bad.

The resumption of sports activity for the first group was, in 2 cases, lower than the pre-lesion level while, in three cases, there was no return to sports activity.

All the other patients resumed at the same or a higher level.

In the secon group 2 patients resumed at a lower level, 2 did not take up sports activity again, while the remaining 26 all resumed at the same or a higher level.

From the lassitometric viewpoint too the differences observed between the two groups was not significant, with 6,6 mm. mean anterior translation of

the tibia on the femur with the Lachman test for the group treated with
the patellar tendon free, and a mean value of 6,5 mm. for the group treated
with the patellar tendon plus augmentation.
No cases of infection or synovitis were observed in the post-operative
period.
In short, the data found on the two patients populations brought out no
significant differences between them.

DISCUSSION

The need to reduce re-education times after reconstruction of the A.C.L.
seems unobtainable meantime through modification of the normal re-educa-
tional protocols.
On the other hand, it appears evident that, at least for the moment, it is
unthinkable to modify biological maturation times of the pro-A.C.L. fitted
tendinous transplant by shortening them.
At present the reduction of re-education times is, in our opinion, obtai-
nable by backing up the normal biological tissue with a synthetic support
that bears the mechanical stresses of the early re-educative period, thus
protecting the biological graft and permitting it to take root and mature
normally.
The preliminary data of our experimentation using SEM or LAD prosthetic
ligaments are highly satisfactory: there are no significant differences
with respect to the sample group treated with traditional technique even
with a return to sports activity 6 months after surgical treatment against
the 12 months of the control group.
There still exist some problems connected with the use of these prostheses
which we intend to face in the second part of our research.
In the first place the mechanical resistance of the prosthesis must be
equal to that of the normal A.C.L. in the initial phases of the implant
but, in time, it is necessary, alongside the progressive attachment and
maturation of the implant as its mechanical resistance increases, to have
simultaneous reduction in the mechanical resistance of the protection
prosthesis.
In this manner, transfer of the mechanical stresses to the biological
graft is progressively achieved without risking the implant's mechanical
hold.
Gradual transfer of the mechanical stresses is necessary to ensure maturation of the biological tissue in the ligamentous sense.
The prostheses used in our sample study, though ensuring mechanical resistance equal to that of the A.C.L; at the moment of implantation, do not,
from the mechanical viewpoint, present a pogrammable reistance curve.
According to our clinical experience it is foreseeable that after an initial period of prosthesis lengthening a sudden mechanical yielding is noted, linked to rupture of the prosthesis itself.

For this reason gradually-yielding prostheses are being studied and are in course of realization, able to ensure progressive, graded transfer of the mechanical stresses to the biological tissue.
At the moment we are not in a position to know the degree of wear resistance this prosthesis must have nor why three types with different yielding times have been foreseen.
The first type foresses the loss of 80% of the mechanical resistance 2 months after the implant, the second an 80% loss after 6 months and the third an 80% loss of mechanical resistance 12 months after the operation.
Besides providing a predetermined yield curve, these prostheses differ from the SEM or LAD types in that they contain about 75% of slow absorption material.
The prostheses used for the first part of our research were made of non-resorbable materials and though clinical research demonstrated their tolerability in time, they stiil constitute foreign intra-articular material.
In this sense, too, the protection prosthesis is placed in a greater biological context of the mixed implant, thus foreseeing the almost complete disappearance of the prosthetic skeleton.

REFERENCES

1. Arnoczky S.P., Tarvin G.B., Marshall J.L.: Anterior Cruciate Ligament Replacement using Patellar Tendon. An evaluation of graft revascularization in the dog. J. Bone Joint Surg. 64A, 2, 217-224, 1982.

2. Clancy W.G., Narechania R.G., Rosenberg T.D., Gmeiner J.G., Wisnefske D. and Lange T.A.:Anterior and posterior cruciate ligament reconstruction in Rhesus monkeys. A histological, microangiographic and biomechanical analysis. J.Bone Joint Surg., 63A, 8, 1270-1284,1981

3. Jones G.K.: Results of use of the central one-third of the Patellar ligament to compensate for Anterior Cruciate Ligament deficiency. Clin. Orthop . and Rel. Res. 147, 39-44, 1980.

4. Marcacci M., Giannini S., Pagani P.A., Buda R., Paladini Molgora A., Catani F.: Proposta di una nuova cartella computerizzabile per la valutazione nella patologia da sport del ginocchio. It. J. Sports Traum. Vol. 11, 29-46, 1989.

5. Noyes F.R., Butler D.L., Paulos L.E., Grood E.S.: Intra-articular Cruciate reconstruction. I: Perspectives on graft strngth, Vascularization, and immediate motion after replacement. Clin Orthop. and Rel.Res. 172, 71-77, 1983.

DENTINE MICROSTRUCTURE IN NORMAL DECIDUOUS TEETH BY COLLAGENASE ETCHING

JANINE MEADS[1], MICHAEL GREEN[2] AND DAVID H ISAAC[3]

[1] Community Dental Service, WGAHA, Orchard Street, Swansea
[2] School of Engineering, University of Wales College of Cardiff, PO Box 917, Cardiff
[3] Department of Materials Engineering, University of Wales, Swansea SA2 8PP

ABSTRACT

Samples of disease-free human deciduous molars were cut, polished and treated with collagenase solution prior to gold coating and observation in the SEM. This treatment dissolved the collagen fibres and led to micrographs of the mineral hydroxyapatite very close to the *in vivo* arrangement. The photographs show the mineral to be composed of small spheroidal units ~10-20nm across which are fused together to form a continuous phase. This picture of the mineral was confirmed by observing a fracture surface which was completely untreated except for gold coating. The intertubular regions of the collagenase etched surfaces contained numerous holes (up to ~50nm across) from which collagen fibres had been removed by the collagenase treatment, whereas the peritubular areas had a smoother surface finish and few such holes, thus clearly indicating the *in vivo* occurrence of collagen.

INTRODUCTION

The mineral component of dentine was first identified as a crystalline form of apatite by Gross (1) using x-ray diffraction. Various other authors have subsequently confirmed these findings using x-ray diffraction (2-4) and selected area electron diffraction (5). The diffraction data have been interpreted variously as resulting from needle-shaped or plate-like crystallites with an average length ~20-30nm (4,6,7). Early electron microscope studies also led to conflicting ideas of the size and shape of the crystallites; again there was support for both plate-like (8,9) and needle-shaped particles (10,11). Johansen and Parks (5) tried to resolve these conflicting views by using a stereoscopic technique to observe individual crystallites from different angles, and they concluded that the dentine crystallites were plate-like with an average thickness of 2-3.5nm and a length of up to 100nm. More recently Takuma *et al* (12) have demonstrated the crystallites from sound dentine as thick hexagonal platelets with mean dimensions of 3.4nm thick by 13.9nm wide and 24.8nm long. Johansen and Parks (13) have also concluded that the mineral crystallites are located within as well as on the surface of the collagen fibres and that their arrangement along the fibres is orderly with their long axes approximately parallel to the collagen fibres. This pattern of mineralisation had previously been described as "anisotropic mineralisation" by Schmidt and Keil (14). They also described a second morphological form, seen in many mammals (including humans) and certain mammal-like reptiles, in which the crystallites were arranged spherically, leading to the formation of discrete globules known as calcospherites. These calcospherites grow by the addition of calcified material to their outer surfaces until they contact each other and fuse together forming a continuous structure. In the areas between the fused calcospherites, interglobular dentine is formed. These features are seen in ground sections of hypomineralized teeth due to the disturbed general metabolism (15) and also in normal teeth in areas of rapid mineralization such as the coronal dentine close to the ameliodentinal junction (16). Boyde *et al* (17,18,19) have also demonstrated the presence of a calcospheritic pattern in circumpulpal dentine following the removal of the organic component of the dentine using cold sodium hypochlorite solution.

It has been argued that many of these techniques used to study the structure of dentine have damaged the mineral component and introduced artifacts. To overcome these criticisms it was decided to use a more subtle biochemical etching procedure originally described by Green *et al* (20) to study the structure of bone mineral. In this technique a polished surface of bone was treated with a collagenase solution to remove the collagen component without significantly affecting the mineral hydroxyapatite. Observations of the resulting inorganic surface in the scanning electron microscope revealed a structural model for the mineral component of bone believed to be closer to the *in vivo* arrangement than previous studies. The similarity between bone and tooth microstructure suggested that such a technique could be used to give some insight into the detailed arrangement of the mineral component of the tooth.

MATERIALS AND METHODS

Sound human deciduous molars from patients with no relevant medical history were used, the teeth having been extracted to balance the extraction of a carious contralateral primary molar. They were carefully examined to ensure that they were caries free.

The teeth were sectioned either longitudinally or transversely using a hand hacksaw and the exposed surfaces were polished using carbide papers and polishing alumina. During the polishing process the samples were examined periodically in a reflection optical microscope and polishing was continued until the surfaces appeared to be scratch free. They were then thoroughly washed in distilled water to remove any adherent alumina and subsequently boiled in distilled water for 30 minutes to denature the collagen, thus enhancing the efficiency of the collagenase. Aliquots (5 ml) of collagenase solution each containing 3000 units were made up in 0.05M Tris Buffer pH 7.4 (Sigma Chemicals Ltd, Type 1A). These solutions also contained 0.1M calcium chloride, the Ca^{++} ions being necessary for activation of the collagenase. The specimens were incubated in the solutions at 37°C for 1 week to ensure adequate etching. Each tooth surface was bathed in 5 ml of enzyme solution which was agitated twice daily. The efficacy of each batch of solution was tested by its ability to disintegrate small pieces of rat tail tendon which were placed in the enzyme solution with at least one of the teeth specimens in each batch. After incubation the samples were coated with gold and examined in the scanning electron microscope. For the low magnification pictures of Figures 1, 2 and 3 a Jeol 35C was used at an accelerating potential of 30kV. The high resolution pictures of Figures 4, 5 and 6 were obtained using a Jeol 120C Temscan operating in the scanning mode at 100kV.

RESULTS

Figure 1 is a scanning electron micrograph of a polished longitudinal section through the dentine of a human deciduous molar. This surface was not treated with collagenase, but simply polished and coated with gold. The tubules are clearly observed as holes ~2-3μm in diameter, against a flat, featureless background. In contrast, the enhanced background detail of Figure 2 shows the effect of the collagenase treatment. To produce this scanning electron micrograph, a polished longitudinal section through the dentine of a human deciduous molar was treated with collagenase, to remove the collagen component, and then coated with gold. The tubules are again evident and now there is a clear distinction between the peritubular and intertubular regions. This distinction is even more evident in the higher magnification micrograph of Figure 3 which shows that the mineral appears to be more densely packed in

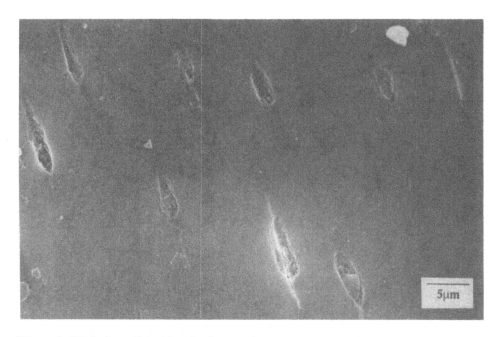

Figure 1. SEM of a polished longitudinal section through the dentine of a human deciduous molar. Although the tubules are clearly seen there is little detail on the surface since this sample was not etched with collagenase.

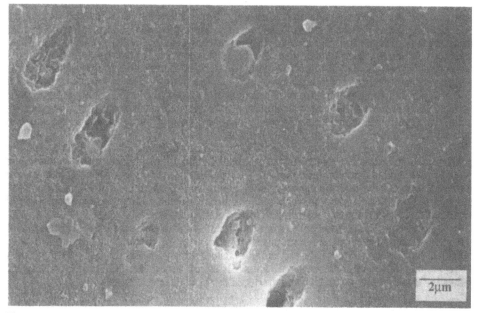

Figure 2. SEM of a polished, collagenase treated longitudinal section through the dentine of a human deciduous molar. The effect of the collagenase solution is evident in the enhanced background detail.

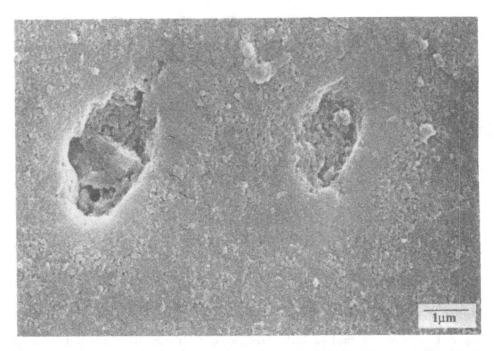

Figure 3. A higher magnification photograph of the surface in Figure 2. Note the apparently more densely packed mineral in the peritubular areas compared with the intertubular regions.

peritubular regions and contains numerous holes, due to the removal of the collagen, in the intertubular regions. The photographs in Figure 4 are higher resolution pictures of each of these regions. The peritubular area of Figure 4a shows a few holes typically ~30-50nm across, whereas the intertubular area of Figure 4b contains numerous holes of similar dimensions. These holes indicate the *in vivo* location and dimensions of the collagen fibrils which were removed by the collagenase treatment. The rougher surface finish that is evident in Figure 4b is thought to result from the removal of collagen fibrils at an oblique angle to the surface. In both peritubular and intertubular regions the mineral is seen to be composed of units ~10-20nm across which appear to coalesce to form larger units ~100nm across which in turn are fused to form a continuous mineral phase.

The photographs in Figure 5 are high resolution scanning electron micrographs of polished and collagenase treated transverse sections through normal deciduous teeth. Figure 5a shows a peritubular region (at bottom left), again with few holes from which collagen was removed, whereas the intertubular regions of Figure 5a (at top right) and Figure 5b exhibit numerous such holes. These transverse sections also reveal the mineral as being composed of small units ~10-20nm across coalescing to form a continuous phase.

It may be argued that the collagenase treatment had itself affected the microstructure of the mineral and so electron micrographs were also taken of fracture surfaces which were prepared completely anhydrously and simply coated with gold. Figure 6 shows that under these preparation conditions the small units ~10-20nm across are still clearly observable indicating that they are not an artifact introduced by the collagenase treatment.

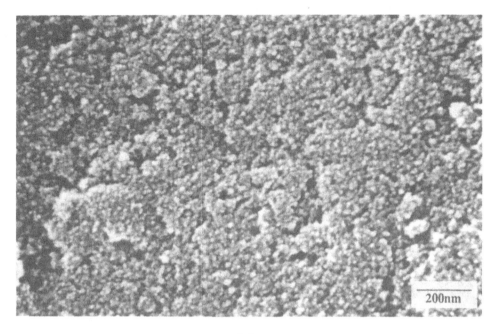

(a)

(b)

Figure 4. High magnification micrographs of polished, collagenase treated longitudinal section through the dentine of a human deciduous molar, showing (a) a peritubular area and (b) an intertubular area. Note the increased proportion of holes and the rougher surface of the intertubular area (b).

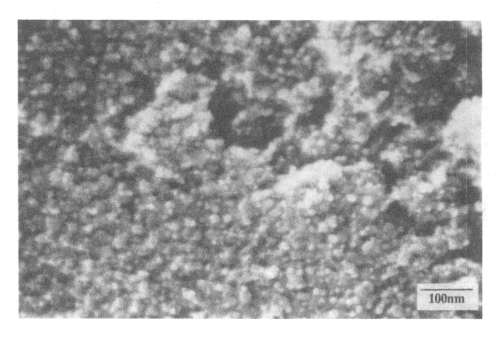

(a)

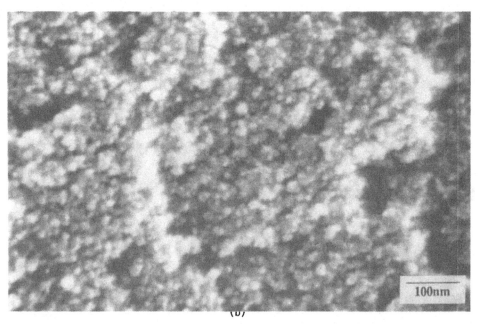

(b)

Figure 5. High resolution micrographs of polished, collagenase treated transverse section through the dentine of a human deciduous molar showing (a) a peritubular area and (b) an intertubular area. A tubule is at the bottom left corner of (a) and the smooth almost hole-free surface immediately adjacent to the tubule contrasts with the rougher, holed intertubular surface of (b). The small spheroidal units ~10-20nm across are resolved and are seen to be fused together to form a continuous mineral phase.

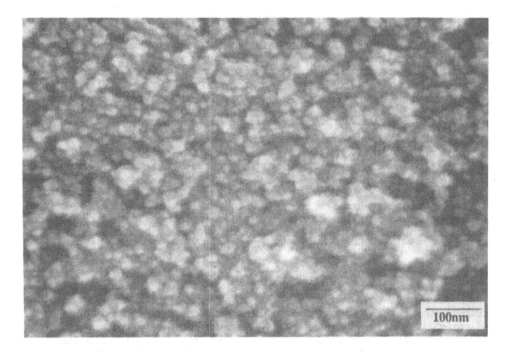

Figure 6. High resolution micrograph of a fracture surface of the dentine of a human deciduous molar. This surface was prepared by simply coating a fracture surface with gold and using no polishing or etching treatment. The small spheroidal units similar to those in Figure 5 are still present.

DISCUSSION AND CONCLUSIONS

It has previously been established that the mineral component of dentine is a form of hydroxyapatite (1-5). The micrographs presented here clearly show that, in both the peritubular (intratubular) and intertubular regions, this mineral is composed of small spheroidal units ~10-20nm across which coalesce to form larger structural entities ~100-200nm across which in turn are fused together to form a continuous mineral phase. This general picture for the morphology of dentine hydroxyapatite is very similar to that of bone (20). The major difference between dentine and bone microstructure is the far greater proportion of collagen in the latter. In bone, the collagenase treatment clearly revealed the orientation of the collagen fibres and showed that, particularly in the case of secondary osteons, a very regular arrangement of the fibrils occurs, giving rise to a lamellar structure (21). The micrographs presented here indicate a very less regular arrangement, more comparable with woven bone, and considerably fewer fibres in dentine. Also the fibres appear to be smaller in dentine (typically up to ~50nm across) than in bone (up to ~200nm across).

The micrographs clearly show that the detailed mineral microstructure is the same in both peritubular and intertubular dentine. The difference between the two regions is due to an increased proportion of collagen passing through the intertubular regions. This effect manifests itself in the micrographs of the collagenase treated surfaces as a higher number of holes in intertubular areas due to the etching away of the collagen fibres. Numerous authors have previously shown that peritubular dentine is hypermineralised compared with intertubular dentine (22,23) and it has been suggested that the matrix of peritubular dentine is not collagenous (16), although its precise composition was not made clear.

There has been some discussion in the literature about 'smear layers' on the surface of dentine (24-27), due to polishing using both hand instruments (25,28) and rotary instruments at low and high speed (29,30). The smear layer is resistant to mechanical removal but can be removed by chemical means (31). Pashley *et al* (32), using an ultrasonic technique to disrupt the smear layer, found that it had a globular substructure ~10-100nm diameter. They also demonstrated that fractured dentine surfaces of specimens, which were free of smear layers, revealed a spheroidal structure consistent with the model discussed above. The appearance of the smear layer is probably due to the presence of dentine dust containing separated spheroidal units and various other components such as microorganisms. It has been shown that using water cooled high speed drills the smear layer is less apparent, presumably because the dentine dust is washed away (29). We are confident that in the micrographs presented here, there is no significant smear layer effect. The photographs of Figures 1-5 were all taken from surfaces which had been water cooled during polishing and very thoroughly washed before collagenasing. Indeed, if any tiny particles forming a smear layer were present we would not have expected to see the holes due to collagen removal so clearly and we would have expected evidence of the smear layer particularly in these regions. Furthermore, the fracture surface of Figure 6 shows a structure totally consistent with the other micrographs and there was no possibility of such a smear layer forming on this unpolished anhydrously prepared surface. This final micrograph also confirms that the spheroidal units observed in the other pictures are not an artifact due to the dissolution and reformation of mineral since the fracture surface received no treatment other than a gold coating.

In summary, we are confident that this new technique of collagenase etching of a polished surface is a powerful method for visualising the microstructure of dentine mineral, and producing an image closer to the *in vivo* situation than previous studies. We have demonstrated that in normal, disease-free dentine, the mineral is in the form of spheroidal units which coalesce to form a continuous mineral phase, through which collagen fibres pass. Our studies are continuing on abnormal dentine samples to see whether the abnormalities are manifest at this microstructural level.

REFERENCES

1. Gross, R., Die Kristalline Structur von Dentin und Zahnschmelz, Festschr 1926, 59, Zahnnarzh Inst. Univ. Greifswald, Berlin.

2. Roseberry, H.H., Hastings, A.B. and Morse, J.K., X-ray analysis of bone and teeth, J. Biol. Chem. 1931, 90, 395-407.

3. Thewlis, J., X-ray analysis of teeth, Brit. J. Radiol. 1932, 5, 353-359.

4. Trautz, O.R., Klein, E., Fessenden, E. and Addelston, H.K., The interpretation of x-ray diffractograms obtained from human enamel, J. Dental Research 1953, 32, 420-431.

5. Johansen, E. and Parks, H.F., Electronmicroscopic observations on the three dimensional morphology of apatite crystallites of human dentine and bone, J. Biophys. Biochem. Cytol. 1960, 7, 743-746.

6. Bale, W.F., Hodge, H.C. and Warren, S.L., Roentgen-ray diffraction studies of enamel and dentine, Amer. J. Roentgenol. 1934, 32, 369-376.

7. Jensen, A.T. and Moeller, A., Determination of size and shape of the apatite particles in different dental enamels and in dentine by the x-ray method, J. Dental Research 1948, 27, 524-531.

8. Watson, M.L. and Avery, J.K., The development of the hamster lower incisor as observed by electron microscopy, Amer. J. Anat., 1954, 95, 109-162.

9. Little, K., Electron microscope studies on teeth, J. Dental Research 1955, 34, 778.

10. Takuma, S., Preliminary report on the mineralisation of human dentine, J. Dental Research 1960, 39, 964-972.

11. Pautard, F.G.E., Calcification in unicellular organisms in Calcification of Biological Tissues Publ. No. 65, 1960, 1-14, Amer. Ass. Advanc. Sci., Washington.

12. Takuma, S., Tohda, H., Watanabe, K. and Yama, S., J. Electron Microsc. 1986, 35, No. 1, 60-65.

13. Johansen, E., and Parks, H.F., Electron microscopic observations on sound human dentine, Arch. Oral. Biol. 1962, 7, 185-193.

14. Schmidt, W.J. and Keil, A., Die gesunden und die erkrankten Zahngewebe des Meschen und der Wirbeltiere im Polarisationsmikroscop, 1958, Carl Hanser Verlag, Munchen.

15. Bradford, E.W., Microanatomy and histochemistry of dentine in Structural and Chemical Organisation of Teeth Vol. II. Ed. A.E.W. Miles 1967, Academic Press.

16. Berkowitz, K.B., Holland, G.R. and Moxham, B.J., A Colour Atlas and Textbook of Oral Anatomy, 1978, Wolfe Medical Publications Ltd.

17. Boyde, A., The contribution of scanning electron microscopy to dental histology, Apex 1970, 7, 9-16.

18. Boyde, A. and Reith, E.J., The pattern of mineralization of rat molar dentine, Z. Zellforsch 1969, 94, 479-486.

19. Boyde, A. and Sela, J., Scanning electron microscope study of separated calcospherites from the matrices of different mineralizing systems, Calcif. Tiss. Res. 1978, 26, 47-49.

20. Green, M., Isaac, D.H. and Jenkins, G.M., Bone microstructure by collagenase etching, Biomaterials 1985, 6, 150-152.

21. Green, M., Isaac, D.H. and Jenkins, G.M., Collagen fibre orientation in bovine secondary osteons, Biomaterials 1987, 8, 427-432.

22. Roullier, C., Huber, L. and Rutishause, E., La structure de la dentine etude comparee de l'os et l'ivoive au microscope electronique, Acta anat. 1952, 16, 16-28.

23. Takuma, S., Electron microscopy of the structure around the dentinal tubule, J. Dental Research 1960, 39, 973-981.

24. Scott, D.B. and O'Neil, R.L., The microstructure of enamel and dentine as related to cavity preparation in Adhesive Restorative Materials, ed. R.W. Philips 1961, 27-31, Indiana University.

25. Provenza, D.V. and Sardana, R.C., Optical and ultrastructural studies of enamel and dentine surfaces as related to cavity preparation, in Adhesive Restorative Dental Materials Vol. II, 1966, 68-97, USPHS Publication No. 1,494 Washington DC Government Printing Office.

26. Eick, J.D., Wilko, R.A., Anderson, C.H. and Sorenson, S.E., Scanning electron microscopy of cut tooth surfaces and identification of debris by use of the electron microprobe, J. Dental Research 1970, 49, 1359-1368.

27. Branstrom, M. and Johnson, G., Effects of various conditioners and cleansing agents on prepared dentine surfaces: a scanning electron microscopic investigation, J. Prosthet. Dent. 1974, 31, 422-430.

28. Branstrom, M., Johnson, G. and Friskopp, J., Microscopic observations of the dentine under carius lesions excavated with the GK101 technique, J. Dent. Child. 1980, 47, 46-49.

29. Friskopp, J. and Larsson, U., Morphology of dentine surfaces of prepared cavities, J. Dent. Child. 1985, 52(3), 177-182.

30. Charbeneau, G.T., Peyton, F.A. and Anthony, D.H., Profile characteristics of cut tooth surfaces developed by rotating instruments, J. Dental Research 1957, 36(6), 957-966.

31. Berry, E.A., Von der Lehr, W.N. and Harris, H.K., Dentine surface treatments for the removal of the smear layer: an SEM study, JADA, 1987, 115, 65-67.

32. Pashley, D.H., Boyde, A., King, G.E., Tao, L. and Homer, J., The substructure of dentine smear layers, J. Dental Research 1987, 66(51), 261.

POROELASTIC FINITE ELEMENT MODELS IN BIOMECHANICS - AN OVERVIEW

BRUCE R. SIMON
Aerospace and Mechanical Engineering
University of Arizona
Tucson, Arizona 85712 USA

ABSTRACT

During the last two decades, biological structures with soft tissue components have been modeled using poroelastic constitutive laws, i.e. the material is viewed as a deformable porous solid matrix that is saturated by mobile tissue fluid. These structures exhibit a highly nonlinear, history-dependent material behavior; undergo finite strains; and may swell or shrink when tissue ionic concentrations are altered. Given the geometric and material complexity of soft tissue structures and that they are subjected to complicated initial and boundary conditions, finite element models (FEMs) have been very useful for quantitative structural analyses. This paper surveys recent applications of poroelasticity and the associated FEMs for the study of the biomechanics of soft tissues and indicates future directions for research in this area.

INTRODUCTION

The mechanics of soft tissues determines the behavior of many biological structures. These soft tissues often contain significant amounts of mobile fluid. This fluid can serve a number of functions including a convection path for transport, heating or cooling, as well as to contribute to the mechanical functions of the structure. Thus it is important to understand and quantify the mechanical response of soft tissues including the role of the free tissue fluid. The development of poroelastic material models for the study of soft tissue mechanics allows a more physically based view of the history-dependent response of these tissues. The approach is based on a continuum model in which the material is considered to be a porous elastic solid that is saturated by a pore fluid which flows relative to the deforming solid.

The development and use of poroelasticity has been made popular by Biot [1,2,3] for the analysis of soils. A number of poroelastic finite element models (FEMs) have been developed for the simulation of soil structural behavior; e.g. see Zienkiewicz, et al. [4,5], Gaboussi and Wilson [6,7], Carter, et al. [8], Prevost [9], and Simon, et al. [10]. Applications of poroelastic ideas for soft tissue mechanics were presented by Mow's group (see Mow, et al. [11,12]) which describe articular cartilage as a "biphasic" material. Kenyon [13] also utilized Biot's poroelastic view to study the mechanics of the arterial wall.

At the outset, when applied to biomechanical studies, the poroelastic and biphasic models are equivalent, the former based on a "u-w" formulation (given here) and the latter based on a "u-U", incompressible formulation. I will use the term poroelastic models to include biphasic models throughout this paper.

Given the complexity of the mechanics of soft tissue structures, there continues to be a need for a finite element procedure that can be used to simulate their behavior. In the last seven years a number of poroelastic FEMs have been developed for soft tissues; e.g. Oomens [14], Huyghe [15], Simon, et al. [16,17,18], and Spilker, et al. [19]. Large strain poroelastic FEMs have been presented recently by Simon, et al. [20,21,22]; Simon [23]; and Suh, et al. [24]. It is the purpose of this paper to comment on the current state of the art of the application of poroelastic FEMs and to indicate some future directions for research and development of such models in the biomechanics of soft tissues.

POROELASTIC FIELD EQUATIONS

Following a notation similar to Simon and Gaballa [22], the Lagrangian rectangular Cartesian field equations are the **kinematic** equations for Green's strain and relative volumetric fluid strain

$$E_{IJ} = \frac{1}{2}(x_{k,I}x_{k,J} - \delta_{IJ}), \quad \tilde{\zeta} = \tilde{w}_{K,K} \tag{1}$$

the **equilibrium** equations

$$(x_{i,K}S_{JK})_{,J} + J\rho(b_i - \ddot{u}_i) - \rho^f x_{i,J}\ddot{\tilde{w}}_J = 0, \tag{2a}$$

$$\pi_{,I} - \tilde{k}_{IJ}^{-1}\dot{\tilde{w}}_J + \rho^f x_{j,I}(b_j - \ddot{u}_j) - n^{-1}\rho^f J^{-1}x_{k,I}x_{k,J}\ddot{\tilde{w}}_J = 0 \tag{2b}$$

and the **constitutive** equations relating total second Piola-Kirchhoff stress, S_{IJ} and pore fluid pressure, π to the conjugate strain measures E_{IJ} and $\tilde{\zeta}$ as

$$\dot{S}_{IJ} = D_{IJKL}\dot{E}_{KL} + J X_{I,k}\dot{\pi}\delta_{km}X_{J,m}, \quad \dot{\pi} = \text{indeterminant} \tag{3}$$

with the incompressibility constraint $X_{I,k}\dot{E}_{IJ}X_{J,k} + J^{-1}\dot{\tilde{\zeta}} = 0$ when *both* the solid and fluid phases in the material are assumed to be incompressible. A "porohyperelastic" material law can be developed where $D_{IJKL} = \partial^2 W/\partial E_{IJ}\partial E_{KL}$ is the tangent material stiffness for the drained, steady state condition when the pore fluid pressure, $\pi = 0$. The necessary material properties are $W = W(E_{KL})$, the drained strain energy density function and $\tilde{k}_{IJ} = \tilde{k}_{IJ}(E_{KL})$, the hydraulic permeability tensor. Other properties include n, the porosity; ρ, the density of the material, and ρ^f, the density of the pore fluid. Equations (1)-(3) describe the boundary value problem to be solved (subject to initial and boundary conditions) for the primary unknowns which are the displacements of the solid, u_I and the (average) relative displacements of the fluid, $\tilde{w}_I$.

THEORETICAL BASIS FOR POROELASTIC FEMs

A principle of virtual velocities and a Galerkine approach provide the basis for the poroelastic FEMs. A number of formulations are possible including the u-w form for compressible solid or fluid and the u-w-penalty form, u-π mixed, and the u-w-π mixed form for incompressible solid and fluid. The u-π mixed approach is theoretically only valid for the quasi-static case. The development of the FEM is carried out using elemental approximations for the the fundamental field variables, e.g. for the u-w forms, both u_I and $\tilde{w}_I$ are interpolated in a typical finite element as

$$u_I = \phi_N^u u_{NI}, \qquad \tilde{w}_I = \phi_N^w w_{NI} \tag{4}$$

where ϕ_N^u and ϕ_N^w are interpolation functions that depend on X_I and u_{NI} and w_{NI} are time dependent nodal displacements. The italicized capital indices are summed and range from 1 to the number of nodes in each finite element. Elemental consistent mass, damping, stiffness, and load matrices can then be developed yielding elemental dynamic equilibrium approximations of the form

$$m \, \ddot{p} + c \, \dot{p} + P^{int} = P^{ext} \tag{5}$$

where the nodal displacment matrix, p contains both the solid and the relative fluid displacements, i.e. $p^T = < u^T \, w^T >$ and P contains corresponding internal and external forces. The coupled response is evident in the elemental matrices, e.g. the consistent mass matrix, m has partitioned entries $m^{uu} = [\int \phi^u{}_N J \rho \delta_{IJ} \phi^u{}_M dV_R]$, $m^{uw} = [\int \phi^u{}_N \rho^f x_{i,J} \phi^w{}_M dV_R]$, and $m^{ww} = [\int \phi^w{}_N n^{-1} \rho^f J^{-1} x_{m,I} x_{m,J} \phi^w{}_M dV_R]$ and the consistent damping matrix, c contains one non-zero submatrix $c^{ww} = \left[\int \phi^w{}_N \tilde{k}_{IJ}^{-1} \phi^w{}_M dV_R \right]$. Note that the subscript R is not summed and denotes the Lagrangian reference configuration. When R denotes the unstressed configuration, the total Lagrangian formulation is used. The global form of the dynamic equilibrium equations for the FEM are obtained using standard assembly procedures as

$$[M(r)] \{\ddot{r}\} + [C(r)] \{\dot{r}\} + \{R^{int}(r)\} = \{R^{ext}(r,t)\} \tag{6}$$

which must be solved (subject to initial and boundary conditions) using suitable time marching algorithms to obtain the unkown displacement vector, $\{r\}$ containing the nodal solid and relative fluid displacements. Similar forms are obtained using the u-π and u-w-π formulations. A penalty method is also useful in the cases where the incompressibility constraint above must be enforced. A u-U form of the field equations has been used by Zienkiewicz, et al. [5]; Prevost [9]; Spilker, et al. [19]; and Suh, et al. [24]. These u-U formulations are equivalent to the u-w approach described in this article. In the u-U formulation, the absolute (average) displacements of the fluid, U_i replace the relative fluid displacements as primary unknowns using $U_i = n^{-1} w_i + u_i$ and $\tilde{w}_I = J X_{I,j} w_j$. The fundamental stress measure is $J(1-n)S_{IJ}^s$, the second Piola-Kirchhoff stress in the solid, that is related to S_{IJ} and π by the relation $S_{IJ} = J(1-n)S_{IJ}^s + JnX_{I,j} \pi \, \delta_{ji} X_{J,i}$.

CURRENT STATE OF THE ART

Poroelastic finite element models for soft tissue structures have been reported in the literature over the last five years. Some of these publications are noted here as selected examples which give an indication of the current level of poroelastic FEA activity in biomechanics.

Researchers in the Netherlands have made use of poroelastic FEMs to study the mechanics of the skin and the myocardium. Oomens [14] developed poroelastic models for the skin and subcutaneous fat. Both experimental and FE modeling were carried out with emphasis on compressive loading associated with pressure sores. Huyghe [15] used poroelastic FEA to model the left ventricle as an axisymmetric structure subject to finite strains. Myocardial fiber orientation was introduced in the FEMs. An attempt was made to include Darcy losses as well as inherent viscoelasticity of the solid phase of the poroelastic model. A quasilinear model with a continuous relaxation spectrum was evaluated as a possible quantitative description of inherent myocardial viscoelasticity. A novel approach was presented in which an equivalent permeability was used to represent the coronary circulation in the ventricular wall.

The extensive fundamental efforts of Mow's research group have lead to the development of biphasic FEMs for the study of articular cartilage. A u-U FEM was developed (see Spilker, et al. [19]) for consideration of cartilage with nonlinear hydraulic permeability and subject to small strains. The experimental configurations of confined compression and unconfined compression were simulated with FEMs and the results compared to analytical solutions. A related biphasic FEA of the repair of articular cartilage was presented by Athanasiou, et al. [25]. A nonlinear biphasic FEM including finite deformations was recently described (see Suh, et al. [24]) which makes use of the mixture formulation for such boundary value problems given by Mow, et al. [12]. Again results were presented for unconfined and confined compression configurations in order demonstrate the need for a large strain analyses of soft hydrated tissue structures.

We have developed a number of poroelastic FEMs for the study of soft tissue mechanics (see Simon [23] for a general presentation). These FEMs have been applied in the study of the spinal motion segment (SMS) and the walls of large arteries. A linear, compressible poroelastic FEM was developed of a lumbar SMS of a rhesus monkey (Simon, et al. [16,17]) and for the human lumbar SMS (Simon, et al. [18]). In these models a SMS was idealized as an axisymmetric structure with a poroelastic intervertebral disc in which the solid was assumed to be compressible and the mobile fluid was assumed incompressible (water). Normal and degenerated discs were simulated under quasi-static and impulse loading. Discal bulge, stresses, pore fluid pressure, and relative pore fluid motions were predicted in the disc and in the cancellous and cortical bone. The presence of pore fluid in the vertebral body produced markedly increased stresses in the boney end plate during impact loading. A three-dimensional, linear u-w poroelastic FEM is now being developed (see Pflaster, et al. [26]) for simulation of human discal motion in excised SMSs. Experimental measurements of internal discal deformation are being used to provide validation of the FEM results. A simple swelling model (based on a linear thermal stress analogy) was used by Simon and Gaballa [18] to predict ionic swelling effects and static intradiscal pressures in rhesus monkey intervertebral discs.

A linear poroelastic FEM was developed to study arterial wall mechanics (Simon and Gaballa [27]). Both solid and fluid materials were assumed to be incompressible and a u-π formulation used to carry out the calculation of arterial wall displacements, stresses, pore fluid pressure, and pore fluid motion. The FEA predicted the characteristic fluid in-flow at the external wall surface for early times after step pressure loading. This in-flow was also predicted analytically by Kenyon [13]. The FEA was extended by Simon and Gaballa [20] to include finite deformations and a nonlinear "porohyperelastic" view of a rabbit aorta. Using the material properties given in Chuong and Fung [28], aortic pressure-radius response was predicted for steady-state drained and intramural flow conditions. Intramural flow versus intraluminal pressure was predicted for various nonlinear hydraulic permeability functions representing aortic tissue.

FUTURE RESEARCH DIRECTIONS

The application of poroelastic FEMs in biomechanics has been limited thus far to a few soft tissue structures. It is anticipated that many other biological structures can and will be studied more effectively using the poroelastic FEA approach. A number of areas deserve additional attention and fundamental research.

1. Fundamental Research Required

One of the most important areas for future research is the development of the necessary constitutive equations for the biological materials in the soft tissue structure. The general poroelastic theory above provides the framework and identifies the necessary material properties. The determination of these properties can be accomplished effectively using both FEMs and experimental models of the test configuration(s) to be analyzed. This is especially important for the structures in which finite straining occurs and the highly nonlinear, compressible drained properties and strain-dependent permeability must be determined.

First, a general comment is warranted regarding the choice of reference configuration, R, for the formulation of FEMs and the poroelastic material laws. When finite strains occur in a soft tissue structure, a Lagrangian view of the boundary value problem is useful-- usually referring the mechanics of the problem to a stress-free reference configuration. In the living state, a biological structure can be altered or remodeled in response to various stimuli including the stresses it must sustain. Such remodeling must be accounted for in the selection of the reference configuration for the Lagrangian view. For example, Fung [29] has presented documentation of remodeling of the arterial wall in response to *in vivo* intraluminal pressure. This remodeling in the arterial wall results in a prestressed condition when the artery is at zero pressure and axial loading. This prestressed condition is evidenced as a significant "opening angle" observed in a transverse section after a longitudinal cut is made through the arterial wall. The opened shape of the arterial segment then is the stress-free configuration which should be used as a reference for total Lagrangian mechanical models. Other soft tissue structures have also been shown to remodel and develop prestresses in what would usually considered to be a stress-free initial configuration. In order to account for prestresses properly, the unstressed configuration must be identified so that Lagrangian poroelastic FEMs can be based on an unstressed reference configuration for the tissue being studied.

The scalar material properties that must be measured include n, ρ, ρ^f. The current density, ρ of the material can be expressed as $\rho = (1 - n)\rho^s + n\rho^f$. Experiments can be developed to determine ρ (or ρ^s) and ρ^f, the latter being the density of water for soft biological tissues. For a poroelastic material with an incompressible solid phase; the current porosity, $n = 1 - J^{-1}(1 - n_R)$ where n_R is the porosity in the reference configuration. Assuming that the porous solid is *saturated* by the pore fluid, the porosity $n_R = n_R(X_I)$ represents the initial distribution of free fluid in the structure which must be determined experimentally.

Considerable experimental effort will be required in order to fully determine the material stiffness and hydraulic permeability for soft tissues. It is instructive to consider the linear poroelastic case for a compressible, anisotropic material where $\sigma_{ij} = C_{ijkl}e_{kl} + M_{ij}Q^{-1}\pi$ and $\pi = M_{kl}e_{kl} + Q\zeta$. Forty material constants (C_{ijkl}, M_{ij}, Q, k_{ij}) must be determined experimentally. If the material is isotropic then $\sigma_{ij} = \lambda e_{kk}\delta_{ij} + 2\mu e_{ij} + \alpha\delta_{ij}\pi$, $\pi = Q(\alpha e_{kk} + \zeta)$, and $k_{ij} = k\delta_{ij}$ so that five material constants (λ, μ, α, Q, k) are required. If, in addition, *both* the solid and fluid are incompressible, then $\alpha = 1$ and $Q = \infty$ and only three material constants (λ, μ, k) remain to be determined. More theoretical and experimental research will be required in order to quantify large deformation nonlinear models where even in the incompressible isotropic case, three material functions must be determined in terms of the three invariants of E_{ij}, i.e. $\tilde{\lambda} = \tilde{\lambda}(I_1, I_2, I_3)$, $\tilde{\mu} = \tilde{\mu}(I_1, I_2, I_3)$ and $\tilde{k} = \tilde{k}(I_1, I_2, I_3)$. An especially useful form for the anisotropic case has been proposed by Chuong and Fung [28] where the strain energy density function $W = (1/2)(C_0)(e^\phi - 1)$. The parameter ϕ must be quadradic in E_{IJ} so that $\phi = B_{IJKL}E_{IJ}E_{KL}$. Then C_0 and B_{IJKL} are ten constants to be determined experimentally for an orthotropic material. This form for W has been used successfully by Simon, et al. [20] for the axisymmetric, plane strain analysis of the rabbit aorta using a porohyperelastic FEM.

In general, additional consideration of finite strain theory and associated experiments is required since the drained response (used to determine W) is *always* compressible, even when both solid and fluid are incompressible. Most data reduction schemes developed for finite strain elastic problems assume an incompressible material and make use of analytical inverse solutions in which a final shape is known and the equilibrium equations are integrated to determine the required tractions. When the material is compressible, as in the case of the drained condition for the poroelastic material, the integrands of the required integrals can *not* be determined analytically leaving solution of such problems to numerical methods such as FEA. Thus a large strain, nonlinear poroelastic FEM is essential for the proper interpretation of experimental results. This difficulty also develops when considering hydraulic permeability and further research is needed in order to determine appropriate forms for the nonlinear functions $\tilde{k}_{IJ} = \tilde{k}_{IJ}(E_{KL})$.

Soft tissues may also exhibit a history-dependence due to inherent viscoelasticity in the porous solid material. In order to include these viscoelastic effects, the second Piola-Kirchhoff solid stress, $J(1 - n)S_{IJ}^s$ is assumed to contain the inherent material history-dependence. One possible form for this dependence is

$$S_{IJ}^s = S_{IJ}^{se}[E_{KL}(t)] + \int_0^t \{S_{MN}^{se}[E_{KL}(t - \tau)](\partial G_{MNIJ}/\partial\tau\}d\tau \tag{7}$$

where S_{IJ}^{se} are nonlinear "elastic response" functions and $G_{MNIJ}(t)$ are anisotropic "reduced relaxation functions" for the porous solid material. Fung [30] has given some possible forms for these material properties for soft tissues. Mak [31] has described a linear "poroviscoelastic" material model of this form. In order to quantify material properties, care must be exercized in developing appropriate experiments in which the effects associated with the 'Darcy losses' of equation (2b) can be separated from the losses associated with inherent solid phase viscoelasticity. The history-dependent term in equation (7) will introduce an additional term c^{uu} in the elemental and global damping matrices of equations (5) and (6) which correspond to the poroviscoelastic FEM.

Further research will be necessary in order to develop efficient dynamic time integration schemes for dynamic FEA. The inertia forces in the field equations are more involved for poroelastic materials than for elastic materials, e.g. consider the complexity of the relative fluid acceleration terms in the field equations and the corresponding entries in m. It should be noted that the u-U formulation has slightly simpler acceleration terms and mass matrix. However, the u-U form of the damping terms are more complicated and the damping matrix, c is full. Time integration is more difficult for both the u-w and the u-U FEMs since, even though a total Lagrangian form is used, the overall density and porosity are not constant and therefore the mass matrix M is dependent on r. Thus the advantage of constant M usually associated with the Lagrangian view of elastic problems is not realized. Time integration may be carried out on the full set of equation (6) or "staggered" time integration algorithms can be considered. For example, the u-w FE equations can be written in staggered form as

$$m^{uu}\ \ddot{u} + P^{u,int} = P^{u,ext} - m^{uw}\ \ddot{w} \tag{8a}$$

$$m^{ww}\ \ddot{w} + c\ \dot{w} + P^{w,int} = P^{w,ext} - (m^{uw})^T\ \ddot{u} \tag{8b}$$

Time integrators can be applied to both equations above and equation (8b), say; solved for a new w. This w and the associated $\ddot{w}$ can be introduced in equation (8a) which can then be solved for u. Iteration can be introduced as necessary until the desired accuracy is achieved for each time step. Similar staggered approaches can be developed for the u-U, u-π, and u-w-π FEMs.

Swelling phenomena can also be studied using a poroelastic FEM. Simon and Gaballa [21] presented a study of linear swelling in the intervertebral disc. This linear model can be extended to include convection in material and/or geometric nonlinear cases. As an illustration of the formulation, consider a simplified anisotropic swelling model based on

Fick's laws including a total Lagrangian view of the large strains and possible convection effects. The poroelastic theory can be extended as follows: Let $\tilde{c}$ be the Lagrangian description of concentration. To the *kinematic* equations, add a concentration strain (concentration gradient) equation

$$\tilde{e}_I^c = \tilde{c}_{,I} \quad \text{and} \quad \tilde{c} = \tilde{c}(X_I, t) = J\, c \tag{9}$$

To the *constitutive* equations, add a "swelling expansion" term, $\tilde{\alpha}_{IJ}^c \dot{\tilde{c}}$ that is analogous to the thermal expansion term in thermal elastic analysis, so that equation (3) becomes

$$\dot{S}_{IJ} = D_{IJKL}\dot{E}_{KL} + J\, X_{I,k}\,\dot{\pi}\,\delta_{km}X_{J,m} + \tilde{\alpha}_{IJ}^c \dot{\tilde{c}} \tag{10}$$

and add a finite strain form for Fick's law

$$\tilde{T}_I^c = \tilde{d}_{IJ}^c \tilde{e}_J^c - n^{-1}\dot{\tilde{w}}_I \tilde{c} \tag{11}$$

Finally, add a finite strain form of Fick's law to the *equilibrium* law so that equation (3) includes

$$\tilde{T}_{I,I}^c = \dot{\tilde{c}} \tag{12}$$

The second term on the right hand side of equation (11) corresponds to convection effects—here referred to the Lagrangian reference configuration. In general *all* of the material property functions in the field equations depend on E_{IJ} and $\tilde{c}$ and include D_{IJKL}, the "drained" elasticity array; $\tilde{k}_{IJ}$, the hydraulic permeability; $\tilde{\alpha}_{IJ}^c$, the coefficients of ionic expansion (swelling or shrinking); and $\tilde{d}_{IJ}^c$, the diffusion coefficients. Additional experimental effort will be required in order to determine these nonlinear material properties. The new field equations can be modified as information becomes available regarding swelling phenomena, e.g. Donnan osmotic pressure ionic effects and electrostatic chemical expansion effects. Mow, et al. [32] presented a "triphasic" theory which includes these effects in an incompressible u-U formulation for articular cartilage. A total Lagrangian view of this triphasic model can be developed by following the ideas outlined in equations (9) - (12). The corresponding formulation for FEMs including swelling behavior with convection and finite strains can be developed by extending the principle of virtual velocities to include the new fields associated with ionic concentration(s). FEMs can then be generated in which the concentration field(s) is interpolated as $\tilde{c} = \phi_N^c c_N$. Various forms of the discretized equations can be developed, e.g. a u-w-c penalty or u-w-π-c mixed formulation. The resulting set of ordinary differential equations must be solved using specialized techniques that allow efficient time integration including the convection terms. A staggered approach should be useful in carrying out the solution for the coupled discretized poroelastic and ionic conservation equations. In this approach the poroelastic problem is staggered with the ionic field problem(s) which is solved first in order to begin the time integration.

2. Future Applications of Porohyperelastic Finite Element Analysis

In the previous section I have mentioned a few research tasks that are fundamental to any biomechanical analysis using poroelastic FEMs. Having looked at the current state of the art in this area, it is appropriate to forecast some specific directions for future research. The new applications for poroelastic FEA in biomechanics will be diverse and will allow realistic simulations of soft tissue structures that includes the role of the mobile tissue fluid. Material response can be determined experimentally using the poroelastic view so that structural analysis methods can be extended to three dimensional FEMs including material and geometric nonlinearity and swelling phenomena. Following are some examples of potential research areas:

(a) *Orthopedics.* The mechanics of entire joints can be studied. FEMs can be developed of articular joints (normal, arthritic, or prosthetic) including synovial fluid-porous solid interaction and lubrication. The intervertebral joint in a SMS should be modeled in three-dimensions including anisotropic effects in the annulus in order to simulate impact, consolidation, swelling, and discal degeneration. Other soft tissue structures that could be modeled include temporary epiphysial hyaline cartilage, tendon, and ligaments. Free tissue fluid effects which may be important in the response of cancellous and cortical bone mechanics could be studied using poroelastic FEMs. Dynamic FEMs could be developed for head injury and trauma research in which the brain is simulated with poroelastic properties. (b) *Orthotics.* FEM studies of contact pressures on soft tissues can be carried out. Pressure and bed sore development could be investigated with consolidation FEMs of prolonged contact between soft tissues and the supporting bed. Determination of stump pressure distributions in prosthetics could be accomplished using detailed three-dimensional poroelastic FEA. (c) *Dental.* Applications here might include FE simulation of the mechanical interactions between hard structures (teeth, bone, fixtures) and normal or diseased soft tissues (gums, ligaments). (d) *Plastic surgery.* FEA could be useful in determining the resulting free tissue fluid motion and solid deformation associated with flap operations and incision closure or healing. (e) *Cardiovascular structures.* Complex processes in the arterial wall such as transmural transport, swelling, remodeling (normal and pathological), smooth muscle activity, etc. can be modeled. Specifically the simulation of diffusive and convective transport associated with nutrition of normal soft tissues or pathological deposition of materials (e.g., LDL cholesterol in the arterial wall) should be considered. The myocardium, valves, and the papillary muscle of the heart are structures which could be analyzed with poroelastic FEMs. The traditional microvessel model of a "tunnel in gel" could be extended by making the gel a poroelastic material.

CONCLUSION

Poroelastic FEA has been shown to be a useful tool for the study of the mechanics of soft tissue structures including the role of the free tissue fluid. The list of potential uses for theory, experiments, and FEMs based on a poroelastic material view is long and is growing. The poroelastic approach provides the basis for a better fundamental understanding of soft tissue mechanics. Poroelastic FEMs (supported by appropriate experiments and implimented on modern computers) will allow new biomechanical discoveries and applications to be made so that normal, pathological, and prosthetic soft tissue structures can be better understood; structurally evaluated; and/or designed in quantitative terms.

ACKNOWLEDGEMENTS

The author acknowledges the continued research interactions with Dr. A. L. Baldwin, Dr. M. A. Gaballa, and Mr. Y. Yuan and the partial support of the American Heart Association (Arizona Affiliate).

REFERENCES

1. Biot, M. A., General theory of three-dimensional consolidation. *J. Appl. Physics*, 1941, *12*, 155-164.

2. Biot, M. A., Mechanics of deformation and acoustic propagation in porous media. *J. Appl. Physics*, 1962, *33*, 1482-1498.

3. Biot, M. A., Theory of finite deformations of porous solids. *Indiana U. Math. J.*, 1972, *21*, 597-620.

4. Zienkiewicz, O. C., Humpheson, C., and Lewis, R. W., A unified approach to soil mechanics problems. In *Finite Elements in Geomechanics*, Wiley publ., 1977, pp. 151-178.

5. Zienkiewicz, O. C. and Shiomi, T. Dynamic behavior of saturated porous media-the generalized Biot formulation and its numerical solution. *Intl. J. Num. Anal. Meth. Geomech.*, 1984, *8*, 71-96.

6. Gaboussi, J. and Wilson, E. L., Variational formulation of dynamics of fluid-saturated porous elastic solids. *J. Engr. Mech. Div. ASCE*, 1972, *EM-4*, 947-963.

7. Gaboussi, J. and Wilson, E. L., Flow of compressible fluid in porous elastic media. *Intl. J. Num. Meth. Engr.*, 1973, *5*, 419-442.

8. Carter, J. P., Small, J. C., and Booker, J. R., A theory of finite elastic consolidation. *Intl. J. Solids Struct.*, 1977, *13*, 467-478.

9. Prevost, J. Nonlinear transient phenomena in soil media. In *Mechanics of Engineering Materials*, C. S. Desai and R. H. Gallagher, eds., Wiley publ., 1984, pp. 515-533.

10. Simon, B. R., Wu, J. S. S., Zienkiewicz, O. C., and Paul, D. K., Evaluation of u-w and u-π finite element methods for the dynamic response of saturated porous media using one-dimensional models. *Intl. J. Num. Anal. Meth. Geomech.*, 1986, *10*, 461-482.

11. Mow, V. C., Kuei, S. C., Lai, W. M. and Armstrong, C. G., Biphasic creep and stress relaxation of articular cartilage in compression: theory and experiments. *J. Biomech. Engr.*, 1980, *102*, 73-84.

12. Mow, V. C., Kwan, M. K., Lai, W. M., and Holmes, M. H., A finite deformation theory for nonlinearly permeable soft hydrated biological tissues. In *Frontiers in Biomechanics*, G. W. Schmid-Schonbein, S. L-Y. Woo, and B. W. Zweifach, eds., Springer-Verlag, publ., 1985, pp. 153-179.

13. Kenyon, D. E., A mathematical model for water flux through aortic tissue. *Bull. Math. Biol.*, 1979, *41*, 79-90.

14. Oomens, C. W. J., A mixture approach to the mechanics of skin and subcutis. Ph.D. Thesis, 1985, Eindhoven University.

15. Huyghe, J. M. R. J., Nonlinear finite element models of the beating left ventricl and the intramyocardial coronary circulation. Ph.D. Thesis, 1986, Eindhoven University.

16. Simon, B. R., Evans, J. H., and Wu, J. S. S., Poroelastic mechanical models for the intervertebral disc. In *Advances in Bioengineering*, ed. D. L. Bartel, ASME publishers, New York, 1983, pp. 106-107.

17. Simon, B. R., Wu, J. S. S., Carlton, M. W., Kazarian, L. E., France, E. P., Evans, J. H., and Zienkiewicz, O. C., Poroelastic dynamic structural models of rhesus spinal motion segments. *Spine*, 1985, *10*, 494-507.

18. Simon B. R., Wu, J. S. S., Carlton, M. W., Evans, J. H., and Kazarian, L. E., Structural models for human spinal motion segments based on a poroelastic view of the intervertebral disk. *J. Biomech. Engr.*, 1985, *107*, 327-335.

19. Spilker, R. L. Suh, J. K., and Mow, V. C., A finite element formulation for the nonlinear biphasic model for articular cartilage and hydrated soft tissues including strain-dependent permeability. In *Computational Methods in Bioengineering*, R. L. Spilker and B. R. Simon, eds., ASME publ., BED-Vol. 9, New York, 1988, pp. 81-92.

20. Simon, B. R. and Gaballa, M. A., Finite strain, poroelastic finite element models for large arterial cross sections. In *Computational Methods in Bioengineering*, R. L. Spilker and B. R. Simon, eds., ASME publ., BED-Vol. 9, New York, 1988, pp. 325-334.

21. Simon, B. R. and Gaballa, M. A., Poroelastic finite element models for the spinal motion segment including ionic swelling. In *Computational Methods in Bioengineering*, R. L. Spilker and B. R. Simon, eds., ASME publ., BED-Vol. 9, New York, 1988, pp. 93-100.

22. Simon, B. R. and Gaballa, M. A., Total Lagrangian 'porohhyperelastic' finite element models of soft tissue undergoing finite strain. In *1989 Advances in Bioengineering*, B. Rubinsky, ed., ASME publ., BED-Vol.15, New York, 1989, pp. 97-98.

23. Simon. B. R., Poroelastic finite element models for soft tissue structures. Chapter 4 in *Connective Tissue Matrix, Part 2*, D. Hukins, ed., MacMillan Press, Ltd., 1990, pp. 66-90.

24. Suh, J. K., Spilker, R. L., Holmes, M. H., and Mow, V. C., A nonlinear biphasic finite element formulation for soft hydrated tissues under finite deformation. In *1989 Advances in Bioengineering*, B. Rubinsky, ed., ASME publ., BED-Vol.15, New York, 1989, pp. 99-100.

25. Athanasiou, K. A., Spilker, R. L., Buckwalter, J. A., Rosenwasser, M. P., and Mow, V. C., Finite element biphasic modeling of repair articular cartilage. In *1989 Advances in Bioengineering*, B. Rubinsky, ed., ASME publ., BED-Vol.15, New York, 1989, pp. 95-96.

26. Pflaster, D., Laible, J., Krag, M. H., Johnson, C., Pope, M. H., and Simon, B. R., A poroelastic model of the intervertebral disc with experimental validation. In *Proceedings of the World Congress on Biomechanics*, San Diego, CA, 1990.

27. Simon, B. R. and M. A. Gaballa, Poroelastic finite element models for large arteries. In *1986 Advances in Bioengineering*, S. A. Lantz and A. I. King, eds., ASME publ., BED-Vol. 2, New York, 1986, pp. 140-141.

28. Chuong, C. J. and Fung, Y. C., Compressibility and constitutive equation of arterial wall in radial compression experiments. *J. Biomech.*, 1984, *17*, 35-40.

29. Fung, Y. C., In search of a biomechanical foundation of tissue engineering. In *Tissue Engineering - 1989*, S. L-Y. Woo and Y. Seguchi, eds., ASME publ., BED-Vol. 14, New York, 1989, pp. 11-14.

30. Fung, Y. C. B., Stress-strain-history relations of soft tissues in simple elongation. Chapter 7 in *Biomechanics, Its Functions and Objectives*, Y. C. Fung, N. Perrone, and M. Anliker, eds., Prentice-Hall publ., Englewood Cliffs, NJ, 1972, pp. 181-208.

31. Mak, A. F., The apparent viscoelastic behavior of articular cartilage-the contributions from the intrinsic matrix viscoelasticity and interstitial fluid flows. *J. Biomech. Engr.*, 1986, *108*, 123-130.

32. Mow, V. C., Lai, W. M., and How, J. S., Triphasic theory for swelling properties of hydrated charged soft biological tissues. *Appl. Mech. Rev., 1990 Suppl.*, 1990, *43*, S134-S141.

STRESS ANALYSES OF BONES AND SIMULATION OF MECHANICALLY INDUCED CORTICAL REMODELING

Richard T. Hart, Nisra Thongpreda, Vincent V. Hennebel
Department of Biomedical Engineering
Tulane University, New Orleans, LA 70118 USA

ABSTRACT

This paper reviews our work in the study of the mechanical adaptation of mature cortical bone. Three topic areas are addressed: the development and analysis of three-dimensional models of the mandible; the application of the RFEM3D (Remodeling Finite Element Method—3-Dimensional) computer program with an in vivo animal experiment performed by Lanyon, Goodship, Pye and McFie [1]; and the review of some ideas for more mechanistic models of net bone remodeling.

INTRODUCTION

The implantation of prosthetic devices is known to disrupt the natural mechanical stress and strain patterns in bone, and it is the adaptive response of bone to these disrupted strain patterns that is thought to contribute to the long term success or failure of implants. Although the triggers for stimulating the net remodeling response is a topic of continuing study, some progress has been made in understanding the mechanics of net remodeling by several groups of researchers that are using phenomenological modeling techniques to relate bone mass changes to the surrounding mechanical environment. Included among these phenomenological models are those proposed by Frost [2, 3], Kummer [4], Gjelsvik [5, 6], Cowin and Hegedus [7], Cowin and Van Buskirk [8], Cowin [9], Pauwels [10], Guzelsu and Saha [11, 12], Huiskes et al. [13] and Fyhrie and Carter [14]. In addition, efforts have been directed toward the development of more mechanistic models that incorporate some of the phenomenological components of previous models but contain parameters based upon observable biological measures such as cellular populations, distributions, and activity [15, 16, 17]. Many of these ideas have, however, not yet been tested with in vivo data.

In order to test the consequences due to any of the proposed phenomenological models, accurate descriptions of the strain range and patterns in bone must be determined both before and after a mechanical loading alteration is made. The changed loading may be the result of the implantation of a prosthesis or — particularly in cases of in vivo animal models — a changed loading regimen. The usual method used to calculate the stress and strain patterns is a numerical approximation technique called the finite element method.

In addition to standard finite element techniques, some researchers have extended the method to incorporate the simulation of remodeling by basing incremental changes in bone mass and geometry upon the results of the finite element model results. Work of this type includes the techniques developed by Gupta, Knoell and Grenoble [18], Umetani and Hori [19, 20, 21], Carter [22], and Huiskes et al. [23]. The most general method appears to be the special three-dimensional finite element based computational method, RFEM3D (Remodeling Finite Element Method—3-Dimensional) [15, 24]. The application of this method has thus far been restricted to simplified tubular models of long bones and for prismatic models of animal bone [25]. A future step in the research involves the use of the RFEM3D program in an attempt to simulate remodeling in complex models of bone such as the mandible and femur. Before that can proceed, knowledge of the natural mechanical strain environment is required.

METHODS

Finite Element Models

Recently, accurate three-dimensional models of a human mandible [26, 27, 28] have been constructed. The geometry was determined from Picker 1200 SX CT scans of the intact bone.

Because the finite element method is an approximate numerical technique, it is essential that some objective measure of the quality of the models be established. The method for establishing the accuracy with which the model is solved is called a convergence test. Because of the steps used in deriving the displacement based finite element method, it will always be true that as more and more degrees-of-freedom (more elements and nodes) are used, the approximate solution will improve, assuming that computer numerical roundoff errors can be controlled. Each added degree-of-freedom represents an added unknown to the problem as well as an added equation to describe that unknown. Modeling the same structure with different numbers of nodes and elements and comparing the solutions constitutes the convergence test, and the appropriate number of degrees-of-freedom can be objectively estimated. This was done for the entire mandible model [26, 27] and the final model geometries required for solution accuracy had over 30,000 degrees-of-freedom as shown in Figure 1.

The bone tissue was assigned mechanical properties based on a series of increasingly realistic assumptions from homogeneous and isotropic to inhomogeneous and transversely isotropic. The analyses were performed using the ABAQUS finite element program (Hibbitt, Karlsson and Sorensen, Inc.,

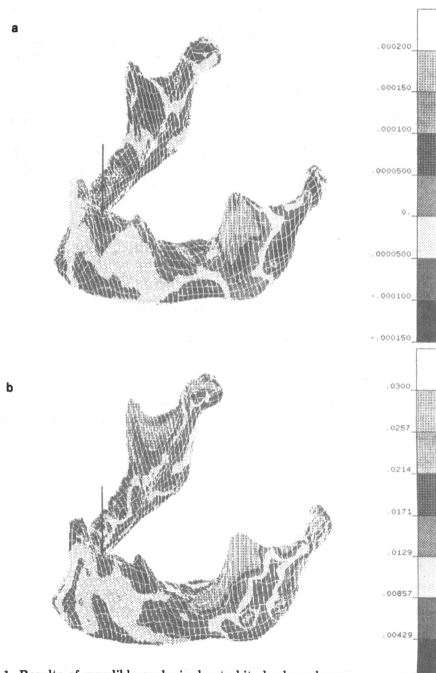

Figure 1. Results of mandible analysis due to bite load as shown.
(a) Dithered gray patterns map the magnitude of one component of the strain,
ε_{33}, with the 3-direction following the bend of the mandible.
(b) Gray patterns map the square root of the strain energy density.

Providence RI) running on the CRAY X-MP/48 at the Pittsburgh Supercomputer Center. The mandible models have been loaded based on an algorithm that assigns muscle forces in accordance with muscle cross-sectional area, while maintaining static equilibrium. It is the description of these natural strain patterns in the mandible that will provide the basis for simulating the evolution of bone architecture if the natural strain patterns are altered.

RFEM3D Remodeling Simulations

A prismatic three-dimensional finite element model of the bones in the sheep forelimb was constructed using the PATRAN finite element pre- and post-processor program. In order to distribute the strains due to a nodal point load, and to provide a loading point that was outside the boundaries of either the radius or the ulna, a "rigid cap" was constructed at one end of the model that had a modulus of elasticity six orders of magnitude larger than that used for the cortical bone properties.

The first order remodeling rate equation describing shape changes associated with altered loading, called a surface remodeling equation, that was developed by Cowin and Van Buskirk [8] was used in this analysis:

$$U = C[\varepsilon(Q) - \varepsilon^o(Q)] \tag{1}$$

where U is the velocity of the bone's external surface, C is a coefficient matrix of remodeling rate constants, ε is the strain tensor at point Q, and ε^o is the remodeling equilibrium strain at the point Q.

To prepare for the subsequent remodeling analysis, finite element calculations were first performed to find the strain distribution in the intact forelimb and these calculated strains were assigned to be the remodeling equilibrium strains, ε^o. Figure 2(a) shows the intact model, with contours corresponding to the axial remodeling equilibrium strain. The remodeling analyses were then run following the removal of the ulna from the model. The loading was left unchanged, and remodeling rate constants that had been empirically found, a posteriori, from in vivo experiments published previously [29] were used for the remodeling analyses.

Mechanistic Model

This section summarizes some ideas that have been previously published regarding a complementary modeling approach to Cowin's phenomenological model that has been used for a cell biology based mechanistic model by Hart [15], Davy and Hart [16], Hart and Davy [17].

This model is intended as an initial attempt to merge some of the phenomenological modeling ideas with observable measures from the field of histomorphometry. This model develops the remodeling rate constants in terms of biological parameters including the number of different cells present and their

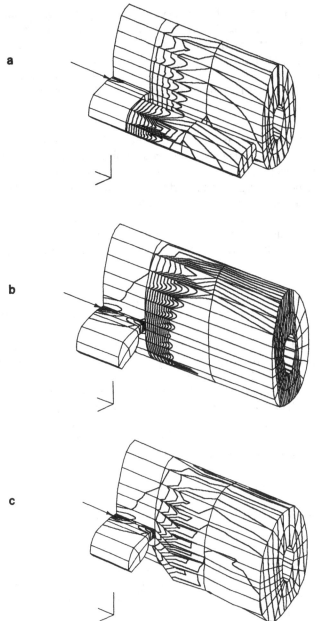

Figure 2. The simulated remodeling sequence in the forelimb of the sheep.
(a) The initial configuration showing the radius and intact ulna. Contours map
the axial strain distribution, $\varepsilon_{zz}{}^{\circ}$.
(b) The forelimb at the start of the remodeling simulation immediately following
the removal of the ulna. Contours map the axial strain distribution, ε_{zz}.
(c) The simulated radius shape and strain distribution at the end of a one year.
The shape change closely mirrors that seen experimentally by Lanyon, et al.
(1982), and the axial strain in the radius is now very close to the initial distribution, $\varepsilon_{zz}{}^{\circ}$, shown in part (a).

average activity. In addition, the theory is developed giving remodeling responses based on the entire strain history with the intent of expressing functional dependence on the strain in the broadest terms possible. When more is known about the nature of the transducing mechanisms and the specific portion of the strain history that may be important, the general theory can be specialized, and the sensitivity of the remodeling response can be adjusted to match experimental results.

The basic premise of the model is that since bone is both resorbed and formed by cells that line the bony surfaces, bone remodeling is the manifestation of surface cellular processes. Based on the concepts of geometric feedback by Martin [30] the net remodeling on the surface of bone is the sum of the osteoblast and osteoclast activity per unit surface area. However, unlike the process described by Martin, the cellular activity is not regulated only by the available surface area per unit volume of bone, but also by cellular response to a proposed strain-induced potential. In addition, the model makes allowances for genetic, hormonal, and metabolic influences.

The theory can be described as follows [17]. The load on the bone, along with the geometric and material properties, completely determines the local strain, assuming elastic material behavior. The strain is then detected by some as yet unspecified transducer that generates a scalar signal called a strain remodeling potential. Modulation of this functional signal is assumed to be possible by genetic, hormonal, and metabolic influences, and is termed a remodeling potential. The history of this potential then determines the rate at which bone is both formed and resorbed. The local magnitudes of bone formation and resorption are determined from the number of cells present, and the average cellular activity. The number of cells present is determined by he level of the cellular recruitment rate. The result of the formation and resorption is the net remodeling.

In general, the loads imparted to the bone will be a function of time. The resulting strain ε at a point x is then dependent on the material properties, C, (that may also depend on position for non-homogeneous material models) and the applied loads.

The theory assumes that the strain remodeling potential, denoted by p(x,t) is a function of the strain history. The cumulative nature of the influence of the strain upon the strain remodeling potential can be expressed as an integral over time in the following form:

$$p(x,t) = \int_{-\infty}^{t} f[\varepsilon(x,t)]dt \quad \text{or} \quad p(x,t) = \int_{0}^{t} f[\varepsilon(x,(t-\tau))]d\tau \quad (2a,b)$$

Because p(x,t) is a function of the strain, and because the strain is, in turn, a function of the position, x, and time, t, the relationship expressed in equation (2b) is a functional. A more concise notation can be achieved by replacing the integral with a symbol denoting the history of strain as follows

$$p(x,t) = H_p(\epsilon,(x,(t - \tau))) \text{ for } 0 < \tau < \infty \qquad (2c)$$

where H_p is a history functional that is dependent upon the strain, the position, and time.

The relationship between remodeling and the remodeling potential is comprised of both the effect of the potential on the average cellular activity rate and the effect of the potential on the cellular recruitment. For the average osteoblast and osteoclast activity rate,

$$a_i = H_{ai}(p(x,(t - \tau))) + a_{iM} + a_{iH} + a_{iG}; \ i = b,c \text{ for } 0 < \tau < \infty, \qquad (3)$$

where the use of the subscript i equal to b or c refers to osteoblasts or osteoclasts and a_i refers to the average volume rate of bone that is deposited (resorbed) per active osteoblast (osteoclast), and H_{ai} is the history functional relating the strain remodeling potential to the cellular activity. Here it is assumed that metabolic, hormonal, and genetic factors act in parallel and are specified by a_{iM}, a_{iH}, a_{iG}. Similar relationships can be written for the number of active osteoblasts (or osteoclasts) per forming area as:

$$n_i = H_{ni}(p(x,(t - \tau))) + n_{iM} + n_{iH} + n_{iG}; \ i = b,c \text{ for } 0 < \tau < \infty. \qquad (4)$$

Finally, since in general the entire surface area will not be available for both formation and resorption, that fraction of the entire surface area that is available for formation is denoted λ_b, and the fraction available for resorption is λ_c. In general, λ_b and λ_c could be functions of time, and could depend on the current magnitude of remodeling. They are related as follows:

$$\lambda_b + \lambda_c = 1. \qquad (5)$$

Thus, on the external surface, or the surfaces of internal voids, the velocity of the surface is written as the following rate equation

$$\dot{d} = n_b \lambda_b a_b - n_c \lambda_c a_c, \qquad (6)$$

where d positive indicates that bone is being deposited, and d negative indicates that bone is being actively resorbed on the surface.

In order to apply the basic concepts embodied in the above equations to specific circumstances, the functional forms must be specified. Several simplified specific relationships have been developed previously [15, 17]. Of interest here is the simplification of assuming constant cell populations which is presented at the end of the following section.

RESULTS

Following the format of the Methods section, the results are divided into three parts.

Standard Finite Element Methods

The standard finite element analyses have yielded several results. The required degree of model refinement has been found using convergence tests with the most refined mandible model requiring 1,836 elements, 10,420 nodes, and thus 31,260 degrees-of-freedom. This result is of interest in terms of the methodology because it objectively shows the required model sizes that are necessary to achieve accurate solutions. These models can only be analyzed with the resources of a supercomputer.

The mandible models, loaded with both symmetric and asymmetric bite loadings, have shown an interesting distribution of strains. A consistent result is the presence of high tensile strains on the anterior portion of the ramus (the portion of the mandible perpendicular to the plane containing the teeth), due to the muscle loading on the thin piece of bone there. Figure 1, shows the results of the indicated biting force. Part (a) shows a strain fringe plot, ε_{33}, while part (b) maps the magnitude of the square root of the strain energy density, chosen to give a more global indication of the mechanical environment.

RFEM3D Remodeling Simulations

The remodeling geometry that simulates the sheep forelimb 9 months following ostectomy is shown as a sequence in Figure 2 (a, b, c), where it is the shape and strain distribution at the non-loaded end that is of interest for this analysis. The initial geometry and axial strain distribution, $\varepsilon_{zz}°$, is shown in Figure 2(a), while the axial strain in the radius immediately post-ostectomy, ε_{zz}, is shown in Figure 2(b). It is the difference in these two strain distributions that drives the surface remodeling as described by equation (1). The simulated geometry and axial strain distribution following one year is shown in Figure 2(c).

Mechanistic Model

Use of the equations presented in the Methods section, coupled with the assumption of a constant cell population and a simple linear relationship between the strain and the strain-induced remodeling potential, $p(x,t) = C_p\varepsilon$, with C_p a constant coefficient matrix, gives the following simplified version of equation (6) [15, 17]:

$$\dot{d} = C_1\varepsilon + C_2 \qquad (7)$$

where $C_1 = C_p(n_{b0}C_{ab}\lambda_b - n_{c0}C_{ac}\lambda_c)$ and $C_2 = n_{b0}a_{b0}\lambda_b - n_{c0}a_{c0}\lambda_c$.

The strain at which no remodeling occurs, $\varepsilon°$, is simply $-C_2/C_1$. This result is the same as the equation developed by Cowin and Van Buskirk [8] except that rather than being entirely phenomenological, the remodeling rate constants are based on observable cellular parameters. The model has not yet been applied to <u>in vivo</u> measurements.

DISCUSSION AND CONCLUSION

Further understanding of the process of functional adaptation in living bone is dependent upon continued research regarding the distribution of mechanical strains in bone; upon in vivo animal experiments and their interpretation, simulation, and ultimately their prediction; and upon development of testable theories. Some progress has been made, but the goal of using these methods and techniques to develop reliable tools for the improvement of prosthesis design and for the betterment of patient care is still in the future.

ACKNOWLEDGMENTS

This work has been partially supported by NIH grants AM33616, DE06859, a BRSG award to Tulane University, and by means of a grant for computational time granted by the Pittsburgh Supercomputer Center (MSM860007P).

REFERENCES

1. Lanyon, L.E., Goodship, A.E., Pye, C.J., and McFie, J.H., Mechanically adaptive bone remodeling. J. Biomechanics, 1982, 15, 141-154.

2. Frost, H.M., The Laws of Bone Structure, Charles C. Thomas, Springfield, 1964.

3. Frost, H.M., Intermediary Organization of the Skeleton, CRC Press, Boca Raton, 1986.

4. Kummer, B.K.F., Biomechanics of bone: mechanical properties, functional structure, functional adaptation, in Biomechanics: Its Foundations and Objectives, eds., Y.C. Fung, N. Perrone, and M. Anliker, Prentice-Hall, Inc., Englewood Cliffs, N.Y., 1972.

5. Gjelsvik, A., Bone remodeling and piezoelectricity — I. J. Biomechanics, 1973a, 6, 69.

6. Gjelsvik, A., Bone remodeling and piezoelectricity — II. J. Biomechanics, 1973b, 6, 187.

7. Cowin, S.C., and Hegedus, D.H., Bone Remodeling I. Theory of adaptive elasticity. J. Elasticity, 1976, 6, 313.

8. Cowin, S.C., and Van Buskirk, W.C., Surface bone remodeling induced by a medullary pin. J. Biomechanics, 1979, 12, 269-276.

9. Cowin, S.C., Wolff's Law of trabecular architecture at remodeling equilibrium. J. Biomechanical Engineering, 1986, 108.

10. Pauwels, F, Biomechanics of the Locomotor Apparatus: Contributions on the Functional Anatomy of the Locomotor Apparatus, Springer-Verlag, Berlin, 1980.

11. Guzelsu, N., and Saha, S., Electro-mechanical behavior of wet bone — part I: Theory. J. Biomechanical Engineering, 1984a, 106, 249.

12. Guzelsu, N., and Saha, S., Electro-mechanical behavior of wet bone — part II: Wave propagation. J. Biomechanical Engineering, 1984b, 106, 262.

13. Huiskes, R., Weinans, H., Dalstra, M., Adaptive bone remodeling and biomechanical design considerations. Orthopedics, 1989, 12, 1255-1267.

14. Fyhrie, D.P., and Carter, D.R., A unifying principle relating stress to trabecular bone morphology. J. Orthopaedic Research, 1986, 4, 304.

15. Hart, R.T. Quantitative Response of Bone to Stress, Ph.D. Dissertation, Dept. of Mechanical and Aerospace Engineering, Case Western Reserve University, Cleveland, Ohio, 1983.

16. Davy, D.T., and Hart, R.T., A theoretical model for mechanically induced bone remodeling, American Society of Biomechanics, Rochester, MN, 1983b.

17. Hart, R.T., and Davy, D.T., Theories of bone modeling and remodeling. Chapter 11 in Bone Mechanics, ed. S.C. Cowin, CRC Press, Boca Raton FL, 1989.

18. Gupta, K.K., Knoell, A.C., and Grenoble, D.E., A bone biomechanical remodeling algorithm. 8th Annual Meeting, Association for Advancement of Medical Instrumentation, Washington D.C., 1973.

19. Umetani, Y., and Hirai, S., An adaptive shape optimization method for structural material using the growing — reforming procedure. Joint JSME-ASME Applied Mechanics Western Conf., 1975, 359.

20. Umetani, Y., and Hirai, S., Adaptive optimal shapes of beam and arch structures by the growing-reforming procedure. Bulletin of the JSME, 1978a, 21, 398.

21. Umetani, Y., and Hirai, S., Shape optimization for beams subject to displacement restrictions on the basis of the growing-reforming procedure. Bulletin of the JSME, 1978b, 21, 1113.

22. Carter, D.R., Mechanical loading history and skeletal biology. J. Biomechanics, 1987, 20, 1095-1109.

23. Huiskes, R., Weinans, H, Grootenboer, H.J., Dalstra, M., Fudala, B., and Sloof, T.J., Adaptive bone-remodeling theory applied to prosthetic-design analysis. J. Biomechanics, 1987, 20, 1135-1150.

24. Hart, R.T., Davy, D.T., Heiple, K.G., A computational method for stress analysis of adaptive elastic materials with a view toward applications in strain-induced bone remodeling. J. of Biomechanical Engineering, 1984, 106, 342-350.

25. Hart, R.T., Computational techniques for bone remodeling. Chapter 12 in Bone Mechanics, ed. S.C. Cowin, CRC Press, Boca Raton FL, 1989.

26. Thongpreda, N. A Three-Dimensional Finite Element Analysis of the Human Mandible, M.S. Thesis, Department of Biomedical Engineering, Tulane University, New Orleans, LA, 1988.

27. Hennebel, V.V. A computational Analysis of a Human Mandible using the Finite Element Method, M.S. Thesis, Department of Biomedical Engineering, Tulane University, New Orleans, LA., 1989.

28. Hart, R.T., Hennebel, V.V., Thongpreda, N., Van Buskirk, W.C., and Anderson, R.C. Modeling the biomechanics of the mandible: a three-dimensional finite element study," manuscript accepted with required revisions the J. Biomechanics, 1990.

29. Cowin, S.C., Hart, R.T., Balser, J.R., and Kohn, D.H., Functional adaptation in long bones: establishing in vivo values for surface remodeling rate coefficients. J. Biomechanics, 1985, 18, 665-684.

30. Martin, R.B., The effects of geometric feedback in the development of osteoporosis, J. Biomechanics, 1972, 5, 447.

BIOMECHANICAL RESPONSES OF CRANIOFACIAL AND ALVEOLAR BONES TO
MECHANICAL FORCES IN ORTHODONTICS

KAZUO TANNE
Department of Orthodontics
Osaka University Faculty of Dentistry
1-8 Yamadaoka, Suita, Osaka 565, Japan

ABSTRACT

Stress patterns and levels were analysed by use of the three-dimensional
finite element method in association with orthodontic and orthopaedic
forces applied to the upper canine and the craniofacial complex in human
beings. Stress distributions in the periodontium and the craniofacial bones
were directly related with actual changes of the tooth and craniofacial
skeleton observed in orthodontic treatments. Thus, it is shown that the
mechanical stress in living structures may be a trigger to induce
biological remodelling of bones.

INTRODUCTION

In clinical orthodontics, various forces are used to correct dental and
skeletal deformities in patients with malocclusion. Therefore, it is of
significance to understand the relationship between bone remodelling and
applied forces. Since the most important factor for bone remodelling is
supposed stress and/or strain in bony structures (1), bone remodelling
should be investigated in relation to biomechanical changes such as
stresses and/or strains in the bones produced by therapeutic forces.

Various analytical techniques have been applied to studies for the
biomechanical response of bones and their surrounding tissues to mechanical
forces. Among these techniques, the finite element method (FEM) has been
recently used for biomechanical studies in the field of medicine and
dentistry (2), (3), (4), (5), (6). The FEM was developed for structural
analysis in engineering, however, its basic principle and procedure are

applicable to biological problems. Thus, various studies by the FEM have indicated possibility and availability of the technique in elucidating biomechanical components.

In this paper, biomechanical changes of bones investigated by use of finite element analysis are shown. Further, the relation of biomechanical changes in bones and their supporting tissues to clinical and experimental findings is investigated.

MATERIALS AND METHODS

Stress distributions in the periodontal tissues by an orthodontic force

A three-dimensional finite element model was made for the upper canine. The model consists of 240 isoparametric solid elements, comprising the tooth, PDL and alveolar bone. Table 1 shows the mechanical properties of the model.

TABLE 1
Mechanical properties of the tooth, PDL and bones

Material	Young's modulus (kgf/mm^2)	Poisson's ratio
Tooth	2.0×10^3	0.30
PDL	6.8×10^{-2}	0.49
Compact bone	1.4×10^3	0.30
Alveolar bone	1.4×10^3	0.30
Cancellous bone	8.0×10^2	0.30

A lingually directed horizontal force of 100 gf was applied at the midpoint of the labial surface of the crown, simulating clinical orthodontic treatment to correct malposition of tooth. As a boundary condition, the model was restrained at the maxillary basal bone area to eliminate the rigid body motion.

The program for the presentanalysis was SUPERB. Three principal stresses were determined midway in the PDL and at the surface of the alveolar bone. These stresses were evaluated for the labial, labio-proximal, linguo-proximal and lingual sides in the horizontal plane at four apicogingival levels of the root, i.e. the cervix, one third root length to the cervix, one third root length to the apex and apex.

Biomechanical changes in the craniofacial bones by an orthopaedic force

A three-dimensional finite element model of the craniofacial skeleton was developed from a young, human dry skull (Fig. 1). The model consists of 2918 nodes and 1776 solid elements. Details of the modelling procedures are described in a previous article (6).

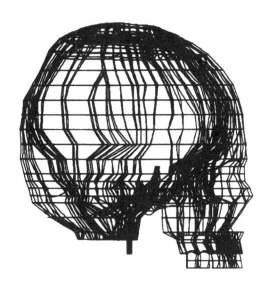

Figure 1. A three-dimensional finite element model of the craniofacial skeleton excluding the mandible.

A 1.0 kgf horizontal force was applied to the upper first molars in the anterior direction parallel to the functional occlusal plane. Material constants of three components of the model are shown in Table 1. The model was restrained at the peripheral region of the foramen magnum to simulate orthopaedic teatment to correct skeletal deformities in Class III patients.

The program for the present analysis was NASTRAN. Displacement pattern of the entire craniofacial complex was evaluated. Further, the absolute maximum principal stress among three principal stresses were determined in various craniofacial bones and their surrounding sutures to evaluate biomechanical responses of the craniofacial complex to the applied force.

In order to comprare the analytical findings with actual changes of the craniofacial skeleton during treatment, lateral cephalograms of a patient who underwent maxillary protraction therapy were investigated.

RESULTS

Fig. 2 shows the stress distributions in the PDL and at the surface of the alveolar bone for the upper canine. In the PDL, three principal stresses were very similar. The highest stresses, approximately 60 g/cm^2 , were observed at the cervix on the labial and lingual sides, with a clear transition of stresses near the centre of the root. Stress levels on the proximal points were substantially less than those on the labial and lingual aspects. At the surface of the alveolar bone, the minimum (compressive) stress on the labial aspect and the maximum (tensile) stress on the lingual aspect produced bending of the alveolar bone, whereas bending stresses were very slight on the proximal points. The remaining two principal stresses exhibited similar changes to those in the PDL.

In orthodontic clinic, tipping movement of a tooth is observed in the direction of force applied to the tooth and the root shifted to the opposite direction. This actual phenomenom is easily understood based on the findings on stress distributions at the PDL-alveolar bone interface. Thus, the relationship between stress and tissue remodelling in the periodontium was supported by this analysis.

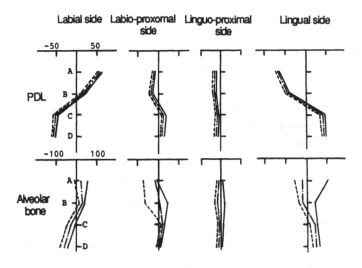

Figure 2. Stress distributions (gf/cm^2) in the PDL and at the alveolar bone of the upper canine. Maximum (——), intermediate (—·—) and minimum (------) stresses are shown for four horizontal points at four apicogingival levels denoted by letters A through D.

Fig. 3 shows displacement pattern of the entire craniofacial complex. The nasomaxillary complex among the entire comlex experienced a substantial repositioning in the forward and upward directions. Centre of rotation of the entire complex located at the lower posterior region of the skeleton near the occipital bone.

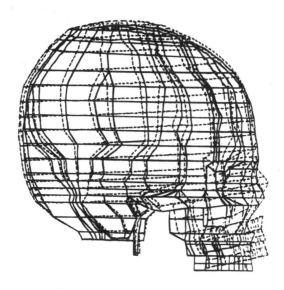

Figure 3. Displacement (——— ;unloaded,- - - - ;loaded) of the craniofacial skeleton. Forward repositioning of the complex was induced with concomitant rotation in the upward direction.

Figs. 4 and 5 show stress distributions in the entire craniofacial complex and in three different planes parallel to the Frankfort horizontal plane. High stress levels were induced in the nasomaxillary complex and its surrounding bones such as the zygomatic corpus and arch (Fig. 4-A).

In the plane passing through the centre of the orbital cavity, stress concentration was induced in the nasal and maxillary bones. These stress values ranged from 0.018 to 0.020 kgf/mm^2 (Fig. 4-B). At the level for the superior ridge of the nasal cavity, high stress levels were produced in the zygomatic bone as well as in the maxillary bone. The maximum stress level approached 0.030 kgf/mm^2 (Fig. 5-A). At the level for the maxillary basal bone, stress uniformly distributed without concentration at the region close to the force application point. The maximum stress level in this plane was 0.014 kgf/mm^2 (Fig. 5-B).

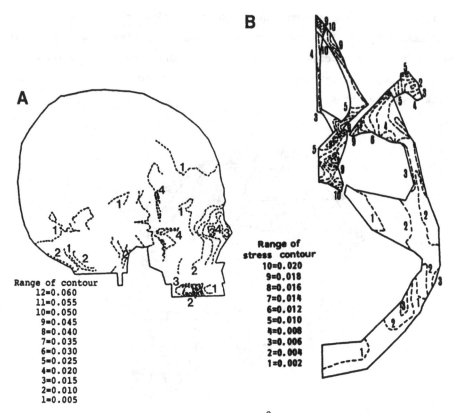

Figure 4. Stress distributions (kgf/mm^2) in the entire craniofacial skeleton (A) and in the horizontal plane passing through the centre of the orbital cavity (B).

Fig. 6 shows cephalometric profilograms of a patient who underwent maxillary protraction for the correction of mandibular protrusion. In this case, an anteriorly directed force of approximately 500 gf was applied to the upper first molars which were supporting teeth of the lingual arch appliance used as an anchorage of maxillary protraction. Protraction force was produced by elastics placed between the hooks on the upper first molars and on the vertical bars in front of patient's face (Fig. 6-A). An acceleration of maxillary growth and/or a foward repositioning of the maxilla was obviously observed (Fig. 6-B). These findings are well understood, if compared with analytical results by the FEM described above.

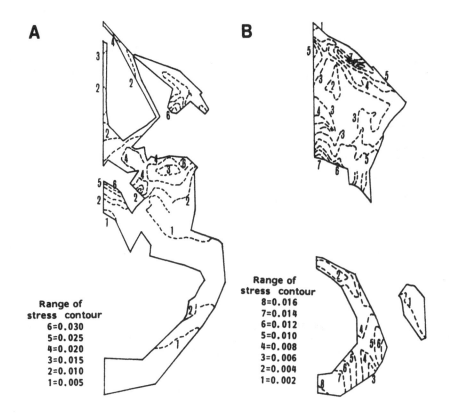

Figure 5. Stress distributions (kgf/mm^2) in the planes traversing the superior ridge of the nasal cavity (A) and the maxillary basal bone (B).

DISCUSSION

Bone remodelling is a well-known phenomenon in human body. Fortunately, remodelling of the cranial, facial and alveolar bones is frequently experienced in daily orthodontic treatment. Therefore, it has been of importance to understand how these bones are remodelled. For solving this question, histological, biochemical and biomechanical studies have been conducted (1), (3), (4), (6), (7), (8). These studies have suggested that a key factor to bone remodelling is mechanical stress or strain.

In this study, a few examples are presented to elucidate the nature of mechanical stress in the periodontal tissues and craniofacial bones. It was indicated that stress distributions in the PDL and the alveolar bone were related with resorption and apposition of the alveolar bone based on a hypothesis that compressive and tensile stresses induce resorption

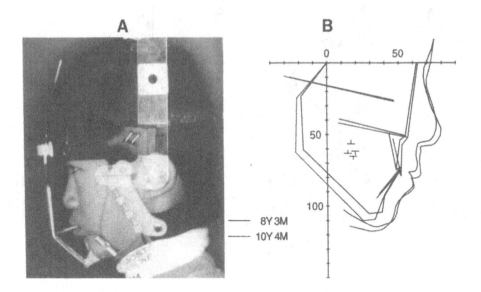

Figure 6. Maxillary protractor for Class III patients (A) and lateral cephalometric profilograms of a patient before (———) and after (- - -) maxillary protraction therapy (B).

and apposition, respectively. The hypothesis is discussed below. Further, it was shown that stresses in the craniofacial bones induced bone remodelling noted by growth changes and repositioning of the complex observed in a patient. In addition, sutural modifications may be related with bone remodelling (6).

With respect to the relation of mechanical stress to bone remodelling, two different analyses were conducted in addition to the present study to discuss the preceding problem.

A three-dimensional model of the metacarpus from a young adult sheep was developed based on the study by Churches and associates (7). A 5.5 MPa stress simulating the experiment was loaded at the mid-diaphyseal point of the bone. Mechanical properties of the compact and cancellous bones are defined as shown in Table 1. Three principal stresses were determined at the mid-diaphyseal region of the bone and were compared with dimensional changes of the bone.

Fig. 7 shows three principal stresses in the mid-diaphyseal region of the bone and ratio of dimensional changes at the same area where the stresses are evaluated. Three principal stresses in the dorsal compact bone

(A) were greatest, and then exhibited a decrease in the dorsal and volar cancellous bones (B and C) and in the volar compact bone (D). The thickness of the compact and cancellous bones measured at four points (A through D) increased substantially in the dorsal compact bone (A), whereas those of the cancellous bone (B and C) and the volar compact bone (D) did not greatly changed or slightly decreased. From these results, two major findings were derived. One is that compressive and tensile stresses induce resorption and apposition as bone remodelling. The other point is that the magnitude of stresses is almost proportional to the amount of bone remodelling shown by changes in the thickness of the bone. These findings strongly emphasise that principal stresses directly pertain to bone remodelling.

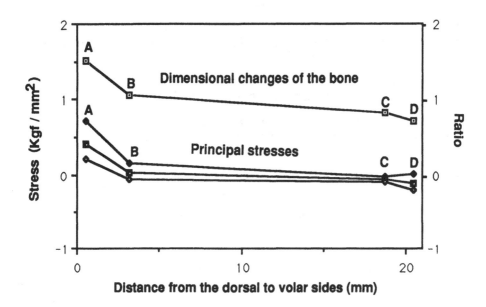

Figure 7. Maximum (■—■), intermediate (□—□) and minimum (◇—◇) stresses and dimensional changes of the bone.

As a next, this relationship was further investigated by use of a numerical simulation based on finite element analysis (9). A two-dimensional finite element model of the upper canine was developed. Basic principle was that restructuring of the model was produced according to the magnitude and direction of principal stresses in the PDL. It was found that analytical tooth movements equivalent to actual tooth movements were achieved by use of the simulation technique. The results indicated that compressive and tensile stresses in the PDL played a role in initiating resorption and apposition of the alveolar bone, respectively. Further, the threshold level of stress in the PDL for initiating bone remodelling was 0.4 gf/mm^2 . It is interesting that the stresses in the PDL, not at the surface of the alveolar bone, trigger bone remodelling facing to the PDL. It is shown that remodelling of the periodontal tissues, in particular the alveolar bone, is induced directly by stresses ,i.e. its magnitude and direction have an important key to induce such biological changes in living structures.

REFERENCES

1. Burstone, C.J., The biophysics of bone remodeling-optimal force considerations. In The Biology of Tooth Movement, eds. L. A. Norton and C.J. Burstone, CRC Press Inc., Boca Raton, 1989, pp. 321-33.

2. Dale, P. J., Matthews, F. L. and Schroter, R.C., Finite element analysis of lung alveolus. J. Biomech., 1980, 13, 865-74.

3. Orr, T.E. and Carter, D.R., Stress analyses of joint arthroplasty in the proximal humerus. J. Orthop. Res., 1985, 3, 360-71.

4. Tanne, K., Sakuda, M. and Burstone, C. J., Three-dimensional finite element analysis for stress in the periodontal tissue by orthodontic forces. Am. J. Orthod. Dentofac. Orthop., 1987, 92, 499-505.

5. Williams, K. R., Edmundson, J. T., Morgan, G., Jones, M. L. and Richmond, S., Orthodontic movement of a canine into an adjoining extraction site. J. Biomed. Eng., 1986, 8, 115-20.

6. Tanne, K., Miyasaka J., Yamagata, Y., Sachdeva R., Tsutsumi, S., Sakuda, M., Three-dimensional model of the human craniofacial skeleton: method and preliminary results using finite element analysis. J. Biomed. Eng., 1988, 10: 246-52.

7. Churches, A. E., Howlett, C. R., Waldron, K. J. and Ward, G. W., The response of living bone to controlled time-varying loading: method and preliminary results. J. Biomech., 1979, 12, 33-5.

8. Rodan, G. A., Mensi, T. and Harvey, A., A quantitative method for the application of compressive forces to bone in tissue culture. Calcif. Tissue Res., 1975, 18, 125-31.

9. Inoue, Y., Tsutsumi, S., Tanne, K., Ida, K. and Sakuda, M., Effects of principal stresses in periodontal tissues on canine retraction. J. Dent. Res., 1988, 67, 168.

NONLINEAR BEHAVIOUR OF THE PERIODONTAL LIGAMENT A NUMERICAL APPROACH

A. N. WILSON*, J. MIDDLETON*, G. N. PANDE*, M. L. JONES**
*Department of Civil Engineering, University College of Swansea,
Singleton Park, Swansea
**Dental School, University of Wales College of Medicine,
The Heath, Cardiff

ABSTRACT

A numerical model is used to investigate the non-linear behaviour of the periodontal ligament which surrounds an upper canine tooth. Comparisons are made with experimental results and it is shown that a finite element elastic-viscous overlay model can be used to simulate the time dependent displacements of an upper canine tooth. The investigation provides quantitative information on tooth movement which can subsequently be used to provide data on the material parameters which describe the periodontal ligament.

INTRODUCTION

Where numerical models have been applied to problems relating to dental mechanics, the behaviour of the periodontal ligament has often been assumed to be linear elastic [1]-[5]. It is however well known that the displacements observed experimentally [6],[7] are much greater than those reported in numerical models developed by Tanne [2,3] and Middleton [4]. Two of the main reasons for these observations are firstly that the material behaviour of the ligament is nonlinear [8-11] and secondly, material parameters obtained from experimental data are limited [12]. These two problems lead immediately to gross variations in results and furthermore compound the problem of developing accurate numerical models to predict displacements of teeth and the surrounding tissue. In order to overcome these difficulties and provide a more realistic model a visco-elastic overlay element will be introduced which can adequately describe the elastic and viscous deformation of the periodontal ligament.

When a tooth is initially loaded the periodontal ligament acts as an hydrodynamic buffer by utilisation of the interstitial fluids [10],[11]. As the fluid moves from the ligament into the surrounding tissue, stress is transferred to the ligament fibres which gives rise to a time dependent response. The maximum displacment is considered to be reached when a balance occurs between the interstitial fluid and the stresses in the ligament fibres.

This behaviour gives rise to a time dependent response in which the displacements of the supporting bone and tooth are significantly less than the viscous behaviour of the periodontal ligament. Ross [7] and Burstone [6] have provided experimental data showing this behaviour which has been obtained from in-vivo measurements. It has also been suggested by Wills, Kurashima, Picton, Ryder and Pryputniewicz [8,13,14,11,6] that the periodontal ligament exhibits properties which are visco-elastic although no material parameters for this behaviour are given.

VISCO-ELASTIC MODEL OF THE PERIODONTUM

In order to model the periodontal response described above a visco-elastic overlay element will be introduced to discretise the periodontal ligament. The tooth and surrounding tissue will be considered to behave elastically. Experimental data published by Ross [7] will be used as a means of obtaining the visco-elastic parameters for the overlay model such that the displacement time response obtained from the numerical solution matches the experimental results. If agreement can be established in matching these results then it can be assumed that the material parameters which describe the numerical model are indicative of those of the in-vivo ligament behaviour.

Visco-elasticity using the overlay model has been previously used in the finite element context by Pande, Zienkiewicz, Owen [17,19,20]. Here the visco-elastic response is modelled by considering two overlay elements which have identical nodal coordinates but different material properties. This allows instantaneous elastic response followed by viscous behaviour, the onset of which is governed by the specified yield limit prescribed to the viscous element. The general visco-elastic stress-strain relationship [15,16,18] can be written by adding the appropriate elastic and viscous components as follows:

$$\{\dot{\epsilon}\} = \{\dot{\epsilon}_e\} + \{\dot{\epsilon}_\mu\} \tag{1}$$

where $\dot{\epsilon}$, $\dot{\epsilon}_e$ and $\dot{\epsilon}_\mu$ are the rates of total elastic strain and viscous strains respectively. In solving equation (1) the material properties which describe the visco-elastic behaviour have to be assigned to the various elements which form the mesh. In the overlay model the material is assumed to be made up of several different layers or overlays as shown in Figure 1. Here each layer can have a different thickness and prescribed material behaviour which gives considerable flexibility in modelling complex material behaviour.

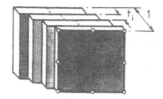

FIGURE 1   THE OVERLAY MODEL

Since the nodal coordinates in each overlay are coincidental the strain produced in each component is identical. This results in different stresses $\{\sigma\}_j$ in each layer which subsequently contribute to the overall stress field $\{\sigma\}$ according to the overlay thickness t_i. This may be written

$$\{\sigma\} = \sum_{j=1}^{k} \{\sigma\}_j \, t_i \quad \text{where} \quad \sum_{j=1}^{k} t_i = 1 \qquad (2)$$

The equilibrium conditions which must be satisfied at each stage of the solution procedure can be written as

$$\int_v [B]^t \sum_{j=1}^{k} \{\sigma\}_j \, t_i \, dv - \{f_i\} = 0 \qquad (3)$$

where $\{f_i\}$ is the vector of nodal forces and $[B_i]$ is the strain matrix. The element stiffnesses are given as the sum of each overlay contribution and can be written as

$$K_T = \sum_{j=1}^{n} \int_v [B]^T [D^n]_j [B] \, dv_j \qquad (4)$$

where $[D^n]$ denotes the operator which models the viscous behaviour of the material.

The material described above can be represented in spring and dashpot form as shown in Figure 2. A typical displacement versus time response provided by this model is shown in Figure 3.

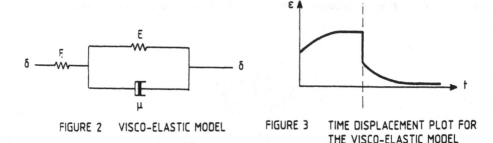

FIGURE 2   VISCO-ELASTIC MODEL

FIGURE 3   TIME DISPLACEMENT PLOT FOR
THE VISCO-ELASTIC MODEL

BIOMECHANICAL FRAMEWORK

The framework described in the previous section exhibits the typical time dependent behaviour reproduced in the experimental work referenced in [6]][7].   It is noted however from these references that the material properties of the periodontal ligament are patient dependent and therefore this behaviour must be considered in the numerical model.   To account for this it was decided to consider the experimental results of Ross [7] where upper and lower bounds are established for a prescribed point load of 10 grammes   which were applied to different patients.

The finite element mesh used to model the upper canine is shown in Figure 4.   The mesh consists of 638 eight-noded isoparametric 2-D elements and the geometrical dimensions and elastic material properties are taken from reference [4].   The periodontal ligament is discretised into 136 elements with three elements across the ligament thickness and only these elements are prescribed with viscoelastic behaviour.   The material properties used in the numerical model are given in Tables 1-3.

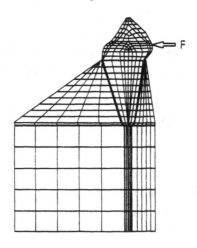

FIGURE 4   FINITE ELEMENT MESH WITH LOAD

For comparison with the experimental results a load of 10 grm was applied to the mid point of the tooth in the buccal palatal direction as shown in Figure 4 and allows direct comparison with the experimental results of reference [7]. The material parameters obtained from fitting the numerical results to the experimental are given in Table 2 where the yield tolerance controls the accuracy of the converged solution and the viscosity parameter controls the rate at which viscous flow occurs.

TABLE 1 – MATERIAL PROPERTIES

	E	ν
Enamel	45000	0.3
Dentine	18600	0.31
Pulp	2	0.49
Bone	4000	0.3
Periodontal L	See tables 2/3	0.45

No. of elements = 638

No. in periodontal L = 68 x 2 = 136

TABLE 2 – LOWER BOUND SOLUTIONS

$E_1 = 4.1$ $E_2 = 2.05$ $\mu = 0.4$ $\nu = 0.45$

Force	δ initial		δ final Before load removed	
	Experi.	Numerical	Experi.	Numerical
5g	0.75	1.0	1.8	1.2
8g	1.0	1.5	2.8	2.2
10g	2.0	2.0	3.1	3.0

TABLE 3 – UPPER BOUND I

$E_1 = 1.1$ $E_2 = 0.55$ $\mu = 1.3$ $\nu = 0.45$

Force	δ initial		δ final	
	Experi.	Numerical	Experi.	Numerical
5g	1.0	2.5	4.1	3.0
8g	1.8	3.8	5.3	5.5
10g	3.0	4.9	8.05	7.8

RESULTS

Figures 5 and 6 show the results obtained from the visco-elastic model and compares these with the measured data of Ross et al [7]. The parameters used to describe the overlay material response are given in Tables 2 and 3. These parameters were established by undertaking a series of computer analyses in which the elastic and viscous components of the numerical model were updated to obtain the best match with the experimental data for an applied force of 10 grm. From these results it is clearly shown that the behaviour provided by visco-elastic model fits closely with the general trend provided by the experimental results for various loading cases [7]. In particular the 8 grm and 10 grm loading cases show good agreement throughout the complete loading history, however the 5 grm loading cases deviates considerably when compared with the experimental results.

Table 4 gives the results obtained from a different patient, as described in reference [7], although these results were obtained using the same material parameters as in Table 3. From the results of Table 4 and Figure 7 it is shown that for the 10 grm load case that a good correlation is obtained. Again as discussed previously the 8 grm and 5 grm load cases do not produce such a good correlation although the trend of the time dependent behaviour is similar.

It can be concluded from these results that the visco-elastic model can provide quantitative information on the time dependent behaviour of the periodontal ligament. In particular the results for the higher loading cases show that a good approximation can be obtained for the non-linear time dependent behaviour. Furthermore, using this numerical model, it is possible to produce the details of the time dependent stress-strain response of the periodontal ligament which up to the present time is unavilable in experimental form.

One of the difficulties encountered in this study is the inclusion of patient dependency in the model and it should be noted that the material parameters used have been obtained from a back analysis for the single loading case of 10 grms. Also it should be mentioned that the loads applied in the experimental work of Ross and Lear [7] are considered to be comparatively small and hence restrict the numerical prediction of the behaviour of the periodontal ligament to small forces only.

TABLE 4 - UPPER BOUND II

$E_1 = 1.1 \quad E_2 = 0.55 \quad \mu = 1.3 \quad \nu = 0.45$

Force	δ initial Experi.	Numerical	δ final Experi.	Numerical
5g	2.0	2.5	6.0	3.0
8g	2.9	3.8	8.0	5.5
10g	4.5	4.9	8.0	7.8

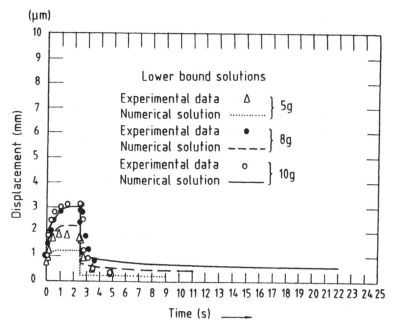

FIGURE 5   A PLOT OF THE EXPERIMENTAL DATA AND THE NUMERICAL
            SOLUTION USING A VISCO-ELASTIC MODEL

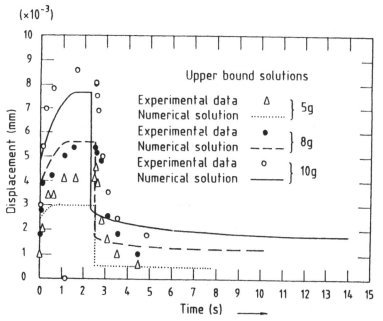

FIGURE 6   A PLOT OF THE EXPERIMENTAL DATA AND THE NUMERICAL
            SOLUTION USING A VISCO-ELASTIC MODEL

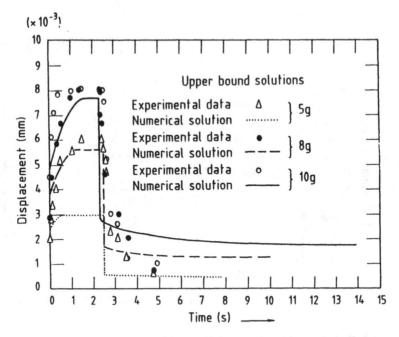

FIGURE 7   A PLOT OF THE EXPERIMENTAL DATA AND THE NUMERICAL
SOLUTION USING A VISCO-ELASTIC MODEL

DISCUSSION

A finite element model based on a visco-elastic framework has been used to predict the
non-linear response, that is displacement against time, for point loadings applied to an
upper canine tooth.   The material parameters obtained for the periodontal ligament have
been established by fitting the numerical time-displacement curve to experimental data
collected from in-vivo tests.   The material parameters provide adequate information to
enable numerical prediction of time dependent ligament displacement to be obtained for a
range of other loading conditions.   Although it has been shown that material response is
patient dependent the visco-elastic framework however does provide the typical nonlinear
response of teeth subject to loading conditions such as those applied in orthodontic
treatment.

Further development of the model in full three-dimensional form is considered
esential to provide greater accuracy in obtaining material parameters which can be verified
against in-vivo displacements.   This would provide not only an accurate means of assessing
material parameters but would also give a quantitative prediction of the time dependent
behaviour of the periodontal ligament and surrounding bone which could be subsequently
developed and applied to bone remoddeling.

ACKNOWLEDGEMENT

The authors wish to thank the Welsh Scheme for the Development of Health and Social Research for providing the funds to undertake this work.

REFERENCES

1. Tanne, S. and Burstone, C., Am. J. Orthod. Dentofac Orthop. Dec. 1987, 92, 439–505.

2. Tanne, S., Stress distributions in the periodontal membrane associated with various moment to force ratios. J. Osaka Dent. Sch., 1987, 27, 1–9.

3. Tanne, K. and Sakuda, M., Initial stress induced in the periodontal tissue at the time of the application of various types of orthodontic force: Three dimensional analysis by means of the finite element method. J. Osaka Univ. Dent. School, 1983, 23, 143–71.

4. Middleton, J., Jones, M. L. and Wilson, A. N., Three dimensional analysis of orthodontic tooth movement. J. Biomed. Eng., 1990, 12, 319–327.

5. Yettram, Wright and Houston, Centre of rotation of a maxillary central incisor under orthodontic loading. Br. J. Orthod., 1977, 4, 223–7.

6. Pryputniewicz, R. J. and Burstone, C., The effect of time and force magnitude on orthodontic tooth movement. J. Dent. Res., 1979, 58 No.8, 1754–65.

7. Ross, G. G., Lear, C. S. and De Cou., Modelling the lateral movement of teeth. J. Biomechanics, 1976, 9, 723–734.

8. Wills, D. J., Picton, D. C. A. and Davies, W. I. R. An investigation of the viscoelastic properties of the periodontium in monkeys. J. Periodont. Res., 1972, 7, 42–51.

9. Parfitt, C. J., Measurement of the physiological mobility of individual teeth in an axial direction. J. Dent. Res., 1960, 39, 608–618.

10. Bien, S. M., Hydrodynamic damping of tooth movement. J. Dent. Res., 1966, 45, 907–914.

11. Ryden, H., Bjelkhon, H. and Soden, P.O., Movements of healthy and periodontaly involved teeth measured with laser reflection technique. J. Periodontal, 1982, 53(7), 439–45.

12. Wright, K. W. S. On the mechanical behaviour of human tooth structures: An application of the finite element method of stress analysis. Ph.D. Thesis, Brunel University, Uxbridge.

13. Karashima, Visco–elastic properties of periodontal tissue, Bulletin Tokyo Medical and Dental University, 1965, 12, 240.

14. Picton, D. C. A. On the part played by the socket in tooth support. Arch. Oral Biol., 1965, 10, 945–955.

15. Hunter, S. C., <u>Mechanics of Continuous Media</u>, Ellis Horwood Publishers, Chichester, Sussex, 1976.

16. Christensen, R. M. <u>Theory of viscoelasticity. An Introduction.</u> Academic Press Inc. (London) Ltd., 1971.

17. Pande, G. N., Owen, D. R. J. and Zienkiewicz, O. C., Overlay models in time dependent non-linear material analysis. <u>Comp. and Structures</u>, 1977, **7**, 435-443.

18. Valliappan, S., <u>Continuum Mechanics</u>, Balkema Academic and Technical Publications, 1981.

19. Zienkiewicz, O. C., Nayak, G. C. and Owen, D. R. J., Composite and overlay models in numerical analysis of elastoplastic continua. <u>Int. Symp. on Foundationa of Plasticity</u>, Warsaw, 1972.

20. Owen, D. R. J., Prakash, A. and Zienkiewicz, O. C., Finite element analysis of non-linear composite materials by the use of overlay systems. <u>Computer & Structures</u>, 1974, **4**, 1251-1261.

SOLID MODELING AND FINITE ELEMENT MODELING OF BONES AND IMPLANT-BONE SYSTEMS

Beat Merz*, Ralph Müller+, Peter Rüegsegger+

* Sulzer Medical Engineering, CH-8401 Winterthur, Switzerland
+ Institute of Biomedical Engineering of the Swiss Federal Institute of Technology and the University of Zürich

ABSTRACT

The use of Finite Element Modeling (FEM) is a valuable tool in the field of orthopaedic implants. In the present projects we emphasized the use of geometrically realistic models and easy understandable results to improve the dialogue with the orthopaedists. We built up models of femurs with and without prosthesis. The bone geometry and material properties were first obtained by dissection and literatur, later on by using a quantitative CT scanner (QCT). The implementation of an interface between QCT and pre/postprocessing program increased the speed of modeling a long bone dramatically. Bevore modeling an implant-bone system, the relative positioning of prosthesis and bone was determined by performing an implantation simulation with the help of solid modeling. Based on the use of three dimensional displays, the discussions with the surgeon during modeling and after the calculation enhance the understanding of the modeled system, as well as the interpretation of changes found in histologic examinations.

INTRODUCTION

The use of FEM for the study of implant-bone systems has been shown to be a valuable method for improving the design of new implants and understanding the involved mechanisms [1]. However, the time consuming calculations and especially the cumbersome modeling of the anatomic geometries have been limiting the accuracy of the models, regarding the geometry as well as the interface conditions.

In order to create detailed models of long bones with and without implants in reasonable time, we use the general purpose pre/postprocessing software I-DEAS Supertab, that allows the use of automatic meshing. By partitioning the model into subvolumes we are able to control the meshing process achieving thus a reasonable mesh. Furthermore the subvolumes can be used as modules to build up different versions of a system. For the first femur we modeled with our system, we dissected a cadaveric femur and digitized the slices [2]. Building up a 3D model this way took us months. Concerning the material properties, we had to use some of the few average values available in the literature, although large scale mechanical testing has shown the great variability of the mechanical properties within the species as well as within one individual [3].

We therefore built up an interface between our FEM pre/postprocessor and a quantitative CT scanner. It was our goal to extract inner and outer contours of long bones, especially femurs and to bring them in the specific vector-format into our software. There subvolumes are formed, which allow automatic meshing and solid models of the bones are created, which are useful for simulation of implantations.

Besides the geometry extraction, the interface is also responsible for the determination of the mechanical constants of each element, corresponding with the local material properties in the measured bone.

METHODS

The base for out FEM studies is consisting from subvolumes that we may mesh directly and solid models of the bones. A subvolume is defined by two following slices and their vertical junctions.The solids are defined by computing B-spline surfaces defined through stacked up profiles. This definition leads to very realistic representations of the bonegeometries (Fig.1).

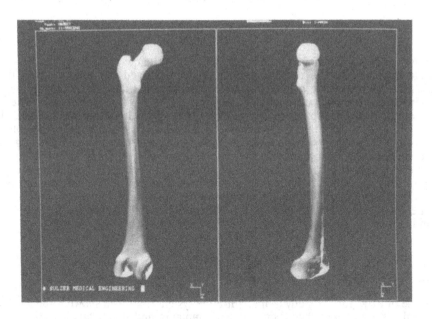

Figure 1. A solid of a scanned human femur.

As mentioned in the introduction, we first obtained the slices by dissection of a cadaveric femur and digitizing each slice. As this turned out to be far to complicated, we decided to use a CT scanner

QCT Scanning

We used the quantitative computed tomography scanner (QCT) at the Institute of Biomedical Engineering [4], which is a translation-rotation scanner, constructed to scan appendicular parts of the body (Fig.2). The measurements were taken with a field of view of 155 mm. Slice thickness was 1mm and the lateral resolution 0.3mm. After image reconstruction the tomograms were stored as 256x256 pixel images with a resolution of 605 μm per pixel. The distance between two slices was 5 mm in regions with strongly varying contour geometries i.e. the epiphyses and metaphyses. In the diaphyseal area the distance was increased to 10 mm and in the transition regions we chose 8 mm. A total number of 66 cross-sectional slices were taken over the full specimen bone length of 450 mm.

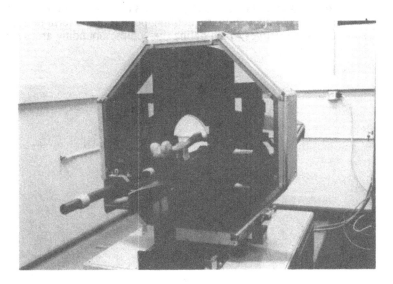

Figure 2. QCT scanner with a femur fixed for measurment.

Contour Segmentation

For the reconstruction of the three dimensional bone geometries, we first had to extract the inner and outer contours out of the stack of QCT images. This was realized by an automatic tracing-algorithm [4,5]. After the segmentation the inner and outer contours are stored separately. The outer contour belongs to each cross-section. This is not the case with the inner contour, which is only present in those regions where the cortical bone has to be separated from trabecular bone. As the corticalis close to the articulations is often very thin and this would lead to serious problems, representing it with three dimensional elements, we decided to use a simplified bone model. In this model cortical bone does only exist if it forms a continuous outer shell with a specific thickness. In all other cases cortical bone is in the segmentation neglected and in the FE model represented with shell elements. Because of this simplification, the totality of cross-sections can be reduced to three different types: Type I) characterizes one single object in a cross-section without cortical bone. Type II) characterizes two objects in a cross-section without cortical bone. Type III) characterizes one single object in a cross-section enclosing cortical bone (Fig.3).

Subvolume Generation

The elementation of a bone is realized with automatic meshing, yet it is not possible to mesh a bone in one part. On the one hand such a volume would have geometries too complex to successfully control an automatic net-generation. On the other hand it would not be possible to discriminate sharply between the different bone materials like trabecular and cortical bone or medullary cavity. Therefore it is necessary to divide the bone previously into smaller significant subvolumes.

322

In consideration of these facts a concept of subvolume generation was developped, where the elementary process is to divide a cross-section into four quadrants relative to its centre of gravity and in sagital and coronal direction.
In the proximal region the division is taken in the direction of the principal axes. In a first step the centres of gravity, the axes and the points of intersection between the axes and the inner, respectively outer contours (defining points) are determined as shown in Fig.4. In a second step the curves which define the volume bounding areas are generated.

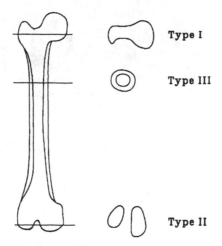

Type I

Type III

Type II

Figure 3. Femur with examples of the three different cross-section types.

Mainly they consist of sectored inner and outer contour segments, which occur when objects are subdivided into four single quadrants, or of the connecting lines between the centre of gravity and the defining points, or of the vertical connecting lines between corresponding points of two neighbouring cross-sections (Fig.4).

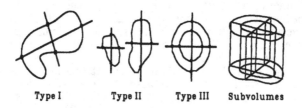

Type I   Type II   Type III   Subvolumes

Figure 4. Cross-sections with centre of gravity and axes. In the neck area the axes are principal axes, in the other areas they lay in coronal and sagital direction. On the right the resulting subvolumes between two sections.

Automatic Meshing

In the automatic meshing process tetrahedrons are generated within the subvolumes. Fist the surfaces of the subvolumes are triangulated in a recursive subdivision process,

then this subdivision is expanded into space. The chosen average length of the element edges is in every subvolume the same as the distance between the two defining slices. Therefore, the corners of the tetrahedrons lie in the cross-sections defined by the measurements.
Figure 5 shows the meshed specimen bone with 20977 elements.

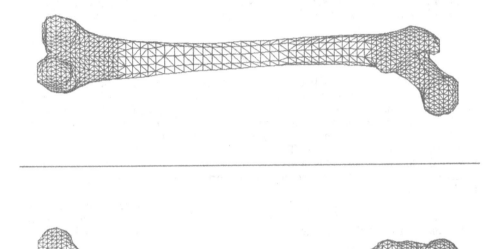

Figure 5. Finite Element Model of a human femur with 20977 elements.

Material Properties

As shown in different investigations over the last two decades [6,7,8,9,10], a highly significant relationship between apparent density and material properties like compressive strength or Young's modulus could be derived from clinical investigations. In essence QCT images are maps of the linear attenuation coefficients. The apparent density ρ can be predicted by evaluating the pixel values of the image matrix [5] :

$$\rho = (\mu/\mu_c) * \rho_c$$

μ_c = attenuation coefficient corticalis
ρ_c = density of the corticalis
(valid for macerated bone)

For the relation between apparent density ρ and Young's modulus E Carter and coworkers [7] found a cubic relationship (i). Rice and coworkers [10] found a quadratic relation (ii).

Gibson [11] found a structure dependent relation between Young's modulus and density. Based on the theory of cellular materials and on the observation, that spongiosa shows a open-cell rod-like structure up to a density of about 0.36 g/cm³ and that it consists over that density of a network of plates they predict the following relations :

$$E \ \alpha \ \rho^2 \quad for \ \rho < 0.36 \, g/cm^3$$
$$E \ \alpha \ \rho^3 \quad for \ \rho > 0.36 \, g/cm^3$$

In our calculation we use the formulas of Carter and Rice. As the function curves $E = f(\rho)$ intersect at $\rho = 0.43$ g/cm³ and this is close to the relation of Gibson, we take the formula of Rice (ii) for lower densities and that of Carter (i) for the denser areas.

$$E = 60 + 900 \cdot \rho^2 \, [MPa] \quad for \quad \rho < 0.43 \, g/cm^3 \qquad (i)$$
$$E = 2865 \cdot \rho^3 \, [MPa] \quad for \quad \rho > 0.43 \, g/cm^3 \qquad (ii)$$

The density for bone is varying between 0 and 2 g/cm³. The poisson ratio is set to 0.33. For the following FE analysis one hundred material tables are defined which are distinguishable only in apparent density and in the Young's modulus. If the mean density of a single tetrahedron is known, its related material table can be predicted. Such a Material Table ID Number can be attached to each tetrahedron as a further attribute. With a spread of attenuation coefficients from 0 to 2 cm⁻¹ the density takes values between 0 and 3.46 g/cm³. This interval is covered by the 100 material tables. As bone density normaly stays below 2.2 g/cm³, only material tables 1 to about 60 are actually attached to elements.

To calculate the density in a single tetrahedron from the QCT images the tetrahedron is projected onto the bounding CT-planes. The mean density is then calculated according to the weights of the projected area sections. Figure 6' shows a coronal section of the meshed proximal femur.

Models of Bones with Implant

With the goal of realistic relativ positioning of implant and bone, we used the possibility of solid modeling for preparing the FE model.

Solids of a femur and of the stem were built up and combined in a way that simulates the implantation of the prosthesis into the bone. With the help of displaying bodies transparently it is easy to judge the correct relative positioning with the help of a physician.

In the example case a cementless stem of the type CLS was 'implanted'. The CLS stem is characterized through proximal ribs that enlarge the contact area and a slim distal part. It's philsosphy consists in a proximal fixation and load transfer in the intertrochanteric region and the avoiding of a distal wedging and overstiffening [12].

To represent such an implant and its philosophy, it is very important to position it correctly in the model. In three dimensional displays (Fig.7) we may control the situation for cortical contact of the stem and for a physiological positioning. In the example it is important to see, that the stem is not largely prominent in the distal part and therefore not wedged into the cortical bone.

After finding the correct position and 'resecting' the femoral head, the implant boundaries in the bone-defining cross-sections were calculated. A series of new core subvolumes representing the stem had to be added and the subvolumes in the neighbourhood were modified.

Calculation

Models of a natural femur and of the same femur with implant were calculated.

The loading case was the single leg stance phase of gait, according to Pauwels, with a weight resultant of 1750 N and an abducting force of 1275 N.

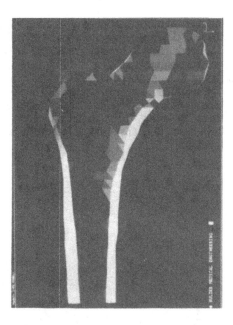

Figure 6. Coronal section through the FE model of a human femur. The greylevels correspond with the assigned Young's modulus of each element.

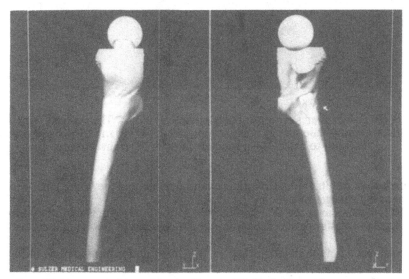

Figure 7. A CLS stem positioned in the femur. Of the femur only spongy bone and intramedullary channel is displayed. Prominent parts of the stem (arrow) signalize direct contact of stem and corticalis.

RESULTS

By developping an interface between QCT and pre/postprocessing for FEM we have made QCT a valuable tool for FE modeling of long bones.

The time needed to build up a detailed model of a specific bone was reduced from 2-3 months (manual method) to 2-3 days (using the present interface). Even more important then this considerable reduction in time is the use of actually measured material properties. This allows to investigate special cases like osteoporotic bones, dysplastic bones or bones with posttraumatic density changes.

Further more solids of the scanned bones are available. The solids help to answer questions on the design and the fit of a prosthesis in a bone (Fig.8) and they are also needed if the complete mechanical system is to be modeled, consisting of a bone with implanted prosthesis. In both cases they are a valuable help to ease the interdisciplinary conversation with the physician.

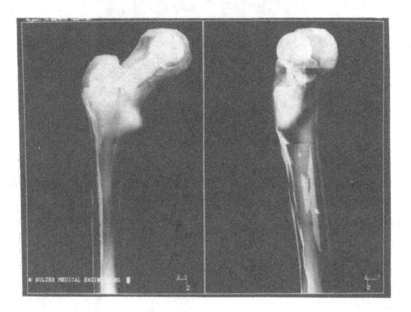

Figure 8   A Zweymüller stem positioned within the femur. The cortical bone is displayed transparently while spongy bone and intramedullary channel are opaque making it easier to detect the prominent parts of the stem (arrow).

A comparison of results from calculating a natural femur and the same femur with CLS stem shows the changes induced by the implant.

The natural femur shows the Ward's triangle and the loading of the primary compressive trabeculae transfering the load from the articulation to the medial cortical neck (Fig. 9).

After an implantation the situation changes. The loadtransfer from implant to corticalis is found in the distal metaphysis. The stresslevels in the medial corticalis below the resection plane are lowered but histologies and clinical results suggest that they still remain big enough to avoid the consequences of a stress shielding that may be found in cases of distally fixed prostheses wedged into the cortical bone.

In the area of the greater trochanter the stresses are increased, probably due to the weakening by resection of the head. In the clinical results we find the shoulder of the stem lying in this area often covered with a thin layer of connective tissue. This tissue has probably a damping function against relativ movements between stem and greater trochanter, which is periodically deformed by strong abductor muscles [12].

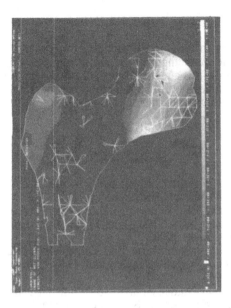

Figure 9. Coronal section through the proximal femur showing the von Mises stress. The white lines and dark spots are artifacts caused by cutting right trough the elements.

DISCUSSION

Bentzen and coworkers showed QCT to be a valuable method for measuring mechanical properties of bone [13]. Being noninvasive and nondestructive it is a method applicable in vitro and in vivo.

Keyak and coworkers [14] described an automated mesh-generation technique, where they created a three dimensional FE model from CT scans. While they generated a mesh consisting of cubic elements directly from the scans, they developped a model that reminds a voxel-display known from 3-D medical imaging, with somewhat coarse and jagged surfaces, making it thus difficult to obtain results in boundary areas.

We chose a different approach. Defining first only the boundaries of subvolumes, cortical and spongy bone, we then let a general purpose pre/postprocessor mesh the model. Afterwards the created elements are backprojected to the CT scans, to assign them the mechanical constants. This way we are able to create a model with smooth boundaries, needing transverse scans only every five to ten mm.

In addition we may define solids of measured bones. By combining them with solids of prostheses simulating this way an implantation, we can modify the subvolumes, including subvolumes that model the implant in correct position. After meshing we may again backproject the elements representing remaining bone and attach the corresponding mechanical properties to them. Doing this, we define a model of an implant and bone short after surgery. It is obvious that remodeling will change the situation after some time, relativating the validity of calculation results for later phases.

While in the last two decades almost all FE models of bones used one average Young's modulus for the cortical bone and typically less than six average values for different spongy areas, we are now able to represent the gradual changes in a single specimen bone. As a disadvantage the significant anisotropy of bone is not detectable in our application of QCT and we are therefore restricted to isotropic models.

CONCLUSION

The use of QCT helps to develop FE models of bones in an effective way. With the ability to speed up the modeling process and to improve the material property and geometry representation, it is a method worth of further improvements.
From scanning long bones in vitro the method can be extended to measurements in vivo and to scanning other bones like e.g. the pelvis bones.

REFERENCES

1. Flemming, M., Scholten, Engelhardt, Kraftflussberechnungen in Knochenstrukturen und Prothesen, BMwF-1A 7a-7291 NTÖ 4, 1972.
2. Merz, B., Schenk, R.K., Wintermantel, E., Finite element model of a femur with a proximally fixed, cementless stem, Proceedings European Biomechanics. 7th Meeting ESB Aarhus, 1990.
3. Knauss, P., Materialkennwerte und Festigkeitsverhalten des spongiösen und kompakten Knochengewebes am coxalen Human-Femur, PhD Thesis, Universität Stuttgart , 1980.
4. Durand, E.P., Quantitative Analyse von Knochenstrukturen aus Computertomogrammen von gelenknahen Skelettabschnitten, PhD Thesis, Eidgenössische Technische Hochschule Zürich (ETH), 1989.
5. Müller A., Quantitative Computertomographie: Ein risikoarmes Verfahren zur Identifikation der osteoporosegefährdeten Frau, PhD. Thesis ETH 8102, 1986.
6. Galante, J., Rostoker, W., Ray, R.D., Physical properties of trabecular bone, Calcif. Tissues Res., 1970.
7. Carter, D.R., Hayes, W.C., The compressive behavior of bone as a two-phased porous structure, J. Bone Jt. Surg., 59A(7):954-962, 1977.
8. Carter, D.R., Spengler, D.M., Mechanical properties and composition of cortical bone, Clin. Orthop., 1978.
9. Gibson, L.J., The mechanical behavior of cancellous bone, J. Biomechanics, 18(5):317-328, 1985.
10. Rice, J.C., Cowin, S.C., Burr, D.B., On the dependence of the elasticity and strength of cancellous bone on apparent density, J. Biomechanics, 21(2):155-168, 1988.
11. Gibson, L.J., The mechanical behaviour of cancellous bone, J. Biomechanics, 18 (5): 317-328, 1985.
12. Spotorno, L., Schenk, R.K., Dietschi, C., Romagnoli, S., Mumenthaler, A., Unsere Erfahrungen mit nicht-zementierten Prothesen, Orthopäde 16: 225-238, 1987.
13. Bentzen, S.M., Hvid, I., Jörgensen, J., Mechanical strength of tibial trabecular bone evaluated by x-ray computed tomography, J. Biomechanics, 20 (8):743-752, 1987.
14. Keyak, J.H., Meagher, J.M., Skinner, H.B., Mote, C.D., Three-dimensional Finite Element Modeling of a proximal femur from CT scan data, Trans. Orthopaedic Research Society (35), p. 492, 1989.

The Effect of Prosthesis Orientation on 'Stress-Shielding' using Finite Element Analysis - Indications as to Bone Remodeling.

P.J. Prendergast[1], B. McCormack[2], T. Gunawardhana[3], D. Taylor[1]
Dublin Biomaterials Research Centre,
(1) Department of Mechanical Engineering, Trinity College, Dublin 2,
(2) Department of Mechanical Engineering, University College Dublin,
(3) EOLAS-The Irish Science & Technology Agency, Glasnevin, Dublin 9.
IRELAND

ABSTRACT

'Stress shielding' in the proximo-medial femur after hip replacement is studied for different prosthesis stem orientations (valgus - neutral - varus). Both axial and hoop stresses are reported for each case. Results are shown to compare favourably with strain gauge studies. It is predicted that valgus orientated stems create greater axial stress changes and varus orientated stems create greater hoop stress changes. The clinical practice of prosthesis insertion is discussed. Finally the results are analysed in the context of a proposed damage model for bone adaption. It is proposed that the hoop stress has greater significance for the 'stress shielding' concept than its smaller magnitude would suggest.

INTRODUCTION

Bone loss from the medial femoral neck has been identified as a factor leading to loosening and subsequent failure in the artificial hip joint (AHJ). Some possible causes for bone loss are (i) wear debris from the acetabular cup being caught between the bone and the cement,(ii) thermal trauma due to recission of the femur neck, (iii) abrasion between the collar and the femur surface with collared prostheses, (iv) disuse antropy resulting from 'stress shielding' caused by the stiff femoral stem.

'Stress shielding' occurs when stress, normally transfered to the femoral neck in the intact femur, 'bypasses' the femoral neck after insertion of the prosthesis. Huiskes et al [1] describe the phenomenon and they simulate the subsequent bone loss that it causes using a 2D finite element model. They use a side plate to represent the 3D structural stiffness of the bone tube and they hypothesise that the bone remodels to changes in strain energy density. Prendergast and Taylor [2] used a fully 3D finite element model to calculate the stress distribution in the medial femoral neck for different prosthesis designs. It was concluded that less stiff (isoelastic) prostheses can transfer axial stresses to the bone that approach the intact femur stress level. However, the value of the hoop stress was predicted to depend on the presence of a prosthesis collar. With a collar the value of the hoop stress was predicted to be compressive and without a collar was predicted to be tensile. In each case the absolute magnitude was greater for less stiff prostheses.

In this paper the authors consider the effect of prosthesis orientation on the stress pattern in the region where bone loss is observed after hip replacement. Prosthesis orientation is considered to be of importance because it is directly under the control of the surgeon during operation. (Hence it is a variable on the INTERFACE between medicine and mechanics.) In the next section we will review the clinical evidence that prosthesis orientation contributes to bone loss and loosening.

BRIEF REVIEW OF CLINICAL LITERATURE

From a study of the literature [3-5,7-10], loosening of the femoral stem and the cement mantle can be attributed to a number of factors. These can be devided into two groups: A) external; B) internal or surgeon/patient interface. External factors include i) body weight; ii) usage (i.e. loading pattern); iii) effects of revision surgery; iv) geometry and surface finish of stem. Internal factors are;

    i)   canal reaming and preparation;

    ii)   cement insertion, packing, and insertion into bone;

    iii) prosthesis insertion and positioning (valgus-neutral-varus);

    iv) matching of stem size to the medulary canal;

v) uniformity of cement distribution around the prosthesis and the thickness of the cement mantle;

vi) infection;

Muller[3] reported a 10% occurance of long term loosening due to complications other than infection. All cases of stem breakage in the 1967-68 group were associated with varus inclination of the prosthesis and very thin cement mantle in the calcar region. This was attributed to high bending stresses in the varus position leading to movement at interfaces and subsequent bone resorption. Amstutz et al[4] reported inadaquate surgical technique as the main cause of loosening in a 2-5 year follow-up of 389 hip replacements. They recommended new procedures such as trial fits, templates and plugging to ensure a uniform thin cement mantle and a non-varus placement.

McBeath and Foltz[5] reported on a follow-up series of 106 implants after 3 years or more. Resorption was present in most cases of stem loosening, mainly for Charnley-Muller type, and tended to be coupled with varus orientation. The implication is that varus implantation is more likely to cause loosening, however the sample is not big enough to have a statistically significant result. An extension of this work[6] reports on a study of the stresses existing in the bone implanted into cadavers in different positions. Valgus positioning gave rise to compressive hoop stresses while the hoop stress for varus positioning was tensile. The conclusion was for improved surgical technique incorporating a gun pressuriser and a holder to insert the prosthesis and ensure valgus orientation.

Moreland et al[7] conducted a study of 444 hips in a 2-9 year follow-up. Calcar resorption > 3mm occurred in 10.5% of those reviewed, but did not necessarily preceed other signs of loosening. Varus positioning was not correlated with failure, but a thin layer of cement between the calcar and the proximal part of the stem was considered undesirable.

Carlsson et al[8] reviewed 288 hips in an 8-12 year follow-up. Of 139 implanted into varus 53(38%) were classified as being loose; 28(27.5%) out of 102 implanted into neutral loosened; and 13(37%) of 35 implanted into valgus loosened. They concluded that stem position does not have a direct effect on loosening. However, the presance of calcar resorption was significantly (p < 0.001) correlated to loosening. Dispite the lack of correlation between loosening and stem positioning they recommend a non-varus positioning to avoid excessive loading and to ensure a uniform thickness cement mantle. Bocco et al [9] study the effects of prosthesis orientation in a follow-up study of 197 hip replacements. In contrast to Carlsson et al[8], they reported a marked reduction (up to 10 times) in 'calcar' bone destruction when the stem was implanted into valgus.

The conclusion from the literature is that varus positioning of the stem is to be avoided for differing reasons among which are its tendancy to cause bone loss and that it creates a thin, easily fractured, cement layer. Interestingly, it has been further observed (Fowler et al [10]) that prosthesis orientation can alter after implantation due to cement creep. They consider that the movement of the prosthesis into valgus is benificial to the long term stability of their prosthesis design.

Hence prosthesis orientation is one design variable which has been found to have distinct relationship to loosening by bone loss. For this reason the authors study it further and use it to assess a previously proposed model for bone adaption (Taylor and Prendergast[11]).

METHOD

The method of stress analysis was 3D finite element modelling (FEM). The mesh consisted of 886 brick and wedge shaped elements resulting in 12159 degrees-of-freedom (figure 1). These were isoparametric

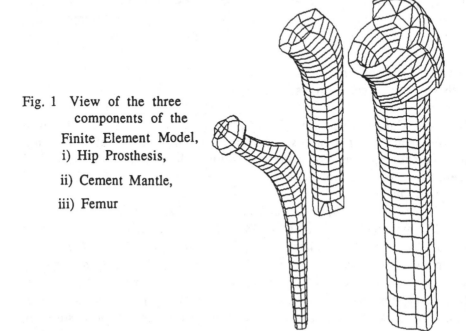

Fig. 1 View of the three
components of the
Finite Element Model,
i) Hip Prosthesis,

ii) Cement Mantle,

iii) Femur

elements consisting of 20 and 15 nodes respectively. Hence they have a quadratic displacement feild. The software used was PAFEC and the CPU time per run was about 1 hr 44 min on the VAX-750.

Geometry

The prosthesis modelled was collarless, with the Sheehan design stem geometry. This prosthesis is fixed into the femur using cement. The bone geometry was not taken as specific to any femur. Instead a characteristic geometry was meshed that incorporated the important structural features of the femur; the linea aspera, the greater and lesser trochanters and the isotrochanteric ridge. This was considered to be a better modelling procedure than direct 'anatomical' modelling since anatomical modelling would necessitate distorted elements. Such distortion of elements would reduce the accuracy of the model and would thus offset any possible gain in model validity. Also, as Huiskes and Chao[12] noted in their review paper, stress around structural details may obscure the more important general stress patterns. With this approach to geometry modelling, there were no distorted elements (see ref. [13]) in the mesh. Figure 1 shows the prosthesis, cement and femur separated. A more complete view of the mesh is given in a previous publication[14]. Care was taken to ensure that no distorted elements arose on rotation of the prosthesis into varus and valgus.

Materials

The effect of materials properties on stress patterns in the AHJ was studied in detail by the authors in a previous publication (Prendergast et al [14]). The baseline values used in that study are used here and are shown in table 1 below.

	Young's Modulus
Prosthesis	200.0GPa
Cement	2.3GPa
Cortical Bone	17.6GPa
Cancellous Bone	32MPa
	(100MPa in the transition region)

Table 1: Young's modulii of the components of the F.E. model

Bone was taken to be isotropic and this seems to be acceptable assuming that bone is deformed primarily in bending (Huiskes [15]).

Boundary Conditions

The hip joint load was taken to be 3kN which aproximates the peak load during gait (Paul[16]). The angle of the applied load was taken to be 20 degrees to the femur shaft. Other loads are experienced by the joint during daily activity but, since most fatigue cycles are incured during walking, it is suffcient to consider this load only in a first analysis. Muscle forces were taken as 1.25kN at 20 degrees to the shaft of the femur for the abductor muscle group and 0.25kN parallel to the femur shaft for the illio-tibial tract. Interfaces were assumed to be perfect bonds. Whilst this is not true for the cement-to-metal bond it is a necessary simplifying assumption for a 3D finite element model with many elements. Iterative solution with 'gap' elements would be prohibitive both in terms of computer time and storage space. The cement-to-bone bond is assumed to be intact which seems to be true initially. The model was restrained at 7 elements distal to the prosthesis tip. Restraints were seen to balance with the applied loads indicating that the finite element model was well conditioned. Further checking of ill-conditioning in the solution is required in this case since rotating the prosthesis could create low stiffnesses within the stiffness matrix. The diagonal decay co-efficient is one measure of ill-conditioning and it is given for all three orientation positions in table 2 below.

Varus and Valgus Positioning
Figure 2 shows schematically valgus and varus positioning and shows also the direction of the axial and hoop stresses. Rotation into valgus and varus was achieved by creating a new axis set for the prosthesis nodes and by rotating through 2 degrees using the distal end as the origon. This generated geometries in the proximal region as shown in figure 4 for varus and figure 5 for valgus. Table 2 shows the largest diagonal decay coeffients encountered during solution. It can be seen that varus and valgus models give results of comparable accuracy to those of the neutrally alligned prosthesis.

Orientation	Diagonal Decay Coefficent
Varus	36503.0
Neutral	36564.0
Valgus	36531.0

Table 2: Measures of ill-conditioning in the 3 F.E. models

RESULTS

Figures 3, 4 and 5 show the axial and hoop stress contours, for neutral, varus and valgus orientation respectively, in the region where bone loss is observed after hip replacement. To aid in assessing the validity of the results a 2D slice of the proximal region of the 3D finite element mesh is shown alongside the stress contours.

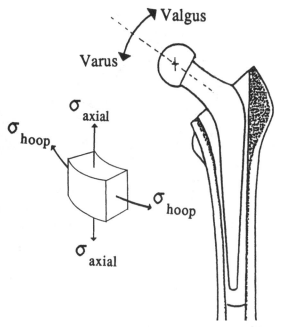

Fig.2   Schematic Diagram of the proximal femur showing valgus and varus positioning and the axial and hoop stresses

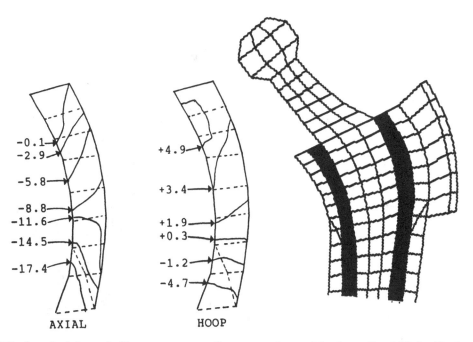

AXIAL

HOOP

Fig.3   Axial and Hoop stresses for neutral positioning (in MPa). Dark region indicates cement

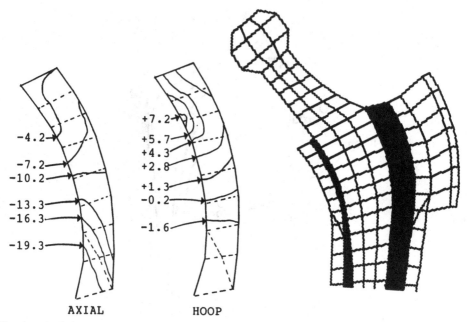

Fig.4 Axial and Hoop stresses for varus positioning (in MPa). Dark region indicates cement

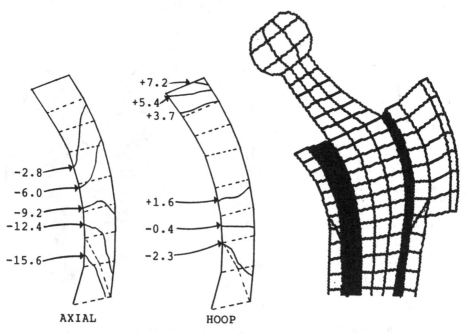

Fig.5 Axial and Hoop stresses for valgus positioning (in MPa). Dark region indicates cement

DISCUSSION

This discussion is divided into two parts. The first part is a comparison of the F.E. stress data in figures 3-5 and a comparison of it with stress data obtained experimentally. Such a comparison is frought with difficulty because the F.E. model neglects interface decohesion and bone anisotropy. The second part is an attempt to reconcile the f.e data with the observed bone remodeling phenomena - namely that valgus implanted stems are less prone to bone remodeling[9].

Analysis of F.E. stress results

Taking the Young's modulus of bone to be 17.6GPa and adjusting to a3kN load, the strain gauge results of Oh and Harris[17] give aproximately 49MPa for the axial stress in the proximo-medial intact femur. The uncollared prosthesis never achieves this stress magnitude no matter what orientation. (A collared prosthesis of Young's modulus equal to 25GPa can achieve this stress level, at least in principle[2].)

The varus orientated stem develops a higher axial stress than the neutral stem; a valgus orientation decreases this stress. This is in accord with the results of McBeath et al[6] who report varus and valgus axial stresses that are 1.5 and 0.3 times the neutral axial stress respectively.

McBeath et al[6] report a tensile hoop stress for the intact proximo-medial femur. Adjusting to the results of Oh and Harris[17] gives a hoop stress of magnitude about +5MPa. However, in a finite element analysis Fagan and Lee[18] calculate the hoop stress to be compressive of magnitude -5MPa for the intact femur. If the latter is the case, i.e. that the hoop stress is compressive prior to implantation, then a considerable change in hoop stress (from compressive to tensile) has occured due to implantation of the prosthesis. The tensile hoop stress is attributable to the wedging effect of the tapered prosthesis stem.

Positioning into varus increases the magnitude of the hoop stress by about a factor of 1.4. It also increases the extent of the area under tensile stress. Implantation into valgus reduces the tensile hoop stress in most of the section, except close to the neck where it is highly increased. This rapid rise in the most proximal element would probably not be sustained in practice, where interface debonding can occur.

Varus-valgus implantation and remodeling

The follow-up studies of Bocco et al[9] and Fowler et al[10] indicate that valgus implantation reduces bone loss. This means that valgus implantation generates a smaller bone remodeling stimulus. However, as we noted above, the varus orientated stem creates a higher axial stress and would therefore be predicted to be beneficial according to the traditional 'stress shielding' concept. Hence we have evidence that the contribution of the hoop stress to the remodeling stimulus is an important, if not governing, contribution in this case.

This is suprising because (as we have predicted above) the axial stress changes by more than 40MPa wheras the change in hoop stress is of the order of 10-13MPa (i.e. -5MPa before to +7MPa after). One explanation for this is that microcracking in the bone is the stimulus for bone remodeling. Evidence for such a theory has been produced by Burr et al[19] who find a correlation between microcracks and resorption cavities. If we take microdamage contributions in the axial and hoop directions to be in the ratio of fracture stresses, then we can suggest ratios for the ratio of contributions (per MPa) to the remodeling stimulus as in table 3.

	tensile	compressive
axial	132 (1.4)	187 (1)
hoop	58 (3.2)	132(1.4)

Table 3: Fracture stresses, in MPa, (from Reilly and Burnstein[20]) with suggested ratios for remodeling stimulus contributions shown in brackets.

Hence, compared to the axial stress (to which 'stress-shielding' is usualy attributed) the remodeling stimulus contribution of the hoop stress is more than three times greater per unit of stress. Applying this ratio to the stress magnitudes of this problem indicates that the hoop stress magnitude is nearly equal to the axial stress magnitude in terms of its contribution to the remodeling stimulus.

CONCLUSIONS

1) Varus orientated stems create smaller axial stress changes and greater hoop stress changes.
2) Clinical studies indicate that valgus implantation is preferable to varus. However, the reason is not entirely clear. One possibility is that the stress state generated in valgus has a reduced stimulus to cause adverse bone remodeling.
3) If this is so then we have evidence that the hoop stress may be of key importance in proximo-medial bone loss in the AHJ.
4) A simple mechanistic basis for this is proposed based on fracture strength and a damage stimulus.

REFERENCES

1. Huiskes R., Weinans H., Grootenboer H.J., Dalstra M., Fudala B., Slooff T.J., Adaptive Bone Remodeling Theory Applied to Prosthetic Design Analysis, J.Biomechanics. 20, 1135-1150, 1987
2. Prendergast P.J., Taylor D., A Stress Analysis of the Proximo-Medial femur after Total Hip Replacement, J.Biomed.Eng. in press
3. Muller M.E., Late Complications in total hip replacement, Proc.3rd Sci.Mtg. Hip Soc., C.V. Mosby Co. 319-327, 1974
4. Amstutz H.C., Marklof K.L., McNeice G.M., Gruen T.A., Loosening of Total Hip Components : cause and prevention, Proc. 4th Sci.Mtg. of the Hip Soc., St. Loius, C.V., Mosby Co., 102-116, 1976
5. McBeath A.A., Foltz R.N., Femoral Component loosenind after Total Hip Arthroplasty, Clin.Orthop.Rel.Res., 141, 66-70, 1979
6. McBeath A.A., Schopler S.A., Seireg A.A., Circumferential and Longitudnal Strain in the Proximal Femur as determined by Prosthesis Type and Position, p.36, Trans. 25th O.R.S., San Francisco, California, 1979
7. Moreland J.R., Gruen T.A., Mai L., Amstutz H.C., Aseptic loosening of Total Hip Replacement: incidence and significance, Proc.6th Sci.Mtg. Hip Soc., 281-291, 1980
8. Carlsson A.S., Gentz C-F., Mechanical Looseninf of the Femoral Head Prosthesis in Charnley Total Hip Arthroplasty, Clin..Orthop.Rel.Res., 147, 262-270, 1980
9. Bocco F., Langan P., Charnley J., Changes in the Calcar Femoris in Relation to Cement Technology in Total Hip Replacement, Clin. Orthop. Rel. Res. 128, 287-295, 1977
10. Fowler J.L., Gie G.A., Lee A.J.C., Ling R.S.M., Experience with the Exeter Total Hip Replacement since 1970, Orthopaedic Clinics of North America, 19, 477-489, 1989
11. Taylor D., Prendergast P.J., Mathematical Modeling in Implant Design: Stress Analysis and Material Selection, in Current Perspectives in Implantable Devices, 2, JAI Press, (Ed D.F.Williams) in press
12. Huiskes R., Chao E.Y.S., A Survey of Finite Element Analysis in Orthopaedic Biomechanics, J.Biomechanics. 16, 385-409, 1983
13. PAFEC User Manual, PAFEC Ltd., Strelly Hall, Nottingham, England.
14. Prendergast P.J., Monaghan J., Taylor D., Materials Selection in the Artificial Hip Joint using Finite Element Analysis, Clinical Materials. 4, 361-376, 1989
15. Huiskes R., Some Fundamantal Aspects of Human Joint Replacement, Acta Orthop. Scand. Suplement no. 185, 1979
16. Paul J.P., Approaches to Design- Forces Transmitted by Joints in the Human Body, Proc.R.Soc.Lond. B ,192, 163-172, 1976
17. Oh I., Harris W.H., Proximal Strain Distribution in the Loaded Femur, J. Bone Joint Surg., 60A, 75-85, 1978
18. Fagan M.J., Lee A.J.C., Role of the Collar on the Femoral Stem of Total Hip Replacements, J. Biomed. Eng., 8, 295-304, 1986
19. Burr D.B., Martin R.B., Schaffler M.B., Radin E.L., Bone remodeling in response to *in vivo* fatigue microdamage, J.Biomechanics, 18, 189-200, 1985
20. Reilly D.T., Burnstein A.H., The elastic and ultimate properties of bone tissue, J.Biomechanics, 8, 393-405, 1975

LE LESIONI DA CASCO MOTOCICLISTICO

ANDREA RICCI
Istituto di Medicina Legale e delle Assicurazioni
della Università di Bologna

GIOVANNI PALLOTTI
Dipartimento di Fisica
della Università di Bologna

RIASSUNTO

Si considerano i vantaggi dell' uso del casco obbligatorio per i motociclisti certamente ragguardevoli per la forte riduzione della mortalità e della lesività, che erano incentrate sul distretto cranio-encefalico e sul viso. Però sono comparse, seppur rare, lesioni della colonna vertebrale e del midollo spinale dovute alla violenta sollecitazione all' indietro del casco in tal caso non integrale. Ed anche eccezionali riscontri di traumi sulle parti anteriori del collo, capaci di determinare pure lo schiacciamento della laringe con inondazione ematica delle vie respiratorie rapidamente letale, legati allo sfilamento del casco, in genere troppo largo e sempre poco esteso posteriormente, per urto frontale con conseguente stiramento trasverso del soggolo.

ABSTRACT

There is no doubt that legislation making it compulsory for motorcyclists to wear crash helmets has led to a sharp drop in the numbers of fatal accidents and injuries involving the skull, head and face. However, there is growing evidence of rare injuries to the vertebral column and spinal cord due to the violent impact of helmets without a chin-piece slipping backwards. Crash helmets are often too big and high at the back of the neck. In some exceptional cases these helmets provoke severe injuries to the front of the neck when they slip off after frontal impact and the chin-strap crushes the larynx so that blood floods into the airways which is rapidly fatal.

INTRODUZIONE

Sono ormai passati oltre quattro anni da quando la legge 11 Gennaio 1986, n° 3, annunciava l' obbligo dell' uso del casco protettivo per gli utenti dei motocicli, ciclomotori e motocarrozzette. Quivi si annunciava pure l' ulteriore impegno del Ministero dei Trasporti a stabilirne entro 60 giorni le caratteristiche tecniche. Così avvenne, infatti, con Decreto Ministeriale 18 marzo, chiedendosi rispondenza al regolamento ECE/ONU n° 22/02. Con successivo Decreto del 9 Agosto 1988, n° 572, si stabiliva, infine, all' art. 2, che con decorrenza ormai attuale, cioè dal 19/7/90, sarebbe cessata la validità di ogni precedente omologazione, ammettendosi al commercio soltanto i caschi conformi al regolamento ECE/ONU n ° 22/03, pur concedendosi l'esaurimento di quelli già in uso, se conformi al precedente regolamento ECE/ONU n° 22/02.

DISCUSSIONE E CONCLUSIONE

Mentre si rimanda direttamente alla lettera della relativa omologazione, per ciò che riguarda i particolari, ai nostri fini pratici è sufficiente ricordare che ai ciclomotori può essere riservata anche una sorta di caschetto assai leggero, munito del relativo sottogola, o soggolo. Lo stesso vale per le motociclette, dove però il caschetto è molto più robusto e più o meno discendente sui lati e posteriormente, oltre che opportunamente foderato, e quindi ammortizzato (commercialmente, modelli "jet" e "demi-jet"). Anche il soggolo è più robusto, più largo e con la fibbia, di foggia spesso diversificata, a perfetta tenuta. A ciò aggiungasi un' altra sorta di casco discendente sui lati fino a chiudersi davanti al mento, mentre posteriormente supera di poco la regione nucale. In tal caso manca il soggolo. Viene da ultimo il cosiddetto casco integrale che ricopre tutto, capo e collo, lasciando praticamente aperta la sola visuale, la prominenza nasale e la fessura buccale e "articolandosi" verso il basso, ovviamente senza entrare in contrasto con le spalle.
I benefici di tale provvedimento (obbligo al casco) sono stati immediati, tanto che con molto cinismo, giustificato soltanto dalla statistica, si è detto che si era venuta a perdere una importante fonte di organi utili per il trapianto.

Però la lesività specifica, che come sempre accompagna qualsiasi forma di artifizio meccanico, è stata incredibilmente grave, ancorché contenuta.

Valga il confronto.

Prima del casco si avevano:

tutte le possibili lesioni traumatiche del cuoio capelluto, dei tessuti molli del viso, specie in corrispondenza delle sporgenze ossee; assai meno sulla regione nucale; quasi indifferenti in corrispondeza del collo;

tutte le possibili lesioni fratturative della volta cranica, spesso trasmesse alla base, ovverosia indotte sulla base dall'elasticità della volta, nonché sfondamenti e schiacciamenti. Lo stesso vale per le ossa del viso, quasi sempre con linee di frattura multiple ed affondate. Assai meno sulla regione nucale. Pressoché nulla sul collo, se non in combinazione con altre lesioni molto gravi e per particolari modalità dell' avvenimento (proiezione, capitombolo, ecc.).

Andando all' interno, serve la semplice enumerazione scolastica e spesso combinata degli:

ematomi epidurali, ematomi sottodurali, emorragie subaracnoidee, contusione diretta cerebrale e da contraccolpo, microemorragie periventricolari di tipo commotivo. Rare, come si è detto, le lesioni del collo e quindi del midollo, che però, quando risultava colpito, era completamente tranciato, soprattutto per lussazione dei corpi vertebrali.

Dopo l' uso del casco:

cranio e cervello risultano praticamente preservati, salvo modeste lesioni escoriative o contusive del cuoio capelluto. Anche il viso ha tratto un notevole giovamento, ovviamente meno del capo, che è completamente coperto. Non si vedono più le notevoli dilacerazioni e soprattutto le fratture con affondamento e sfondamento, eccezion fatta per quelle del mento e quindi della mandibola che sembrano anzi aumentate, o relativamente tali, in rapporto al sensibile calo delle altre consimili e adiacenti, che in ogni caso sono direttamente riconducibili all' infrazione e allo sfondamento della rigida chiusura sovrammentoniera del II tipo di casco chiuso davanti e sui lati, privo di soggolo e alquanto aperto di dietro. Ma non è soltanto a quest' ultima caratteristica (apertura posteriore), poichè si vede anche per il casco integrale, che va riportata un' altra lesione praticamente nuova, almeno come frequenza statistica: cioè la frattura, e soprattutto la lussazione, delle vertebre del collo con tranciamento immediato del midollo spinale. Avviene a varie altezze a seconda dello spazio libero lasciato posteriormente dal casco. E' inversamente proporzionale all' estensione di questo medesimo spazio, siccome al grado di mobilità rimasto complessivamente al collo.

Infatti la spiegazione sembra essere del tutto ovvia.
L' energia cinetica scaricatasi prevalentemente sul casco
mentoniero, ma anche sul relativo cercine frontale e assai meno
in laterale, supera la resistenza del collo, che è l'unico e
l'ultimo ad opporsi. Di per sé molto robusto, ma estremamente
vulnerabile se viene colto di sorpresa in una fase di riposo,
ancorché tensiva, come avviene per il casco.
Infatti si tratta di tensione obbligata, che ritarda l' eventuale
reazione, peraltro impossibile per la rapidità dei tempi.
Veniamo da ultimo all' esclusiva, cioé alla lesione da casco vera
e propria, ancorché indiretta, mai descritta prima d' ora quando
non si usava il casco.
Pretende un casco aperto davanti e poco esteso di dietro, quindi
dotato di soggolo, casco che deve essere preferibilmente più largo
del necessario, come spesso capita ai più giovani accompagnati dai
genitori che acquistano anche in previsione della crescita.
In tal caso il meccanismo lesivo è addirittura elementare;
nellimpatto frontale a capo reclinato in avanti, il casco si sfila
venendo proiettato all' indietro e il soggolo discende sul collo,
mentre il capo, ormai scoperto, continua la sua corsa in senso
fronto-parietale. Perciò il casco stira violentemente verso
l'alto il soggolo, ormai disposto in senso trasversale, determinando
una sorta di strangolamento o di garrottamento. Lo strangolamento
vero e proprio consiste infatti nell' avvolgimento e nello
stringimento di tutto il collo, con mezzo morbido flessibile e
ovviamente resistente; la garrotta è un triste strumento per
l'esecuzione giudiziaria spagnola, fatto da una lamina
opportunamente sagomata, anch' essa disposta all' avvolgimento
del collo, e ristretta di lato mediante una vite a lungo passo,
per risultare più rapida.
Nel caso di specie l' effetto si arresta ovviamente davanti e sui
lati, più simile alla garrotta, data l' altezza sempre superiore
ai 2 cm. del soggolo, la sua sottigliezza e addirittura
proverbiale resistenza. Ma è certamente sufficiente. Infatti alla
rapida compressione che, fatti i debiti calcoli, si estende quasi
sempre fino alla colonna vertebrale, può conseguire, nella
migliore delle ipotesi, un' ischemia e una stasi cerebrale. Ciò
in quanto l' arteria carotide che irrora e la vena giugulare, che
trasporta il deflusso, scorrono in coppia sui due lati del collo,
strettamente ravvicinate tra loro.
L' effetto anatomico è l' edema, o rigonfiamento cerebrale con
perdita della conoscenza, che però, data la rapidità dell' evento
provocatore, si risolve ancora prima di conseguirne l' impegno dei
centri cardio-respiratori, e quindi la morte. Il che esclude
l'intervento dell' altro effetto contemporaneo, o anche isolato,
che invece può dare la morte, pur se durato un solo istante,

rappresentato dal riflesso vaso-vagale, promosso dall' apposito
ganglio che si trova alla biforcazione carotidea, donde lo stimolo
risale fino al nucleo dorsale del vago situato nel bulbo, per poi
discendere nel cuore, che subito si arresta in diastole. L' assenza
di tale riflesso conferma che lo stimolo efficace sul collo non
può essere dato da una striscia larga e robusta, com' è il soggolo
violentemente affondato, bensì da un corpo solido in genere smusso
e affusolato, comunque affondato di poco, in senso vibratorio.
Tutto ciò lo abbiamo appreso dai Clinici, che appunto hanno
valutato giustamente, in senso soltanto vascolare, alcuni di
questi casi giunti in Ospedale privi di conoscenza e subito
risoltisi felicemente, senza complicazioni.
Invece noi abbiamo visto al tavolo anatomico un altro effetto
drammatico e letale. Lo schiacciamento del collo contro la colonna
vertebrale aveva determinato la frattura della laringe con
comprensibile ostacolo respiratorio. Ma a ciò si sarebbe potuto
anche ovviare,per l' ulteriore e altrimenti irreversibile
progresso, se non fosse sopraggiunta la contemporanea e logica
emorragia della stessa laringe e degli altri vasi del collo, che
sono molto abbondanti e ricchi. Emorragia che inevitabilmente
viene "assorbita" all' interno della laringe ad ogni atto
inspiratorio, donde deriva il relativo intasamento della trachea
e dei bronchi che, molto figurativamente, è stato denominato
annegamento o sommersione interna. La cui conseguenza asfittica
è invece irreversibile all' istante, potendo addirittura trarre
nocumento dalle stesse manovre rianimatorie convenzionali, come
la ventilazione forzata.
In definitiva, l' uso del casco ha ridotto sensibilmente i casi
di morte, ma ha aperto altre due strade pressoché specifiche che
sono: la frattura-lussazione delle vertebre del collo con
interessamento quasi sempre letale, la "garrotta" operata dal
soggolo per il casco che si sfila, con interessamento vascolare
e quindi cerebrale (svenimento), che peraltro si risolvono
felicemente, ovvero lo schiacciamento fratturativo con difetto
asfittico meccanico, che di solito "attende" l' arrivo dei
soccorsi, oppure quanto precede con emorragia inspirata nelle vie
respiratorie, che è invece inesorabilmente letale.
La risoluzione è facile:caschi che non permettano grande mobilità
sul perno inevitabile del collo durante il ballottamento del capo
(peraltro rispettato); caschi che non si sfilino, stirando il
soggolo e che, secondariamente, non siano mai troppo larghi.

LOCAL STRAIN DEVIATION (L.S.D.) AND BONE REMODELING: A COMPARATIVE STUDY ON CEMENTLESS TOTAL HIP ARTHROPLASTY.

M. VICECONTI*, A. TONI°, A. SUDANESE°, D. DALLARI° AND A. GIUNTI°

* Laboratory for Biomaterials technology, Istituto Rizzoli, Bologna, Italy.
° Orthopaedic clinic, University of Bologna, Bologna, Italy.

ABSTRACT

In the last years many theories correlating mechanical load and bone remodelling have been proposed. The majority of them indicates the strain as activating factor of the bone adaptivity. This mechanism is still unclear. However, many authors suggest that it has to be a "local" mechanism. If the strain induced by external loads assumes in any point of bony tissue a persistent value which is different from the average physiological pattern in that point, there we have a biological response. This response is probably provided by a small unit (as the Basic Multicellular Unit proposed by Frost) placed close to that point.

In this study we attempt a "local" correlation between bone remodelling occurred after total hip arthroplasty and the strain alterations induced by the cementless prosthesis.

To determine these alterations we used both experimental and computer aided simulation approaches. Strain gauges were applied on cadaver femurs and strain history was recorded during a standard cyclic load test under an Instron test device. The strain history was measured again after a cementless arthroplasty of the femur and the two results were compared. To consider the effect of the force applied to press-fit the stem into the medullary canal, in some samples the prosthesis was applied through a controlled load; in others we implant with the same technique used in vivo. In both cases we controlled through the gauges also the peak strain induced by the implantation. With similar purposes a parametric finite element (F.E.M.) computer model was used. Such analysis was carried out on the Digital VAX 8800 of the "Formula 1" racing car department of Ferrari S.E.F.A.C., Maranello. The general geometry and the elastic module distribution were modified to model the real conditions of some cases where we observed an adaptive behavior after total hip arthroplasty. The assessment of remodeling and its evaluation is reported in a correlated paper (Toni A. *et. al.* "The role of bone remodelling in cementless total hip arthroplasty").

INTRODUCTION

Although numerous careful studies were undertaken since J.Wolff published his law in 1884, bone behavior under stress is still far from being fully understood. The suitability for a specialized function of a complicated structure as the trabecular bone was pointed out in the works of ancient scientists as, for example, Galileo Galilei [5]. Although it was Bourgery who first observed an architectural structure in cancellous bone in 1832, the origin of bone studies is often reported as "The Law of Bone Transformation" published by J. Wolff in 1892 [7].

During bed-rest experiments, the bones which in physiological conditions are more loaded react to the new environment with a more pronounced mineral loss than those less loaded [10][12][4]. This supports the hypothesis that the bone behavior is related to the deviation of the functional strain pattern from the physiological strain pattern, which will be referred to as Strain Deviation [3] [6] [8] [9]. This study is undertaken to investigate the feasibility of applying these observations to a specific problem of the modern orthopaedic surgery: the bone remodeling induced by a cementless total hip arthroplasty. After experimental and numerical assessments of the strain alterations induced by an AN.C.A. cementless total hip arthroplasty [11], we try to relate these results to the bone remodeling observed in clinical cases reported in a correlated paper (Toni A. et. al. "The role of bone remodelling in cementless total hip arthroplasty"). As last step, we apply a prediction model developed by M. Viceconti and A. Seireg [13] to this particular case.

MATERIALS AND METHODS

The determination of the mechanical strain alterations due to cementless total hip arthroplasty was realized with a strain gauges *in vitro* study on cadaver femurs loaded by an Instron test device. According with the finding of Rubin et. al. [8], we assumed as standard load pattern a cyclical load with a 1 Hz. frequency. The load intensity was fixed on the base of a maximum strain level of 2000 microstrains. The loading test was repeated before and after hip arthroplasty; strain measurements were conducted also during the press-fit of the prosthesis.

The detailed method and the preliminary results have already been presented [1]. To investigate the internal strain alterations and the influence of the natural geometry differences on the results, we realized also a three dimensional F.E.M. analysis of the experimental model. A MARC software of MARC Corporation running on a Digital VAX 8800 was used. A particular effort was put on the interface modeling between bone and prosthesis, using "Gap" elements. Also in this case the detailed method and the preliminary results have already been presented [2]. The *in vivo* assessment of bone remodeling, made by image analysis software, is described in the above mentioned correlated paper.

The last step was to compare the clinical observations with the prediction obtained from the model [13]. We used as input data the strain deviation as calculated on the base of the experimental measures.

RESULTS

As observed in the above mentioned correlated paper, we found that AN.C.A. total hip arthroplasty induces, during the follow-up, in some patients two different phenomena: reduction of bone density at the level of great trochanter and cortical hypertrophy at the level of the stem, either distally and medially. We are interested to the possible correlation between these phenomena and the alterations of the mechanical bone environment due to the presence of the prosthesis. A first, simple observation is to compare the diagram of fig. 1 describing in term of frequency of incidence the resorption sites observed clinically, with that in fig. 2 reporting the average longitudinal strain deviation in each site as measured by strain gauges during the in vitro test. It is possible to see a good correlation between the two factors either on the medial side and on the distal side.

Another aspect we tested during the in vitro experiments was the evaluation of the strain induced by the press-fit of the prosthesis into the femur. We have seen a circumferential pre-deformation of the femur at the level of the trochanter. The local intensity of this strain and its location is probably strictly connected with the extension of the prosthesis-bone contact. In this case also the FEM study didn't help because of the influence of the real contact area, which is very hard to model in that kind of problems.

However, some fiber of the bony tissue were tensioned or compressed permanently for the presence of the press-fitted prosthesis, with an intensity varying between 500 and 1000 microstrains.

The FEM analysis substantially confirmed the data from strain gauges. The method gave us more information on the contact between bone and prosthesis. Using the gap element we evaluated the contact area under load and the displacement between the two surfaces. As aspected the lower part of the stem was the site which presents the higher micromotion respect the endosteal wall. This can explain the hypertrophy observed to the level of the stem in some patients. During the loading cycle the stem probably present a micro motion orthogonal to the bone axis which induce cyclical load in the cortical diaphyseal wall. However, the basic mechanism of this hypertrophy is still not clear; it will take a specific study in the future.

On the base of the strain reduction data, we try to predict the bone mass reaction to that alteration. For this purpose we used a program developed by M. Viceconti and A. Seireg [13] in University of Wisconsin-Madison. We explored the range of strain deviation 0.2-0.8 (which means reduction of physiological strain between 20% and 80%) considering two hour of walking per day (7200 cycle per day on the base of 1 Hz. step frequency). The time horizon was three years. The results are reported in the diagram of fig. 4. The model predict for the worst observed case a reduction of the trochanteric trabecular density almost 30% of the initial density.

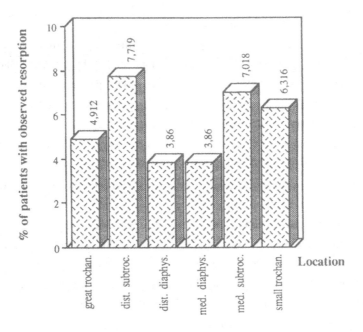

Fig. 1: The percentage of patients, for each site, who presented on x-ray an appreciable resorption of the trabecular bone. (average follow-up 17,4 months).

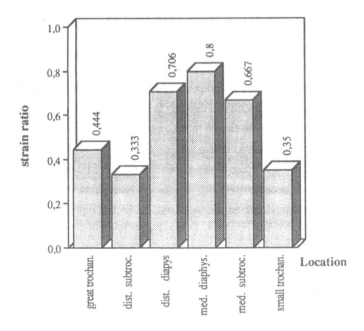

Fig. 2: Average percentual deviation from physiological values of the cortical mechanical strain intensity after hip arthroplasty.

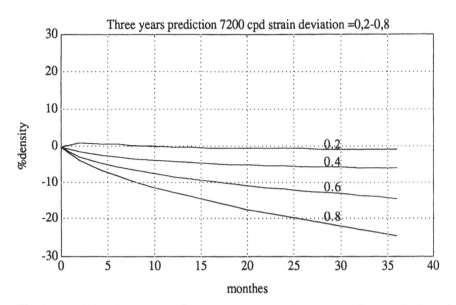

Fig. 3: Prediction diagram; the four curves represent the percent of trabecular bone density reduction for 0.2, 0.4, 0.6 and 0.8 strain deviation after 36 months under two hour/day walk.

CONCLUSIONS

It seems to be confirmed that the qualitative theory of the stress shielding is translatable in a much more quantitative theory of the local strain deviation.

On the other hand, we are still far from to understand the complete mechanism. In each point the strain is described by a tensor which is a three by three matrix; when we speak of strain deviation, what we really mean? The simple intensity of principal components may be not enough to describe the physiological phenomena.

However the exposed criteria can drive bio-engineer and orthopaedic surgeon through the design or the evaluation of a cementless hip prosthesis. Higher is the alteration of the physiological biomechanical environment, stronger will be the reaction of the bone tissue to alloplasty.

REFERENCES

1.   Andrisano A. O., Cassese F., Viceconti M., Toni A. and Giunti A. (1990) A strain-gauge study on alterations of the proximal femur strain pattern induced by a cementless hip arthroplasty. *Presented at the First World Congress of Biomechanics, S. Diego USA September 1990*

2.   Cassese F., Mazzinghi A., Stefani G., Viceconti M., Toni A. and Sudanese A. (1990) A parametric F.E.M. study of the femoral strain pattern modifications due to cementless hip arthroplasty. *Presented at the First World Congress of Biomechanics, S. Diego USA September 1990*

3.   Cowin S. (1984) Mechanical modeling of the stress adaptation process in bone. *Calcif. Tissue Int.* **36** , S98-S103.

4.   Currey J. (1987) The evolution of the mechanical properties of amniote bone. *J. Biomechanics* **20** , 1035-1044.

5.   Galilei G. (1981) Discorsi e dimostrazioni mathematiche intorno a due nuove scienze. In: Treharne R., Reviw of Wolff's law and its proposed means of operation. *Orthopaed. Rev.* **10**, 35-40.

6.   Lanyon L. (1987) Functional strain in bone tissue as an objective, and controlling stimulus for adaptive bone remodelling. *J. Biomechanics* **20** , 1083-1093.

7. Roesler H. (1987) The history of some fundamental concepts in bone biomechanics. *J. Biomechanics* **20**, 1025-1034.

8. Rubin C. and Lanyon L. (1987) Osteoregulatory nature of mechanical stimuli: function as a determinant for adaptive remodelling in bone. *J. Orthop. Res.* **5**, 300-310.

9. Skerry T., Bitensky L., Chayen J. and Lanyon L. (1988) Loading-related reorientation of bone proteoglycan in vivo. Strain memory in bone tissue? *J. Orthop. Res.* **6**, 547-551.

10. Smith M. , Rambaut P. , Vogel J. and Whittle M. (1977) Bone mineral measurement - Experiment M078. Biomedical results from Skylab. NASA SP-377, US Government Printing Office, Washington DC, 183-191.

11. Toni A., Sudanese A., Ciaroni D., Dallari D., Greggi T. and Giunti A. (1990) Anatomical ceramic arthroplasty (AN.C.A.): preliminary experience with a new cementless prosthesis. *Chir. Organi Mov. LXXV, 81-97, 1990.*

12. Uhthoff H., Sekaly G. and Jaworski Z. (1985) Effect of long-term nontraumatic immobilization on metaphyseal spongiosa in young adult and old beagle dogs. *Clin. Ortho. Rel. Res.* **192** , 278-283.

13. Viceconti M., Seireg A. (1989) A generalized procedure for predicting bone mass regulation by mechanical strain. *Accepted by Calcif. Tissue Int.*

DEVELOPMENT OF BONE-IMPLANT INTERFACE DURING HEALING PHASE OF DENTAL IMPLANTS

I. BINDERMAN, *D. GAZIT, Y. CHAIT, R. YAHAV, S. WEISMAN,
Y. EILON AND N. FINE
Department of Dentistry and Hard Tissues Laboratory,
Ichilov Hospital, Tel Aviv University and
*Department of Pathology, Hadassah School of Dental Medicine,
Hebrew University, Jerusalem

INTRODUCTION

Dental implants serve as artificial abutments to support a prosthetic appliance. Occlusal function and load is usually transmitted from the prosthetic apliance through the implant, to hard tissues touching the implant. It is therefore crucial to achieve an intimate interface of the implant with most possible dense bone. The structure of the mandible and maxilla consists of cortical bone which has considerable strength because of its high density and of trabecular bone which is much less dense or strong. The amount of cortical and trabecular bone varies due to sex, age, anterior versus posterior part of the jaw and mandible versus maxilla. It has also individual variations. Anatomically the peripheral part of the jaws consists usually of cortical bone and the inner part is trabecular. It is therefore proper to expect that implant modalities are designed to achieve maximal initial anchorage in the cortical bone and prevent if possible irreversible resorption of this bone.

Most of the published studies which describe development of interface using different implant modalities and different implant materials are based on the long bone model system [1, 2], rather than the mandible or maxilla bones. The high osteogenic response of the marrow to injury and implant surgery is well established for the long bone model of young and mature animals, and not in the mandible. Differences in the osteogenic capacity of marrow of long bone in comparison to mandible or maxilla emphasize the importance in establishing more data on the marrow responsiveness of the jaw bones to implant surgery.

In this study we tested the hypothesis that surgical placement of endosseous implants into the jaw bone are enhancing repair mechanisms to preferentially deposit bone on the implant material. We compared two implant modalities, namely Blade type and Cylinder-Screw type, both fabricated from pure titanium metal. We tested whether the osteogenic capacity of the extraction socket will develop more bone at implant interface and its density. The study was peformed on six local strain dogs. New bone formation was evaluated by tetracycline labelling and bone strength at interface was estimated from microradiographs of plastic embedded section of mandibles with the implant in place.

MATERIALS AND METHODS

The present study was performed on 6 dogs of local strain which were raised in the Ichilov Animal House. Each dog was anasthesized using Combelen (1ml/20kg) and Nembutal (1ml/5kg) and local injection of 2% Lidocain prior to surgery. Premolar teeth were extracted on one side of the lower jaw. Three months later the same procedure was repeated on the other side of the mandible. In addition, surgical placement of one Screw and one Blade type Universal titanium implants was performed on both sides of the mandible. One set of implants placed into the residual ridge and the other set (same implants) into the fresh extraction site. The soft tissue was sutured to cover the exposed mandible bone. The implant posts were exposed into the oral cavity and not submerged. The sutures were removed after one week and the tissues cleaned of food debris and plaque. Each week similar tissue treatment was given under light sedation. After 2-3 weeks the soft tissue healing proceeded with no signs of swelling or inflammation of soft surrounding tissues. In this study, no loosening of implants was observed. After 7 weeks Tetracycline tablets were given orally during 3 days, a pause of 6 days and again Tetracycline tablets during 5 days, to achieve double lable of the newly formed bone. A week later the dogs were sacrificed and the mandibles removed and fixed in 10% Buffered Formalin. The fixed mandibles were cut into blocks which contained the implants, x-rayed for orientation, embedded in plastic and sectioned into 3-5 mm blocks including the titanium implant. Each of the sections was microradiographed by Faxitron x-ray apparatus and screened under phase contrast microscope with ultraviolet attachment to visualize the Tetracycline label. The sections were also stained with Toluidine Blue to visualize the cellular and matrix components of the bone. The x-rays obtaind were further analysed by videocamera attached to IBM personal computer for image analysis. This type of analysis was performed on 8-10 different areas in the cortical bone, trabecular bone and along the interface of bone-implant. Every time 1 mm x 2 mm surface area was analysed. It is assumed that the x-ray radioopacity is in direct relationship to the bone density in individual x-rays. The purpose of this analysis was to compare, in each radiograph, the relative bone density at the implant interface to that of cortical and trabecular bone.

RESULTS

In the present study a comparison of implants placed immediately into fresh extraction sites and in residual ridges (3 month after extraction) with regard to development of bone-implant interface was investigated. Tetracycline label was a measure for the amount and pattern of new bone formation surrounding the implants. Microradiographs were analysed for bone density at implant interface, in comparison to cortical and trabecular bone densities. A comparison between Blade type and Screw type Universal Titanium implants was done in order to understand the effect of different implant modalities on the development of bone at implant interface.

Figure 1 shows tetracycline labeling pattern in the bone surrounding implants placed in fresh extraction sites and in residual ridge. A striking difference in the tetracycline uptake between the two groups is seen. It is almost impossible to oberve discrete lines of labeling in the fresh extraction group and the entire extraction site is heavily labelled showing very active bone formation. It seems that the pattern of the label is dependent on the root forms. In comparison, the surgical site of the residual ridge is labelled very little and lines of tetracycline label seperated by unlabelled bone is observed as expected (Figure 1). Also, in those implants the pattern of labelling is not consistent and does not follow the geometry of the implants, either Screw or Blade type. Even in those sections where tetracycline label was in close proximity to implant surface, representing newly formed bone, this bone seems to develop from bone periphery approaching the implant surface. This observation may suggest that the titanium material does not inhibit bone formation toward its surface, however, it does not attract bone cells preferentially attaching to its surface to form new bone.

Bone density was analysed from microradiography films prepared from sections used for histology (Figure 2). The microradiographs were analysed through a video camera attached to an IBM personal computer programmed for image analysis. The range of radioopacity was related to bone density. Squares of 1 mm x 2 mm were measured in 8-12 different areas of the cortical bone and trabecular bone and the bone along the implant interface. Table 1 shows the relative bone density of trabecular and interface bone in comparison to cortical bone. The interface bone density is a function of

Figure 1. Tetracycline labeling of bone.

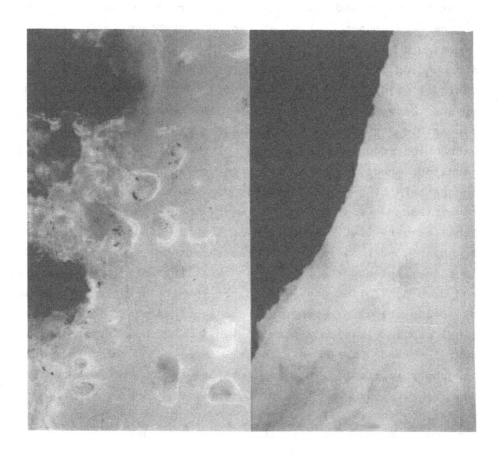

a) Extraction site   b) Residual ridge site

Note, the extensive uptake in the extraction site (a) 3 months
after implant insertion; a) and b) in the same dog.

Figure 2. Microradiography of sections prepared for histology.

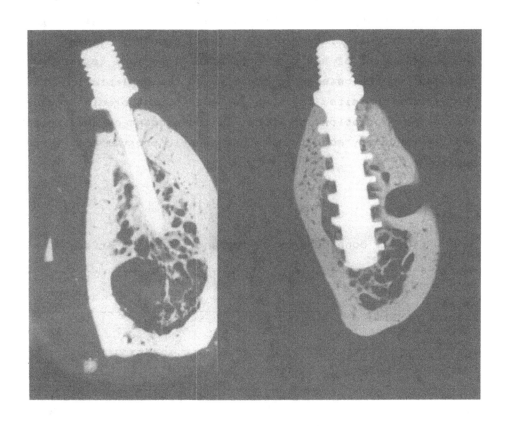

a) Blade type   b) Screw type

how much of the implant surface is in close proximity to
cortical or trabecular bone and therefore never reaches
cortical bone density.   It seems that the geometry of the
implant does not affect bone density at the interface, at this
stage of healing.   Bone density of fresh extraction sites is
however higher than the residual ridge sites.   Therefore, the
interface density of the extraction site group is also higher
than the residual ridge group.   These results suggest that
the development of new bone at implant interface is very much
dependent on the osteogenic capacity of the implantation site.
The extraction site is such a potent osteogenic tissue.
Also, the formation of bone is independent of the implant
modality and very much follows the genetic program of tissue
repair.

TABLE 1

Relative Bone Density of Implant - Bone Interface

	Cortical Bone	Trabecular Bone	Interface Bone
Extraction Site (Screws & Blades)	162 ± 8 100%	94 ± 23 58%	100 ± 18 62%
Residual Ridge Site (Screws & Blades)	147 ± 6 100%	50 ± 22 34%	53 ± 18 36%
Screw Type	156 ± 8 100%	91 ± 23 58%	98 ± 25 63%
Blade Type	135 ± 8 100%	66 ± 22 49%	96 ± 24 71%

All figures are the mean ± S.E. of 8 - 12 estimations.

DISCUSSION

The mandible consists of dense cortical bone which maintains
its shape with varying thickness throughout its structure and
trabecular bone which comprises its internal structure in
close relation to marrow tissue.   The mechanic support of an

endosseous implant is directly related to bone density at the implant interface. It is therefore inevitable that in order to achieve maximal mechanical support the implant surface should be in close proximity to cortical bone. Another possibility will be to stimulate development of cortical bone along the implant interface. Chiarenza [1] pointed out that usually in animal studies and in humans narrow implant modalities like Blades which are placed into jaw bone will usually develop tissue comprising of trabecular bone and fibrous marrow repair tissue, at implant interface. While, wide implant modalities like cylinder type implants will be largely embedded in cortical dense bone, comprising most of the implant interface (1). He observed that cortical bone, Haversian systems and dense lamella, do not form within the marrow cavities around any implant form. Branemark and his coworkers [2] suggested that following a specific implant procedure will produce direct bone implant contact along the entire implant interface, termed by them as osseointegration. However, they were unable to prove that even by following their procedure precisely dense cortical bone will always develop and surround the implant at all its aspects. It is noteworthy to point out that most of their histologic studies were performed on long bones where implant surgery of the marrow produces osteogenic response, independently of the type of implant. However, the marrow cavity of mandible is poor of osteogenic elements. In our studies we were unable to show that pure titanium implants of Screw or Blade type stimulate to form cortical bone where usually trabecular bone develops. It seems that using fresh extraction sites will increase the amount of new bone formed at implant interface. Unless the implant surface will attract specifically osteoprogenitors which upon their attachment with the implant surface will proliferate and differentiate into bone, the bone implant interface will represent at most the typical structure of the bone where the implant is placed.

ACKNOWLEDGEMENT

We would like to thank Dr. J. Kaufman, Orthopedic Research, Mount Sinai Hospital, N.Y.C., for his help in measurements of bone density.

REFERENCES

1. Chiarenza,   A.R.,   Retrospective   observations   on   the influence of bone type in determining the nature of bone implant interface.   Int. J. Oral Implant, 1989, 6:43-48.

2. Branemark, P.I., Osseointegration and its experimental background.   J. Prosth. Dent. 1983, 50: 399-410.

PREDICTION OF THE LONG-TERM EVOLUTION OF THE BONE STOCK AROUND A TAILOR-MADE HIP IMPLANT

J. Vander Sloten[*] and G. Van der Perre
Katholieke Universiteit Leuven
Division Biomechanics and Engineering Design
Celestijnenlaan 200A
B-3030 Heverlee (Belgium)

[*] Research Assistant of the Belgian National Fund for Scientific Research

ABSTRACT

An iterative algorithm has been written to simulate changes in bone geometry (the so-called external remodelling) when changes occur in the stress distribution within the bone. The core of this process is the finite element program ANSYS-PC; the mathematical model for external remodelling was programmed in PASCAL. The research was aimed at prediction of bone stock evolution in the case of cementless hip implants and especially the tailor-made prosthesis IMP (Instant Mulier Prosthesis). We investigated equivalent thickness two-dimensional models of a cylindrical model, a conical model and a realistic geometry of a proximal femur with implant. Phenomena that are clinically observed, such as calcar resorption or distal callus formation, were also observed in our computer simulations.

INTRODUCTION

Basically, Wolff's Law states that most bony structures within the body of humans and animals are optimally adapted to their task and that they have the capability of maintaining this optimality. In this view, we like to use the concepts of 'statical optimality' and 'dynamical optimality'.

It will be shown that the quantitative models that exist for external remodelling can be used to predict changes in bone geometry. This was done for the first time by Hart in 1983, who simulated external remodelling in complex three dimensional finite element models. Dalstra simulated both internal and external remodelling in two dimensional finite element models of a cylindrical model for the proximal femur [6].

Though the quantitative models that are currently used have many uncertainties within them, the evolutions in bone geometry and structure that will be simulated, will show to match phenomena that are observed in clinical practice.

MATERIALS AND METHODS

The algorithm

Given the geometry of a structure, loading upon a structure, material properties and boundary conditions, a finite element program can calculate the resulting stress distribution. Using a mathematical model for external remodelling, the existing geometry can be changed and a new stress distribution can be calculated. This process is continued until equilibrium has been established. Figure 1 illustrates the feed-back loop of the external remodelling.

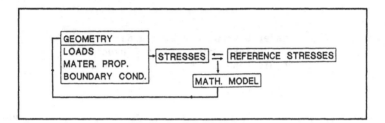

Figure 1. The feed-back loop of external remodelling.

We have investigated the effect of the bending stiffness E.I of a tailor-made cementless hip implant upon external remodelling in the proximal femur. 'Tailor making' refers to a procedure where a cementless hip implant is milled intra-operatively, based upon a mould of the medullary cavity [14]. The geometry was simulated in a cylindrical, a conical and a realistic model. Stresses were calculated by means of equivalent thickness two-dimensional finite element models with a side plate. The equivalent parameters are based upon equivalence of bending stiffness and medio-lateral compressive stiffness [15].

We have programmed the mathematical model of Cowin to predict changes in bone stock around a tailor-made cementless hip implant. The finite element program ANSYS-PC (University Version) was used on an IBM-PS/2 model 80-111.

The algorithm is based upon the formula, proposed by Cowin :

$$U = C_{ij}(n,Q).[E_{ij}(Q)-E_{ij}^{\,o}(Q)]$$

where n is the direction of the normal to the bone surface at a surface
   point Q;
   U is the translation speed of the surface in the n-direction;
   $E_{ij}(Q)$ is the actual strain at a surface point Q (normally the
   dominant strain component is taken for $E_{ij}(Q)$);
   $E_{ij}^{\,o}(Q)$ is a reference value for the strain at a point Q when no
   remodelling occurs;
   $C_{ij}(n,Q)$ is a surface remodelling rate coef-ficient, dependent
   upon the direction n and the point Q under consideration, but not
   dependent upon the strain value.
[3], [4].

In this formula, the strain is used as the variable that controls the adaptation mechanism. This corresponds with the ideas of Lanyon who states that strain is something that is actually present in a material and can therefore be captured by bone cells, whilst stress is something that is mathematically derived from strains [11]. Other investigators have used the elastic energy density as control variable [9]. We have chosen to use the longitudinal stress as control variable, because the component of the stress due to bending is dominant over the other stresses ; because it is proportional to the strain and because stresses are a direct output of the finite element program ANSYS-PC, whilst strains are not.

The algorithm assumed that, at each cross section, only the medial and lateral points were able to move in- or outwards. In our cylindrical and conical models, the cross sections were assumed to remain circular with variable outer diameter. In these models the equivalent thickness of all elements was recalculated at each iteration step.

The 'Lazy Region'

The early mathematical models of Cowin did not take into account the effect of a 'Lazy Region'. According to Carter, there is not just one value of the control variable where there is no remodelling activity, but there is a region where remodelling activity is zero [1]. We have assumed a lazy region of 10% of the local reference stress value. The remodelling activity that is assumed is then as shown in figure 2.

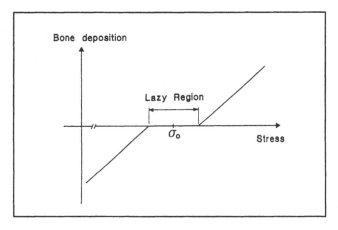

Figure 2. The lazy region around the equilibrium state.

The dimensional change per time unit

The absolute change in dimension in a given point is found by multiplying the surface rate U with a time step δT. In our algorithm, the surface remodelling rate coefficient C_{ij} and δT were taken together in one coefficient. At each iteration step, this coefficient was determined by the demand that, at the location where there was the greatest difference between the actual stress and the reference stress, there would be a change in bone stock (resorption) that was a predetermined fraction of 10 percent of the remaining bone stock. By this means, convergence is assured and the bone thickness cannot become negative.

The finite element models

Three distinct degrees of approximation were considered : a cylindrical model; a conical model and a more realistic model of the proximal femur. The cylindrical model was chosen as starting point, because it contains the basic elements of stress transfer that are present in a proximal femur and allows to understand the mechanics of stress transfer [8]. The conical model was meant as an intermediate step between the very simple cylindrical model and the more realistic model. The results however have shown that its usability is very limited. The realistic model finally took into account most structural details that are present in the proximal femur.

The meshes are shown in figure 3. They consists of quadratic rectangular elements, which are a good choice in a structure which is predominantly loaded in bending. The interface is simulated by means of gap-elements.

The stem of the cylindrical model is loaded by a pure bending moment of 74.95 Nm at its proximal end and the hollow cylinder is clamped at its distal end.

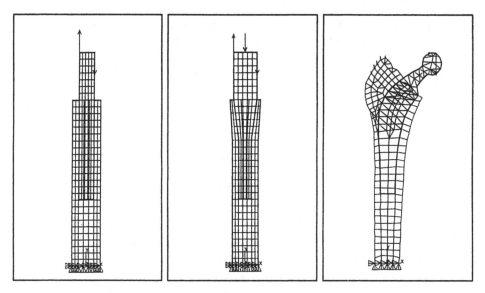

Figure 3. The mesh, the load vectors and the boundary conditions of the cylindrical model, of the conical model and of the realistic model.

The implementation of the feed-back algorithm requires the knowledge of the reference stress at each point. In the cylindrical models this reference stress distribution is calculated by assuming that the total bending moment acts on the proximal side of the hollow cylinder, representing the femur, with no prosthesis being present.

In the conical model, the prosthesis was loaded by a vertical force of 655 N and a bending moment of 74.95 Nm and the distal boundary of the bone was fixed as boundary condition.

Values for the reference stress at each node were obtained by assuming that the normal force and bending moment acted on the proximal end of the bone, without prosthesis.

The geometry and inertial properties of a realistic model for a proximal femur are based upon a specimen that was obtained from the Orthopaedic Hospital St. Barbara, Pellenberg. The geometry of the frontal view was derived from a frontal radiography and the inertial properties were derived from a series of CT-scans, along the height of the femur.

The loading of the realistic model reflected one-legged stance, with load vectors on the prosthetic head and on the trochanter major, caused by the abductor muscles. The reference stress distribution was calculated by completing the bone geometry of the reamed femur with a femoral neck and head.

RESULTS

The cylindrical model
The stem, fitting exactly the shape of the cavity, was assumed to be made of the following materials :

    Stainless steel : E = 210 GPa
    Titanium alloy : E = 110 GPa
    Aluminium   : E = 70 GPa
    Bone   : E = 17 GPa

The results of the iterative process are shown here for a prosthesis, made of a titanium alloy and of an iso-elastic material. They are shown graphically (figures 4 and 5). Two different graphs show the following variables versus the distance along the proximal femur as a function of the iteration step : the longitudinal stress at the medial side (fig. 4a and 5a) and the bone deposition medially (fig. 4b and 5b). Negative values for bone deposition are to be interpreted as bone resorption. The bold line at the bottom of each figure indicates the length of the stem in the bone. The iteration process stopped at 14 iterations, after some five hours of computer time.

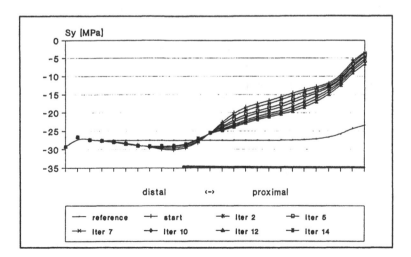

Figure 4a. The evolution of the longitudinal stress at the medial side
throughout the iteration process (E=110 GPa).

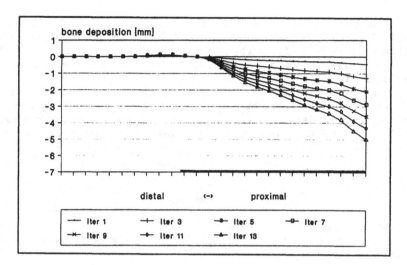

Figure 4b. The evolution of the bone deposition at the medial side throughout the iteration process (E=110 GPa).

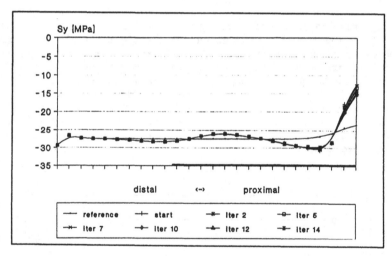

Figure 5a. The evolution of the longitudinal stress at the medial side throughout the iteration process (E=17 GPa).

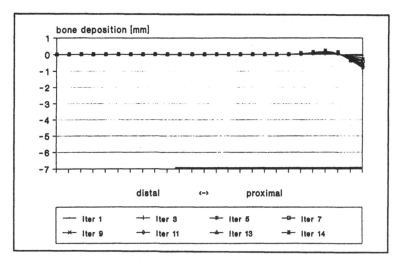

Figure 5b. The evolution of the bone deposition at the medial side
throughout the iteration process(E=17 GPa).

The conical model

The stress concentration in the bone around the transition between the
cylindrical and the conical part dominated the whole stress distribution.
The mentioned stress concentration is not due to the normal force that was
applied in the conical model and not in the cylindrical model; it is due
to the transition between the cylindrical and the conical part of the
geometry, which is clearly not a realistic approximation of the proximal
femur.

The realistic model

The bone deposition throughout the iteration process is shown in figures 6
and 7 for stems made in a Titanium-alloy and an iso-elastic material.

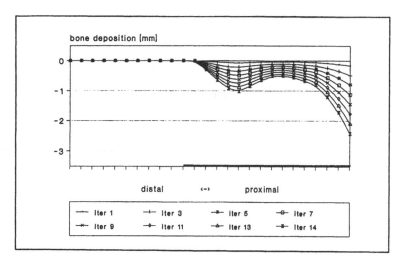

Figure 6. The evolution of the bone deposition at the medial side
throughout the iteration process (E=110 GPa).

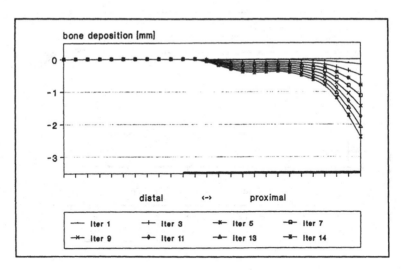

Figure 7. The evolution of the bone deposition at the medial side
throughout the iteration process (E=17 GPa).

DISCUSSION

Cylindrical model
From a comparison of the results of the cylindrical model for the
different materials it is clear that the bone resorption in the proximal
region (the calcar zone) is more massive with the high modulus materials
and is smaller with the isoelastic bone. Though, bone resorption in the
calcar zone exists with all materials, which is clearly a result of the
absence of a collar. Due to stress concentrations in the distal region
near the stem tip, the stress is higher than the reference stress, leading
to a limited amount of bone deposition, which is also observed clinically.
Only in the case of the isoelastic stem there is a second overstress
region in the bone more proximally (figure 5).

The amount of bone resorption in the mid-stem region can also be
predicted by an analytical model : the load transfer from a prosthetic
stem to the bone can be subdivided into three regions.

In the proximal part, the total moment M is gradually divided between
bone and stem. In the mid-stem region bone and prosthesis each carry a
fraction of the bending moment, according to their own flexural rigidity
and finally, in the distal region, the moment is transfered to the bone
alone [8]. The bending stress in the outer fiber of the femoral bone
versus the outer radius of the bone is shown in figure 8. This is done
for different flexural rigidities of the implant and the result holds for
the mid-stem region of the bone.

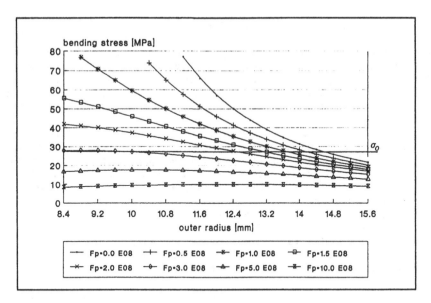

Figure 8. The bending stress in a femur versus the outer radius of the
bone as a function of the flexural rigidity of the implant.

This shows clearly that the bending stress raises when the outer
radius of the bone is decreased, given a certain flexural rigidity of the
stem. Furthermore, for some values of the flexural rigidity of the stem,
it appears that the reference stress (order of magnitude 30 MPa) is never
reached, even not when the bone is totally resorbed.

Realistic model
The results of the realistic model show that, in the mid-stem region, the
bone resorption and deposition are small, compared to the cylindrical and
conical models. The same holds for the longitudinal stress, which shows
also little variation throughout the process. An explanation was found by
considering the influence of the force on the prosthetic head and the
trochanter major separately.

The force on the trochanter major is almost completely carried by the
bone, independent of the presence of an implant, except near the stem tip.
The resulting difference in stress between actual and reference stress is
mainly caused by the influence of the force on the femoral or prosthetic
head. This is in contrast with the cylindrical and conical model where
all forces and bending moments act upon the prosthesis.

The lateral side
The realistic model is the only one that allows us to tell something about
the adaptation that is to be expected at the lateral side when implanting
a proximal femur with an endoprosthesis, because this model is not
symmetrical. The effects on the lateral side are minimal : there is only
minimal bone resorption around the stem, due to stress shielding.

CONCLUSIONS

The implementation of the mathematical model of Cowin for external remodelling has allowed us to simulate the evolution of the bone stock around a hip implant, made in materials with varying elastic properties and implanted without bone cement. Some general tendencies were noticed in the three models we have employed : a cylindrical, a conical and a realistic model.

a. Stress shielding around medullary stems leads to bone resorption in the mid-stem region, which is more pronounced with implants made of high modulus materials. Our simulation predicted that, depending upon the geometry, there exists an upper limit for the elastic modulus of the stem, above which the reference stress can never be reached, independent of the amount of bone resorption.

b. Bone resorption medially and proximally (the so-called calcar resorption) appears with all stems, independent of their elastic modulus. This problem is inherent to the geometry of a stem without a collar that provides a firm support on the calcar region.

c. Bone reaction at the lateral side is minimal.

Bone reactions that are predicted with our simulation correspond qualitatively with phenomena that are observed clinically (mid-stem resorption : [10], [12]; calcar resorption : [2], [7]; distal callus formation : [13]. Quantitative agreement was not checked, but is expected to be limited because of the limitations of the finite element models that were employed; because of the fact that only external remodelling was considered and because the mathematical model for external remodelling that was used is by no means an exact representation of reality. Furthermore a lot of other parameters act upon the bone and determine its behaviour : the surgical procedure as well as genetic, hormonal and nutritional factors also influence bone reaction and, though it will be necessary to introduce them into the simulation programs, this will be very difficult and require further research, including well-defined and controlled animal experiments.

REFERENCES

1. Carter, D.R., Harris, W.H., Vasu, R. and Caler, W.E., 'The mechanical and biological response of cortical bone to in-vivo strain histories.', ASME Symposium, The mechanical Properties of Bone, 1981, pp. 81-93.

2. Charnley, J., 'Low friction arthroplasty of the hip.', Springer Verlag, Chapter 6 : 'Long term radiological results.', 1974.

3. Cowin, S.C. and Van Buskirk, W.C., 'Surface bone remodelling introduced by a medullary pin.', J. Biomechanics, Vol. 12, 1979, pp. 269-276.

4. Cowin, S.C., Hart, R.T., Balser, J.R. and Kohn, D.H., 'Functional adaptation in long bones : establishing in vivo values for surface remodelling rate coef-ficients.', J. Biomechanics, Vol. 18, No. 9, 1985, pp. 665-684.

5. Cowin, S.C., 'Bone remodelling of diaphyseal surfaces by torsional loads.', J. Biomechanics, Vol. 20, No. 11/12, 1987, pp. 1111-1120.

6. Dalstra, M., 'De ontwikkeling van een tweetal computermodellen voor het simuleren van adaptieve botremodellering.', Eng. Thesis, A.Z. Nijmegen, The Netherlands, Afdeling Orthopedie, 1986.

7. Griss, P., Heimke, G., Werner, E., Bleicher, J. and Jentschura, G., 'Was bedeutet die Resorption des Calcar femoris nach der Totalendoprothesenoperation der Hüfte?', Arch. Orthop. Traumat. Surg. Vol. 92, 1978, pp. 225-232.

8. Huiskes, R., 'Some fundamental aspects of human joint replacement.', Ph.D. Thesis, T.H. Eindhoven, The Netherlands, 1979.

9. Huiskes, R., Weinans, H., Grootenboer, H.J., Dalstra, M., Fudala, B. and Slooff, T.J., 'Adaptive bone-remodelling theory applied to prosthetic-design analysis.', J. Biomechanics, Vol. 20, No. 11/12, 1987, pp. 1135-1150.

10. Indong, O. and Harris, W.H., 'Proximal strain distribution in the loaded femur.', J. Bone and Joint Surgery, Vol. 60A, No. 1, 1978, pp. 75-85.

11. Lanyon, L., 'Functional strain in bone tissue as an objective and controlling stimulus for adaptive bone remodelling.', J. Biomechanics, Vol. 20, No. 11/12, 1987, pp. 1083-1093.

12. Lewis, J.L., Askew, M.J., Wixson, R.L., Kramer, G.M. and Tarr, R.R., 'The influence of prosthetic stem stiffness and of a calcar collar on stresses in the proximal end of the femur with a cemented femoral component.', J. Bone and Joint Surgery, Vol. 66A, No. 2, 1984, pp. 280-286.

13. Morscher, E.W. and Dick, W., 'Cementless fixation of 'Isoelastic' hip endoprostheses manufactured from plastic materials.', Clin. Orthop. and Related Research, No. 176, June 1983, pp. 77-87.

14. Sleeckx, E., Mulier, J.C., Mulier, M., Steenhoudt, H., Cauwe, Y., Van Audekercke, R. and Van der Perre, G., 'A new technique for the design and fabrication of custom-made hip prostheses.', Proc. European Congress on Biomaterials, 1986.

15. Vander Sloten, J., 'The functional adaptation of bones in vivo and consequences for prosthesis design.', Ph.D. Thesis, K.U.Leuven, Division Biomechanics, 1990.

A COMBINED ANATOMICAL AND BIOMECHANICAL APPROACH
TO THE SACRO-ILIAC JOINTS

A. VLEEMING*, PhD, R. STOECKART*, PhD, C.J. SNIJDERS**, PhD
J.P. VAN WINGERDEN**
*Department of Anatomy
**Department of Biomedical Physics and Technology
Erasmus University Rotterdam
P.O. Box 1738, 3000 DR Rotterdam, The Netherlands

ABSTRACT

The amount of friction between the articular surfaces of the sacro-iliac joints (SI-joints) was determined and related to the degree of macroscopical roughening. Results show that articular surfaces with both coarse texture and ridges and depressions have high friction coefficients. The influence of ridges and depressions appears to be of greater influence than coarse texture. The data are compatible with the view that roughening of the SI-joint concerns a physiological process.

Movement in the SI-joints was measured in preparation of embalmed cadavers of elder humans and correlated with a radiographic survey. The pelvis was fixed at the fifth lumbar vertebra. The connections between sacrum and fifth lumbar vertebra were spared, as were the surrounding ligaments. To induce movement, forces were directed at the acetabula.

In the sagittal plane both ventral rotation (as part of nutation) and dorsal rotation (as part of contranutation) could be demonstrated. Most SI-joints were mobile, allowing - for the combination of nutation and contranutation - total rotation of up to 4°. Larger rotation can be expected in young people. The SI-joint with the lowest mobility showed radiographically marked arthrosis. The importance of intra-individual differences is emphasized.

Based on studies of embalmed specimens, the sacrotuberous ligaments are considered to be important structures in the kinematic chain between the pelvis and vertebral column. Muscles attached to these ligaments, such as the gluteus maximus, and in some individuals the piriformis and long head of the biceps femoris, may influence movement in the SI-joints. The effect of load application to the sacrotuberous ligament was studied on rotation in the SI-joint. It was shown that load application along the direction of hamstrings and gluteus maximus muscles significantly diminished ventral rotation of the sacrum. The results imply that loading the sacrotuberous ligament restricts nutation of the sacrum. Consequently, muscles attached to the sacrotuberous ligaments, such as gluteus maximus, and in certain individuals the long head of the biceps, can dynamically influence movement and stability of the SI-joints.

INTRODUCTION

Our understanding of the human locomotor system is significantly increased by using biomechanical models. The present study focusses on the relation between biomechanical and anatomical aspects of the pelvis, especially the SI-joints. These joints connect the iliac part of the pelvis with the spinal column, <u>viz.</u> the sacrum (Fig. 1). To enlarge the possibility of communication with professionals mostly working in a non-medical field, emphasis is laid here on three topics: 1. relevant historical data, 2. sexual differences and 3. basic propositions with respect to form-function relationships of the SI-joints.

1. A historical review of the morphology of sacro-iliac mobility

From Hippocrates [1] until Vaesalius [2], it has been suggested that the

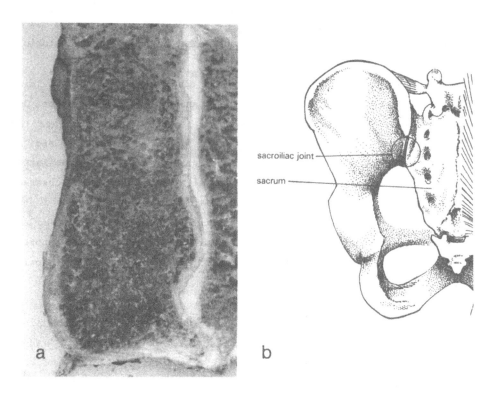

Figure 1. a. Section of the SI-joint. The joint is conspicuous by the presence of two layers of cartilage (white). The cartilage at the sacral side (right) is thicker than the cartilage at the iliac side (left).
b. Dorsal view of the left side of sacrum and pelvis.

SI-joint is mobile during pregnancy only. Paré [3] confirmed this by examination of corpses. Diemerbroek [4], however, claimed that the SI-joint is usually mobile, both in women and men. Albinus [5] observed that the SI-joint has a synovial membrane and concluded that it must be mobile. In 1864, Von Luschka [6] describes the joint as a real diarthrosis, i.e. a mobile joint with a joint cavity between two bony surfaces. Zaglas [7] investigated sacro-iliac mobility in corpses. He concluded that most sacral movement takes place around a transversal axis, situated at the level of the second sacral vertebra.

Upon careful analysis of topographical anatomy, as well as investigation of corpses, Duncan [8] concluded that the generalized pivot of the SI-joint must be localized at the level of the iliac tuberosity. This tuberosity is a bony structure located behind the auricular part of the SI-joint (in 1984 by Bakland and Hansen [9] described as the iliac aspect of the axial joint). Meyer [10] and Klein [11] supported this opinion. Duncan [8] suggested to coin sacro-iliac rotations around a transversal axis as 'nutation' (ventral tilting) and 'contranutation' (dorsal tilting).

In 1949 Testut and Laterjet [12] concluded that the SI-joint contains a freely mobile ventral aspect and an ossified dorsal aspect. One could speak of an 'diarthro-amphiarthrosis', i.e. a joint that has the characteristics both of a freely mobile joint (diarthrosis) and an ossified joint (synarthrosis). Bakland and Hansen [9] described the latter dorsal aspect of the SI-joint as the 'axial joint'. Since they observed the presence of cartilage, they concluded that movement may occur in the axial joint as well. Ehara et al. [13], using CT-scanning, observed the axial joint in not more than 13 out of hundred test persons. They reported that the axial joint may be present at birth, just as a true diarthrosis, but that it also can develop postnatally, in which case the joint is supposed to be fibrocartilaginous.

According to some anatomists (Faraboeuf [14]; Strasser [15]; Fick [16]; Palfrey [17]), the iliac articular surface has a central ridge, corresponding to a groove on the articular surface of the sacrum.

In 1954, Weisl [18] described the joint as consisting of two condyles, forming what may schematically be regarded as a sellar joint. However, several investigators (Brooke [19]; Schuncke [20]; Solonen [21]) have shown that congruence is rare in the SI-joint. Strong reliance on simple models of a joint, therefore, can be misleading for the SI-joint.

Moreover, Fischer et al. [22] demonstrated great intra- and inter-individual sacro-iliac variability. This was confirmed by Bakland and Hansen [9].

Solonen [21] described the articular surface as 'nodular and pitted' and regarded this as pathological. He tried to come to an understanding of sacro-iliac function by conceiving the sacrum as a wedge between the iliac bones. This form, according to Solonen [21], guarantees a strong connection.

2. Sexual differences and the roughening of sacro-iliac surfaces

Female sacro-iliac mobility may be functional in allowing passage for the child during labour. Sacro-iliac curvature is usually less pronounced in women than in men, a fact which is supposed to allow for a higher mobility (Weisl [18]; Solonen [21]).

During pregnancy, the sacro-iliac fibrous apparatus loosens under the influence of relaxine (a hormone produced by the ovary), and some symphysiolysis occurs. Both factors result in increased sacro-iliac mobility. Hypermobility of the joint may lead to complaints (Bonnaire and Bué [23]; Brooke [19]; Hisaw [24]; Chamberlain [25]; Heyman and Lundquist [26]; Abramson et al. [27, 28]; Thorp and Fray [29]; Borell and Fernstrom [30]; Walheim et al. [31, 32]).

Research by Walheim et al. [31, 32] and Snijders and Snijder [33] indicates a positive effect of a pelvic belt ('Trochantergurt') on sacro-iliac hypermobility in pregnant women. Both studies report pain reduction. This observation could serve to explain why in several cultures (e.g. Indonesia, Turkey, Morocco) an elastic corset is worn at the S2-level from the 6th month of pregnancy onwards. In the male pelvis, contrary to the female, there appears to be no functional role for pronounced sacro-iliac mobility. Especially in men, 'blocking' of this joint has often been described (Lynch [34]; Jackson [35]; Haggart [36]; Cazeviel [37]).

Moreover, particularly in males above the age of 35, macroscopic sacro-iliac changes have been described (Stewart [38]). Without justification, terms such as 'arthrotic processes' (British literature, e.g. Brooke [19]) or 'arthritis' (American literature, e.g. McCarty [39], are often used in the sacro-iliac literature in referring to these changes.

In 1892, Braune and Fischer [40] related the position of the centre of gravity to sacro-iliac function. They assumed that a change in the localization of the centre of gravity is related to a change in sacro-iliac function. The larger the distance between the SI-joint and the vertical line through the centre of gravity, the less stable the joint, since rotational torque increases as a function of the length of the lever arm (see Fig. 2).

It is our opinion that the large distance between the assumed rotational pivot of the SI-joint and the vertical line through the centre of gravity is a major cause of the development of the specific form of the SI-joint, which secures its necessary stability.

Several authors indicated a different position of the centre of gravity in men and women. In women the vertical line through the centre of gravity is supposed to pass directly in front of, or through the SI-joint whereas in men its position is further ventral (Braune and Fischer [40]); Tischauer et al. [41]; Bellamy et al. [42]). This would imply a greater lever arm in men than in women, resulting, of course, in higher loads on the joints (Fig. 2).

As a result, the male SI-joint will become stronger, with restricted mobility, if, indeed, the surmised difference in the localization of the centre of gravity exists. More research will be needed before this can be decided. At the moment, the whole discussion is based on rough empirical estimates.

Apart from these differences between men and women, the load carrying surface of the female SI-joint is usually smaller, and the position of the sacrum is usually more horizontal (Derry [42]; Brooke [19]; Sashin [44]; Schuncke [20]; Solonen [21]).

Brooke [19] reported roughening of the articular surfaces in 70% of male preparations above the age of 76; this roughening was ascribed to arthrotic processes. In female preparations, Brooke, however, found no evidence of roughening. Stewart [38] concluded on the basis of an extensive investigation using material from different races that female sacro-iliac ankylosis is extremely rare, even in the high age groups. In the male SI-joints, on the other hand, there is an increasing tendency to roughening of the articular surfaces above the age of 35, a phenomenon which Stewart [38] attributed largely to arthrotic processes.

Sacral and iliac auricular roughening are different. In 1938, Schuncke [20] concluded that differences between the auricular surfaces exist as early as the intra-uterine period. Sacral cartilage was regarded as hyaline cartilage by Schuncke, iliac cartilage, on the other hand, has

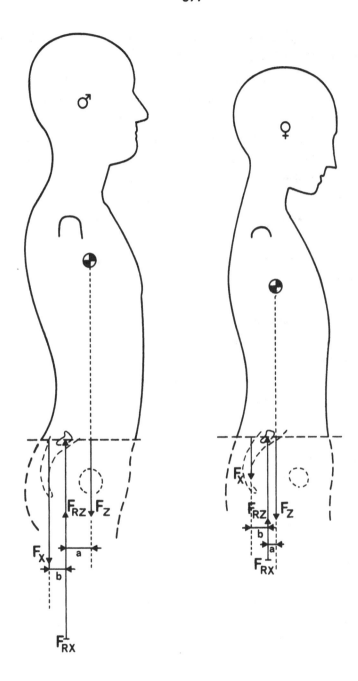

Figure 2. Relationships between the SI-joint and the vertical line through the centre of f gravity in men and women. Note that the lever arm in females is supposed to be smaller.

the macroscopic appearance of fibrocartilage. After birth, Schuncke stated, also sacral cartilage becomes fibrocartilage.

As a rule, auricular roughening starts at the ilium. Although the sacral auricular surface will start to roughen also, it will continue to stay behind the development of the ilium in this respect (Brooke [19]; Sashin [44]; Schuncke [20]; Weisl [18]; Bowen and Cassidy [45]; Paquin [46]; Stewart [38]; Dijkstra et al. [84]). Bowen and Cassidy [45] observed, as did Schuncke [20], that already before birth iliac cartilage is rougher than sacral cartilage. Paquin et al. [46] also observed macroscopic and microscopic differences between iliac and sacral cartilage, iliac cartilage being more fibrous and, hence, rougher; biochemically, however, iliac cartilage resembles hyaline cartilage more than fibrous cartilage.

3. Form-function relationships of the SI-joints

In view of the literature and our own data, the following theory is formulated concerning the form-function relationship in the SI-joint. The form of the healthy SI-joint is specifically related to functional demands, the most important demand in the non-pregnant subject being the stable connection between the spinal column and the pelvic girdle. In preparations of young persons, sacral and iliac articular surfaces are usually flat. Small ridges and grooves, however, can already be observed in frontal sections. This process has been described in the literature but usually attributed to incipient arthrosis (Brooke [19]; Sashin [44]; Schuncke [20]; Weisl [18]; Solonen [21]).

In general, signs of iliac roughening become more pronounced after puberty. This roughening can be divided into two types (Vleeming et al. [48, 49, 50]). The first type is restricted to the cartilage, and mostly to the iliac side of the joint. This roughening, which can hardly be seen macroscopically, is coined 'coarse texture'. The second type involves also the underlying bone; it is classified as 'ridges and grooves'. It occurs on both sides of the joint, ridges being complementary to grooves. Especially this last form of roughening is important since cartilage appears to have the same thickness over ridges and grooves as in the more flat regions. In frontal sections, the ridges resemble osteophytes (if one does not take the cartilaginous covering into account).

We hypothesize that sacro-iliac roughening - coarse texture as well as ridges and grooves - is a physiological process starting at the iliac side of the articular cartilage. We do not think that roughening is a sign of incipient arthrosis, eventually leading to complete ankylosis.

In order to test this hypothesis, the anatomy of the joint is investigated with particular attention to frictional forces (Dijkstra et al. [47]; Vleeming et al. [48, 49, 50, 51]).

The basic assumption is a functional one. If mobility between two bones in a joint would impede normal function, it is to be expected that a bony connection (i.e. a synostosis) will develop. If some mobility is functional, flexible fusion occurs in the form of a syndesmosis, synchondrosis or symphysis. If considerable mobility would be functional, a synovial diarthrosis would emerge.

Since the SI-joint is not a synostosis, without, however, allowing for considerable mobility, we surmise that its major function is to ensure 'flexible stability'.

The following propositions have not or as yet not sufficiently been underpinned by experimental data:

- ridges and grooves on the articular surfaces of the SI-joint, are complementary;
- intra-articular presence of ridges and grooves in the SI-joint, implies a higher friction;
- sacro-iliac stability depends more on the presence of ridges and grooves than on coarse texture.

These propositions call for determining frictional coefficients of a variety of SI-joints, with different architecture and texture. A biomechanical model is required to serve as a framework for closure in terms of form and forces. This model will include frictional values.

REFERENCES

1. Hippocrates, 460-377 B.C., acc. to Lynch.
2. Vaesalius, A., De corpori humanis fabrica, Oporini, Basel, 1543.
3. Paré, A., The works of generation of man, Cotes and Young, London, 1634. Translation T. Johnson.
4. Diemerbroek, I., The anatomy of human bodies, Brewster, London, 1689. Translation W. Salmon.
5. Albinus, 1697-1770, acc. to Lynch.
6. Von Luschka, 1864, acc. to Solonen.
7. Zaglas, 1851, acc. to Weisl.
8. Duncan, J.M., The behaviour of the pelvic articulations in the mechanism of parturition. Dublin Quart. J. Med. Sci., 1854, 18, 60-69.
9. Bakland, O. and Hansen, J.H., The "axial sacroiliac joint". Anat. Clin., 1984, 6, 29-36.

10. Meyer, C.H., Der Mechanisms der Symphysis sacroiliaca. Arch. Anat. U. Physiol., 1878, 1, 1-19.

11. Klein, G., Zur biomechaniek der Iliosacralgelenkes. Ztschr. Geburtsch und Gynak., 1891, 21, 74-118.

12. Testut, L. et Latarjet, A., Traité d'Anatomie Humaine, G. Doine et Cie., Paris, 1949.

13. Ehara, S., El Khoury, G.Y. and Bergman, R.A., The accessory sacroiliac joint: A common anatomic variant. A.J.R., 1988, 150, 857-859.

14. Faraboeuf, L.H., Sur l'anatomie et la physiologie des articulations sacro-iliaques avant et après la symphysectomie. Ann. Gynecol. Obstet., 1894, 41, 407-420.

15. Strasser, H., Lehrbuch der Muskeln und Gelenkmechanik, Springer, Berlin, 1913.

16. Fick, 1911, acc. to Solonen.

17. Palfrey, A., The shape of the sacroiliac joint surface. J. Anat., 1981, 132, 457.

18. Weisl, H., The articular surfaces of the sacroiliac joint and their relation to the movements of the sacrum. Acta Anat., 1954, 22, 1-14.

19. Brooke, R., The sacro-iliac joint. J. Anat., 1924, 58, 299-305.

20. Schuncke, G.B., The anatomy and development of the sacroiliac joint in man. Anat. Rec., 1938, 72, 313-331.

21. Solonen, K.A., The sacroiliac joint in the light of anatomical, roentgenological and clinical studies. Acta Orthoped. Scand., 1957, (suppl.) 27, 1-127.

22. Fischer, L.P., Gonon, G.P., Carret, J.P. et Dimmet, J., Biomécanique articulaire. Ass. Corp. et Med. Tome 2, Lyon, 1976, pp. 33-36.

23. Bonnaire, E. et Bué, V., Influence de la position sur la forme et les dimensions du bassin. Ann. Gynecol. Obstet., 1899, 52, 296.

24. Hisaw, F.L., The influence of the ovary on the resorption of the pubic bones. J. Exper. Zool., 1925, 23, 661.

25. Chamberlain, W.E., The symphysis pubis in the roentgen examination of the sacro-iliac joint. Am. J. Roentgenol., 1930, 24, 621-625.

26. Heyman, J. and Lundqvist, A., The symphysis pubis in pregnancy and parturition. Acta Obstet. and Gynecol. Scand., 1932, 12, 191-197.

27. Abramson, D., Roberts, S.M. and Wilson, P.D., Relaxation of the pelvic joints in pregnancy. Surg. Gynecol. and Obstet., 1934, 58, 595-613.

28. Abramson, D., Hurwitt, E. and Lesnick, G., Relaxin in human serum as a test of pregnancy. Surg. Gynecol. and Obstet., 1937, 65, 355.

29. Thorp, D.J. and Fray, W.E., The pelvic joints during pregnancy and labour. J.A.M.A., 1938, 111, 1162-1166.

30. Borell, U. and Fernstrom, I., The movements at the sacroiliac joints and their importance to changes in the pelvic dimensions during parturition. Acta Obstet. Gynecol. Scand., 1957, 36, 42-57.

31. Walheim, G.G., Olerud, S. and Ribbe, T., Mobility of the pubic symphysis. Acta Orthop. Scand., 1984, 55, 203-208.

32. Walheim, G.G. Stabilization of the pelvis with the Hoffmann frame. Acta Orthop. Scand., 1984, 55, 319-324.

33. Snijders, C.J., Snijder J.G.N. and Hoedt, H.T.E. Biomechanische modellen in het bestek van rugklachten tijdens zwangerschap. Tijdschr. voor Soc. Gen. en Gez. Zorg, 1984, 62, 141-147.

34. Lynch, F.W., The pelvic articulations during pregnancy, labour and puerperium: An X-ray study. Surg. Gynecol. Obstet., 1920, 30, 575-580.

35. Jackson, R.H., Chronic sacroiliac strain with attendent sciatica.

Am. J. Surg., 1934, 24, 456-477.

36. Haggart, G.E., Sciatic pain of unknown origin: An effective method of treatment. J. Bone and Joint Surg., 1938, 20, 851-856.

37. Cazeviel, 1973, acc. to Solonen.

38. Stewart, T.D., Pathologic changes in aging sacroiliac joints. Clin. Orthop. and Rel. Res., 1984, 183, 188-196.

39. McCarty, D.J., Arthritis and Allied Conditions. Lea Febiger, Philadelphia, 1979, 9th edition.

40. Braune, C.W. und Fischer, O., Bestimmung der Trägheitsmomente des menschlichen Körpers und seine Glieder. Abhandl. math. phys. kl. Sächs. Ges. Wiss., 1892, 18, 409.

41. Tischauer, E.R., Miller, M. and Nathan, I.M., Lordosimetry: A new technique for the measurement of postural response to materials handling. Amer. J. Indust. Hyg. Ass., 1973, 1, 1-12.

42. Bellamy, N., Park, W. and Rooney, J.R., What do we know about the sacroiliac joint? Sem. in Arth. and Rheum., 1983, 12, 282-313.

43. Derry, D.E., The influence of sex on the position and composition of the human sacrum. J. Anat. Physiol., 1912, 46, 184-192.

44. Sashin, D., A critical analysis of the anatomy and the pathological changes of the sacroiliac joints. J. Bone and Joint Surg., 1930, 12, 891-910.

45. Bowen, V. and Cassidy, J.D., Macroscopic and microscopic anatomy of the sacroiliac joints from embryonic life until the eight decade. Spine, 1981, 6, 620-628.

46. Paquin, J.D., Van der Rest, M., Marie, P.J., Mort, J.S., Pidoux, I., Poole, A.R. and Roughley, P.J., Biochemical and morphologic studies of cartilage from the adult human sacro-iliac joint. Arthritis and Rheumatism, 1983, 26, 887-894.

47. Dijkstra, P.J., Vleeming, A. and Stoeckart, R., Complex motion tomography of the sacro-iliac joint. RöFo, 1989, 105, 635-642.

48. Vleeming, A., Stoeckart, R. and Snijders, C.J., The sacrotuberous ligament: A conceptual approach to its dynamic role in stabilizing the sacro-iliac joint. Clin. Biomech., 1989, 4, 201-203.

49. Vleeming, A., Van Wingerden, J.P., Snijders, C.J., Stoeckart, R. and Stijnen, T., Load application to the sacrotuberous ligament: Influences on sacro-iliac joint mechanics. Clin. Biomech., 1989, 4, 204-209.

50. Vleeming, A., Stoeckart, R., Volkers, A.C.W. and Snijders, C.J. Relation between form and function in the sacro-iliac joint, part 1: Clinical anatomical aspects. Spine, 1990, 15, 130-132.

51. Vleeming, A., Volkers, A.C.W., Snijders, C.J. and Stoeckart, R. Relation between form and function in the sacro-iliac joint, part 2: Biomechanical aspects. Spine, 1990, 15, 133-135.

The Biological Response to Bone Cement and the influence of incorporated Growth hormone.

S. Downes and M.V. Kayser
Institute of Orthopaedics, Brockley Hill, Stanmore, Middlesex, HA7 4LP, England.

ABSTRACT

The biological response to bone cement, with and without the addition of growth hormone was investigated in an animal model. A new technique has been used to study the intact interface of bone and cement at the ultrastructural level. Five biological responses are reported:
1. The reaction to the stable PMMA implant.
2. The reaction to the unstable implant
3. The formation of a fibrous type tissue in the spaces between the bone and the cement.
4. Osteogenesis at the growth hormone loaded cement interface.
5. The formation of new mineral in the growth hormone loaded cement.
It is clear that there is a spectrum of tissue responses that can occur at the bone-cement interface and that these can be greatly affected by growth hormone released at the interface.

INTRODUCTION

In view of the increasing need for revision after joint replacement surgery there is need for radical improvement in the materials used to fix the prosthesis to bone. Aseptic loosening of the femoral component tends to occur at the bone-cement interface. While, the ideal cementing technique should result in a very good bone-cement bond, unfortunately the interlock is compromised by a number of post-operative changes that occur at the bone-cement interface, including:

1. The "foreign body reaction" used to describe the fibrous cartilage or fibrous connective tissue that develops at the bone-cement interface (1, 2)

2. Necrosis of the bone at the interface with the cement, thought to be caused by leakage of residual methylmethacrylate monomer which is cytotoxic (3)

3. Shrinkage of the cement during polymerisation resulting in the formation of a gap of between 20-250 μm at the bone-cement interface (4).

4. Thermal damage at the bone-cement interface caused by the heat released due to the exothermic reaction of the polymerisation of the cement (5).

Bone is a tissue that is capable of regeneration and is constantly being resorbed and remodelled throughout adult life. Hormonal agents have complex direct and indirect effects on bone, and our understanding of the mechanism by which growth hormone stimulates bone growth is still incomplete.There have been many studies indicating that growth hormone has a direct effect on bone cells, in adult animals. Lindholm (6) and Koskinen (7) observed positive results in the treatment of slow healing fractures. Harris (8) and Heaney (9), in two related papers, observed an increase in endosteal and periosteal new bone formation with a net increase in skeletal mass in adult dogs after treatment with growth hormone. There is also evidence that growth hormone can exert a direct effect on bone cells in culture in an auto- or paracrine fashion (10,11).

In order to stimulate the formation of new bone at the bone-cement interface we have developed a new cement with human growth hormone additive (12). Growth hormone released at the interface stimulates osteogenesis and the formation of bone at the bone-cement interface, thus improving the stabilisation of the implant.

MATERIALS AND METHODS

Preparation of Growth hormone loaded bone cement

All the reagents, mixing bowl and spatula were kept at room temperature for one hour before mixing. Using sterile technique with sterile conditions, 10g of polymethylmethacrylate powder (CMW Densply) were added to a plastic mixing bowl; 12IU Growth hormone (Novo Nordisk, Denmark), as a lyophilised powder, was then added and mixed thoroughly for one minute using a stainless steel spatula. The ampoule of the methacrylate monomer component of the cement was opened and 5 ml added to the powder in a well ventilated area. The cement was mixed for a further minute until in a "dough" state. It was then inserted into a sterile plastic syringe for insertion into the rabbit.

Animal model

Adult Sandy Lop rabbits weighing at least 3.5 Kg were used in the experiments. Access to the knee was gained through a medial parapatellar capsulotomy and using a 3mm diameter drill bit the medullary cavity of the femur was reamed to a depth of 2cm starting at the intercondylar notch. One femur was filled with Growth hormone loaded cement, with plain cement in the contralateral femur as a control. One month after surgery the rabbits were sacrificed and the femora removed.

Processing and Embedding

The adherent tissue was removed from the bone and the undecalcified femora cut into two halves longitudinally starting at the intercondylar notch. The tissue was fixed in 2% Glutaraldehyde in 0.1M Sodium Cacodylate buffer pH 7.2 for two hours. A small area (3mm x 6mm) was sectioned out and further fixed for 24 hours in fresh fixative at 4°C.Secondary fixation was in 1% Osmium Tetroxide in 0.1M Sodium Cacodylate buffer for one hour.

The block was washed in Sodium Cacodylate buffer, dehydrated in graded series of alcohol (70%,90%,100%), impregnated with 1:1 (by volume) alcohol and Spurrs' resin for 6 hours, two hours of which were by vacuum impregnation at 150 mbar using a Anglia Scientific Vacuum embedding chamber.

This was followed by 4 changes of 12 hours each, with Spurrs' resin, alternating every 6 hours with vacuum impregnation. The block was finally embedded and cured at 70°C for 18 hours.

Light Microscopy

For light microscopy 1μm sections were cut using a diamond knife floated onto a drop of water onto a glass slide and dried onto a hotplate. Sections were then stained with Methylene Blue-Azure II and basic Fuschin (Humphrey's stain). Similar areas from both the Growth hormone loaded and the plain cement interface were selected for ultrastructural studies.

Transmission Electron Microscopy

Selected areas were further trimmed down and sectioned with a diamond knife between 60-90nm, picked up onto 0.5% Pioloform support films on copper grids. The sections were then stained with 2% Uranyl acetate and lead citrate (Reynolds) for ten minutes each. Observation was made using a Philips CM 12 Electron microscope with EDAX PV 9800 microanalysis.

RESULTS

THE STABLE PMMA BONE-CEMENT INTERFACE

The characteristic reaction to plain PMMA bone cement was amorphous tissue containing bone dust, marrow and blood components, with little cellular activity and evidence of bone remodelling Fig.1.

Fig.1. (overleaf)

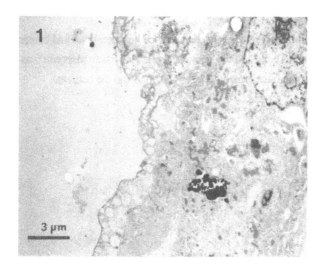

THE UNSTABLE PMMA BONE-CEMENT INTERFACE

At the unstable bone cement interface a sclerotic type tissue (Fig2) was observed, with numerous macrophages containing cement particles (Figs. 3 and 4).

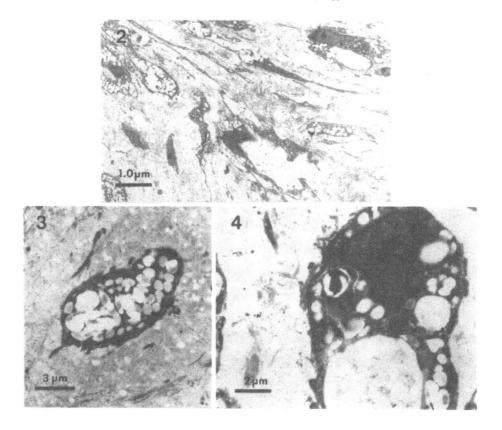

FIBROUS TISSUE

The formation of fibrous type tissue at the interface between bone and cement was observed in the spaces between the bone and the cement (fig5). This tissue response was totally acellular and ocurred after a gap had formed between the bone and the cement.

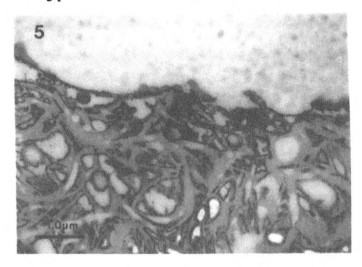

GROWTH HORMONE LOADED BONE CEMENT.

At the interface between bone and growth hormone loaded cement; new bone remodelling was observed with a layer of active osteoblasts along the bone cement (Fig.6). The formation of new collagen and an advancing mineral front, in the direction of the bone cement was observed.

MINERALISATION

There was an overall increase in the new mineral formed at the growth hormone loaded interface. A striking feature of the growth hormone loaded cement was the formation of hydroxyapatite in the cement matrix (Fig.7).

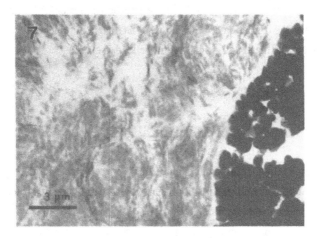

DISCUSSION

There are a number of changes that occur at the bone-cement interface after implantation of the cement and these events can determine the success or failure of the implant. The methods described in this paper enable the intact bone-cement interface to be observed and we report the various tissue responses at the interface.

At best, Polymethylmethacrylate bone cement is an inert material and at the "stable" bone-cement interface, a few fibroblast-like cells with some collagen can be seen, but very little true remodelling of the bone. The bone 'dust' created during the reaming of the bone and some marrow and blood components remained trapped between the bone and the cement. However, there are other biological responses to PMMA bone cement that can lead a poor bone-cement interface.

The formation of a fibrous connective tissue capsule around implanted bone cement, has been described by a number of workers and has been implicated in aseptic loosening. In this work, we were able to observe fibrous-type tissue, only at the plain cement interface and it appeared to occur in the spaces between the bone and the cement. We suggest that, at the plain cement interface various post-operative changes (including the shrinkage of the cement) result in the formation of a slight gap between the bone and the cement, because there is inadequate bone remodelling, this gap becomes filled with fibrous type tissue. This should not be confused with a cellular reaction involving the invasion of the bone-cement interface with macrophages and giant cells and occurs in response to the an unstable bone-cement interface. Both these tissue reactions compromise the bone-cement bond and may eventually lead to failure of the implant. In order to achieve a stable bone-cement interface early stimulation of active bone remodelling and new bone formation is necessary. At the interface with growth hormone loaded cement it was possible to observe osteogenesis; characterised by, active osteoblasts along the bone-cement interface and the formation of collagen with an advancing mineral front. New mineral was also observed in the matrix of growth hormone loaded cement.

CONCLUSIONS

1. There is poor remodelling at the stable plain PMMA bone cement interface.

2. A sclerotic type tissue, containing macrophages which engulf cement particles, occurs at the unstable PMMA bone cement interface.

3. An acellular fibrous-type tissue can form in the spaces between bone and the cement

4. Growth hormone released from bone cement can stimulate osteogenesis and mineral formation at the bone-cement interface.

REFERENCES

1. Charnley J. (1970) The reaction of bone to self-curing acrylic cement: a long term histological study in man.
J.Bone and Joint Surg. 52B: 340-352.

2. Willert H.G. (1983) The unstable bone-cement interface.
In: Revision Arthroplasty 2" p. 6-7. Ed. Elson R.A. Franklin Scientific Publications.

3. Willert H.G., Ludwig J. and Semlitsh M. (1974).
Reaction of bone to methacrylate after hip arthroplasty.
J Bone and Joint Surg 56A: 1368-

4. Draenert T.K. (1981)
Histomorphology of the bone-cement interface: remodelling of the cortex and revascularisation of the medullary canal in animal experiments.
In: Hip ch.7 p71-110

5. Feith R (1975)
Side effects of acrylic cement, implanted into bone.
Acta Orthop Scand 161 (Suppl)

6. Lindholm R.V., Koskinen E.V.S., Puranen R.A., Nieminen M., Kairaluoma and Attila V. (1977)
Human growth hormone in the treatment of fresh fractures.
Hormon Metab Res **9**: 245-246.

7. Koskinen E.V.S., Lindholm R.V., Nieminen R.A., Puranen J.P. and Attila V. (1978)
Human growth hormone in delayed union and non-union of fractures.
Int. Orthop (SICOT) **1**:317-322.

8. Harris W.H., Heaney R.P., Jowsey J., Cockin J., Atkins C., Graham J. and Weinberg E.H. (1972)
Growth hormone: The effect on skeletal renewal in the adult dog.
I Morphometric Studies. Calc. Tiss. Res. **10**:1-13.

9. Heaney R.P., Harris W.H., Cockin J. and Weinberg E.H. (1972)
Growth hormone: The effect on skeletal renewal in the adult dog.
II Mineral Kinetic Studies. Calc. Tiss. Res.**10**:14-22.

10. Stracke H., Schultz A., Moellar D., Rossol S. and Schatz H.(1984)
Effect of growth hormone on osteoblasts and demonstration of Somatomedin C/IGF-I in bone organ culture.
Acta Endocrinol. (Copenhagen) **107**: 16-24.

11. Ernst m. and Froesch E.R. (1988)
Growth hormone dependent stimulation of osteoblast-like cells in serum-free cultures via local synthesis of Insulin-like-Growth Factor-I.
Biomed. and Biophys. Res. Comm. **151**: 14-48.

12. Downes S., Wood D.J., Malcolm A. J. and Ali S.Y. (1990)
Growth hormone in Polymethylmethacrylate Cement
Clin. Orthop. and Rel. Res. **252**:294-298.

EXPERIMENTAL AND THEORETICAL THERMAL EFFECTS DURING
CEMENTATION OF AN ENDOMEDULLARY INFIBULUM

G. Paganetto, S. Mazzullo
HIMONT Italia. Research Centre "G. Natta"
44100 Ferrara Italy

ABSTRACT

A 1-dimensional mathematical model of a femoral prosthesis
implant has been developed. The actual complex geometry has
been idealized as a composite cylinder of infinite length
having a multishell bone/cement/stem structure. Attention is
focused on the cement, which is a polymer that polymerizes "in
situ", inside the medullar channel and generates heat. The
"cementation" phenomenon is described by the heat diffusion
Fourier equation coupled with the polymerization kinetics. The
numerical solution of the model has been accomplished by a
finite difference explicit scheme. The experimental kinetic
data has been obtained through isothermal polymerization tests,
carried out by using a DSC calorimeter and a commercially
available Howmedica Simplex P cement. The two most relevant
results are:
1. The conversion of monomer into polymer is never 100% under
   the imposed initial and boundary conditions.
2. Bone/cement interface temperature is a function of the
   interface heat transfer coefficient.
Depending on such a parameter, ranging from 10. to 1000. (Wm^{-2}
$°K^{-1}$), the temperature may reach the maximum values of 80 °C
and 40°C, respectively. In other words, the lower the heat
transfer coefficient, the higher the interface temperature.
Therefore, the model predicts relatively high temperature
values at the bone/cement inteface. Such values confirm the
criticality of this type of implant. The mathematical model
developed is capable of taking into account both the geometry
of the implant and the chemical-physical and kinetic properties
of the cement. It represents a useful tool for setting-up the
optimal conditions for the new materials developed in this
orthopaedic field.
Heat generated within the polymerizing mixture partly diffuses
through the bone, causing damages. We have therefore tested
bone behaviour before and after thermal treatment in order to
find possible mechanical effects.

INTRODUCTION

Cementation technique of hip prostheses by polymethylmethacrylate (PMMA) and its copolymers started less than 30 years ago [6].
Total hip prostheses generally consist of two pieces: the femoral component and the acetabular component. We shall deal with the cementation thermal effects on the femoral component: the cement is, in fact, a polymer which polymerizes "in situ", inside the medullar channel, and generates heat. It is pretty natural to idealize the complex geometry of a femoral implant as a bone/cement/stem composite cylinder. Assuming that the stem length/medullar channel ratio is generally greater than 5:1, the composite cylinder can be considered of infinite length Thus the study shall be limited to the heat flow in the radial direction only.

MATHEMATICAL MODEL

The cementation phenomenon is described by the heat diffusion Fourier equation coupled with the polymerization kinetics.
If $T(r,t)$ is temperature and $X(t)$ the dimensionless extent of polymerization, we obtain in radial coordinates:

$$\varrho \, c_p \frac{\partial T}{\partial t} = K\left(\frac{\partial^2 T}{\partial r^2} + \frac{1}{r} \frac{\partial T}{\partial r}\right) + \varrho \, \lambda \frac{dX}{dt} \tag{1}$$

$$\frac{dX}{dt} = k(T) \, f(T,X) \tag{2}$$

where ϱ, c_p, K, λ are density, specific heat, thermal conductivity and enthalpy of polymerization, respectively. For simplicity all these quantities are assumed independent of temperature and are given a value in Tab. 1 and 2.

TABLE 1
Thermal properties (Huiskes, [3])

	Density $Kg \ m^{-3}$	Specific heat $J \ Kg^{-1} \ ^\circ K^{-1}$	Thermal conductivity $W \ m^{-1} \ ^\circ K^{-1}$
PMMA	$1.19 \ 10^3$	$1.6 \ \ 10^3$	0.17
Cr-Ni Steel	$7.85 \ 10^3$	$0.46 \ 10^3$	14.0
Cortical bone	$2.10 \ 10^3$	$1.26 \ 10^3$	0.3-0.5
Spongeous bone	$2.1-2.3 \ 10^3$	$1.15-1.73 \ 10^3$	0.5

Table 2
Kinetic contrants of a commercial cement (eq.(2))

Cement	λ J Kg^{-1}	k_o sec^{-1}	E J mol^{-1}
SIMPLEX P Howmedica	193. 10^3	2.640 10^8	62.966 10^3

Table 2
continue

Cement	α -	Tg °K	R J mol^{-1} °K^{-1}
SIMPLEX P Howmedica	9.20	378	8.314

Although polymerization of methyl-methacrylate (MMA) is qualitatively well known, not all kinetic parameters are always available for the more common cements in use [3], [4]. Therefore we have decided to carry out isothermal polymerization experiments (Fig.1) in order to identify all the actual kinetic parameters of a commercial cement.

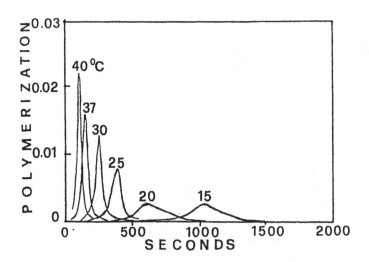

Figure 1. Isotherms of polymerization
of a cement by DSC

The acrylic cement is composed of polymer powder (PMMA) and monomer liquid (MMA). The composition of the mixture is expressed in the ratio gr polymer/ml monomer, which is usually around 2. The main problem in these experiments is to avoid the early polymerization of the mixture before the beginning of the isotherms. To this purpose the mixture has been prepared at the temperature of liquid nitrogen. Experimental data suggest an Arrhenius behaviour for the kinetic constant:

$$k(T) = k_o \exp - E/RT \qquad (3)$$

and the following functional form for the extent of polymerization:

$$f(T,X) = \begin{cases} \dfrac{\alpha}{\eta(T)} X^{1-\frac{1}{\alpha}} (\eta(T)-X)^{1+\frac{1}{\alpha}} & ; \quad X < \eta(T) \\ \\ 0 & ; \quad X \geqslant \langle T \rangle. \end{cases} \qquad (4)$$

where $\alpha>1$ is an empirical parameter. The function (T) is the equilibrium conversion of monomer to polymer. It behaves as follows [1]:

$$\eta(T) = \begin{cases} \dfrac{T}{Tg} & ; \quad 0 \leqq T \leqq Tg \\ \\ 1 & ; \quad T > Tg \end{cases} \qquad (5)$$

where Tg is the glass transition temperature of cement. The actual values of these kinetic constants are summarized in Tab.2 To complete the model, appropriate initial and boundary conditions are required. The temperature of cement, before its introduction into the femoral cavity, is either known or can be determined experimentally. The same applies also to the initial polymerization extent. Therefore, as initial conditions we have:

$$T(r,0) = T_o \qquad (6)$$
$$X(0) = X_o \qquad (7)$$

At the stem/cement boundary, $r = a$, we assume perfect contact between the cement and a perfect heat conductor (the metal infibulum) [2]:

$$\pi a^2 \rho_m Cp_m \frac{\partial T}{\partial t} = 2 \pi a K \frac{\partial T}{\partial r} \qquad (8)$$

where ρ_m, Cp_m, refer to the metal. At the cement/bone boundary, $r = b$, we assume linear heat transfer between the cement and the sorrounding medium (the bone) at body temperature $T_B=37°C$:

$$K \frac{\partial T}{\partial r} = - U(T - T_B) \qquad (9)$$

where $U(W\ m^{-2}\ {}^\circ K^{-1})$ is the surface heat transfer coefficient. The actual value of U strongly depends on the preparation of the femoral cavity by the surgeon [3].

NUMERICAL SOLUTION

On physical grounds, the model formulated previously seems well posed. From a mathematical view point the model belongs to the class of Stefan problems. More formal sufficient conditions of existence and uniqueness of the solution can be found in [5]. The numerical solution of the model has been obtained in three steps as follows: i) transformation of the problem to dimensionless form, ii) change of the coordinate system from hollow cylindrical "r" to rectangular "x", by setting r=exp x, iii) numerical solution of the cartesian problem by explicit finite difference schemes. Use has been made of forward difference operators to approximate time derivatives, and central difference operators to approximate space derivatives. Stability restrictions on time step can be easily obtained.

NUMERICAL RESULTS

Numerical simulation shall be limited to a sensitivity analysis with respect to variations of the heat transfer coefficient, at the cement/bone interface. The U parameter takes values in the range 10 - 1000 ($W\ m^{-2}\ {}^\circ K^{-1}$). The metallic infibulum has a 16 mm diameter and the cement a 6 mm thickness. Temperature at the cement/bone interface, as a function of time, is shown in fig.2.

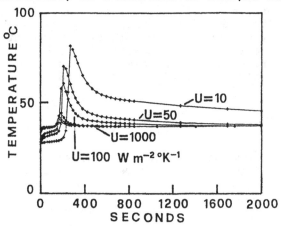

Figure 2. Cement/bone interface temperature

The temperature profile always reaches a maximum, T max, whose intensity increases when U decreases. The maximum conversion, X max, shows a similar trend. The results mentioned above are summarized in Tab.3. It should be mentioned that the temperature behaviour is basically in agreement with the results of Huiskes' model [3] which is numerically solved by finite elements. An important and so far unique feature emphasized by the numerical simulations, is that in no case a 100% conversion is achieved. Therefore, presence of residual unreacted monomer is always noticed in the cement. Fig.3 provides the radial conversion profile at fixed times, in case the U parameter assumes the value of 1000. The maximum radial conversion is 82.1%, while maximum cement/bone interface temperature is 41.7°, (Tab.3).

TABLE 3
Effects of variation of bone-cement
heat transfer coefficient U

U	10.	50.	100.	500.	1000	W m^{-2} °K^{-1}
T max	81.6	70.0	60.6	44.5	41.7	°C
t max	275.2	211.7	211.7	190.5	175.8	sec
X max	94.0	91.2	88.7	82.7	82.1	%

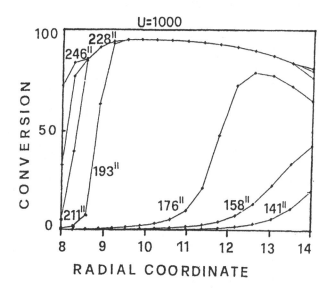

Figure 3. Radial conversions at fixed times, (sec)
U = 1000   W m^{-2} °K^{-1}

THERMOMECHANICAL BEHAVIOUR OF BOVINE CORTICAL BONE

Heat generated within the polymerizing mixture partly diffuses through the bone, causing damages. We have therefore tested bone behaviour before and after thermal treatment in order to study, if any, mechanical effects. For this purpose we have used Dynamical Mechanical Thermal Analyzer (DMTA) equipment, at the frequency of 1 Hz and a scanning rate of 2 °C/min in the temperature range from -140 to 160 °C.

The lower part of Fig.4 shows the real part of the complex dynamic bending modulus (E') as a function of temperature. The lower curve refers to fresh bone. The upper one to the same sample after the thermal treatment due to the first scansion. An increase of the elastic modulus is clearly shown.

The upper part of fig.4 gives the loss factor (tanδ) as a function of temperature. Three relative maxima are clearly distinguishable. We have labeled these peaks as α,β,γ transitions. The γ-transition occurs in the range from -140°C to -100°C and may be attributed to the glass transition of the aliphatic chains of the fatty component of the bone. The β-transition occurs around 0°C and can be explained in terms of water interaction with bone tissue.

Finally, the α-transition occurs in the broad range from 80°C to 160°C and can be associated to removal of bound water or to a solid-solid transition induced by water interaction with collagen [7].

The intensity of the   β-transition increases after the thermal treatment due to the first scansion.

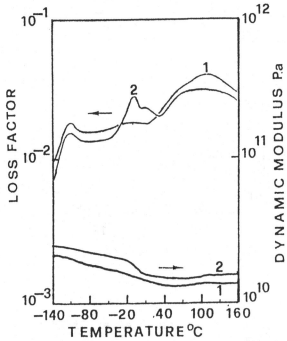

Figure 4. DMTA spectrum of bovine cortical bone
1) before and 2) after thermal treatment

REFERENCES

1. Burnett, G.M., Duncan, G.L., High conversion polymerization of vynyl system. I Methyl-methacrylate. Die Makromoleculare Chem. **51** (1962) 154-170.

2. Carslaw, H.S., Jaeger, J.C., Conduction of heat in solids. Oxford Univ. Press, Oxford (1959), Cap 1.

3. Huiskes, R.. Some fundamental aspects of human joint replacement. Acta Orthop. Scand.; Supp. 185 - Copenhagen (1979) Sect.2 Cap. 7.

4. Jefferis, C.D., Lee, A.J.C., Ling, R.S.M., Thermal aspects of self curing PMMA. J. Bone Joint Surg. **578** (1985) 511-518

5. Verdi, C., Visintin, A., Numerical analysis of the multidimensional Stefan problem with supercooling and superheating. Boll. U.M.I. **1-B** (1987) 795-814.

6. Pipino, F., Il punto sulla cementazione degli impianti protesici. OIC Medical Press, Firenze (1987)

7. Civjan, S., Selting, W.J., De Simon, L.B., Battistone, G.C. and Grower, M.F.. Characterization of Osseus Tissues by Thermogravimetric and Physical Techniques. J. Dent. Res. **51**, (1972) 539-542.

AN INVESTIGATION INTO THE INTERFACE MECHANICS IN FRACTURES, PART 1: THEORETICAL MODELS

J A Brandon   and   O N L Abraham
School of Engineering   Ecole Nationale des Travaux
University of Wales   Publics de l'Etat
    College of Cardiff   Rue Maurice Audin
P O Box 917, Cardiff CF2 1XH, UK   69518 Vaulx en Velin Cedex, France

ABSTRACT

The paper presents a model of the fractured tibia which is based on alternation of two piecewise linear systems. It is shown that the behaviour of the composite system is non-linear and indicators of the behaviour of this system are derived. A companion paper describes the experimental investigation of this model.

INTRODUCTION

In a paper by Brandon and Richards [1] a conjecture was offered concerning the interface mechanics in fractures. This was based on the interpretation of a number of previous papers by this research group (Pocock, Brandon and Richards [2] Pocock et al [3] and Brandon, Richards and Mackie [4]) and other methods described in the literature. In the same issue of the Journal of Engineering in Medicine as the paper cited above other relevant papers presented at the 1989 Leeds Biomechanics Seminar were published.

The current paper presents theoretical analysis which seeks to predict the consequences of the behaviour predicted by Brandon and Richards. A companion paper (Brandon and Macleod [5]) offers experimental evidence which supports the predicted behaviour. As previously, the analysis presented is somewhat simplistic in its theoretical foundation, since it is primarily intended to explore system behaviour from a qualitative point of view, offering pointers to diagnose response characteristics which are unexplainable under linear assumptions. In particular, the mechanism by which the vibration behaviour described in the current paper could be excited is difficult to conceive. As shown by Schmidt and Tondl [6], however, the exact analysis of even very simple non-linear problems can be of daunting complexity and may not even be tractable in the sense generally understood in the field of linear theory.

SUMMARY OF PREVIOUS CONJECTURE

Based on intuitive arguments, Brandon and Richards [1] suggested two phenomena which, they suggested, would tend to undermine the common assumption that fractured bones can be adequately modelled using Modal Analysis, with its fundamental assumption of linearity (see, for example, Van der Perre et al [7]). These were

(i) an impulsive effect caused by dynamic crack closure,
(ii) a stiffness discontinuity caused by the crack transmitting compressive loading, during half of a cycle, but not transmitting loads when the crack is in tension, during the other half cycle.

MODELLING CONSTRAINTS

Although the authors accept that damping is significant in the orthopaedic context, and may be used as a clinical indicator, it is neglected in this study. This is by no means an unusual assumption, for example most finite element codes discount damping completely.

As discussed by Van der Perre et al [7], the definition of boundary conditions in orthopaedic studies is an extremely difficult problem. The authors have followed Van der Perre et al [7] in the choice of boundary conditions.

The emphasis in the current paper will be on the second of the two parts of the Brandon-Richards conjecture.

Whilst the use of sinusoidal excitation has some dangers (see Pocock et al [3]), it is recognised that such methods are widely used in engineering orthopaedics (see, for example, Rosenstein et al [8]) and this assumption simplifies the current analysis.

In terms of traditional engineering idealisations and approximations, for example symmetry, homogeneous properties, etc, bones are difficult to characterise. For the purposes of the current study the bone is simplified as a prismatic bar whose shape is a best fit to the genuine structure. Two types of bone have been considered: cortical bone and cancellous bone. However, in the present study the only difference is between their density. Consequently, the modulus of elasticity is constant in the model and is taken equal to that of the cortical bone. All these hypotheses must be taken into account in the use of the results.

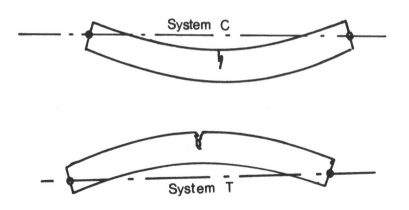

System C

System T

COMPRESSION-TENSION SYSTEM MODEL

The model separates the two assumed states of the system, as shown in figure 1, and assumes each has a complete set of modal characteristics. The general assumption for this mixed model is that the initial conditions and subsequent excitation are such that the system executes a smooth transition from each state to the other alternately. The systems are denoted System C (for the compressive half cycle) and System T (for the tensile half cycle). System C has the properties of the intact member whereas system T has a weakened section whose mechanical properties are unknown.

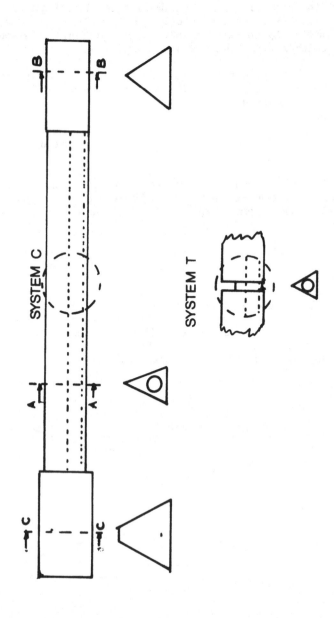

Each is assumed to behave linearly within its own half cycle. Associated with the former are natural frequencies $\omega_{c,i}$ and periods $T_{c,i}$ and with the latter frequencies $\omega_{t,i}$ and periods $T_{t,i}$. Because of the reduction in stiffness, the natural frequencies in the tensile half cycle are less than (or at most equal to) those for the compressive half cycle (see, for example, Brandon [9]), pages 28-29).

System C is modelled as a simple beam, as shown in figure 2, whereas system T contains a reduction in section to simulate the change of stiffness at the location of the fracture.

For the model given, The first and second natural frequencies for system C were computed to be 252Hz and 1092Hz respectively. The corresponding properties for the weakened section were 243Hz and 1078Hz. The computed mode shapes are shown in figure 3. As can be seen the computed mode shapes are almost indistinguishable.

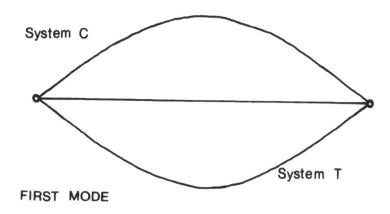

FIRST MODE

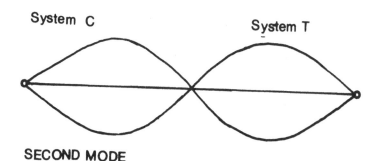

SECOND MODE

402

FOURIER TRANSFORM CONSIDERATIONS

The spectra of periodic functions can be regarded from the point of view of Fourier Tranforms rather than Fourier Series.

1 The Fourier Transform of a periodic function consists of a series of equally spaced delta functions apart in frequency.

2 The Fourier transform of a sinusoidal function comprises a single delta function.

Assuming such initial and excitation conditions are realisable, a composite cycle comprising a compressive half cycle period $T_{c,i}$ and a tensile half cycle of period $T_{t,i}$ will give rise to periodic behaviour of period

$$T = \frac{T_{c,i} + T_{t,i}}{2}$$

In the event of the absence of a fracture, the periods of the two half cycles will coincide, the response will be strictly sinusoidal, satisfying condition 2, and lead to a spectrum containing a single delta function corresponding to the response frequency, as shown in figure 4.

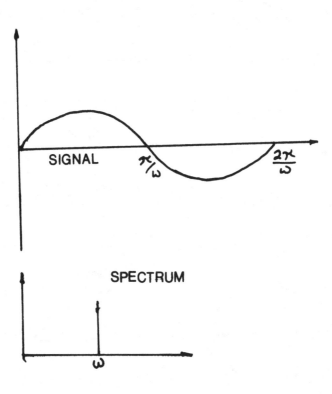

If the compression/tension model applies, ie a fracture of the type conjectured is present, the periods of the two half cycles will no longer coincide, the response will no longer be strictly sinusoidal, but will satisfy condition 1, because the response is still periodic, leading to a a discrete spectrum of delta functions which correspond to integer multiples of the response frequency, as shown in figure 5.

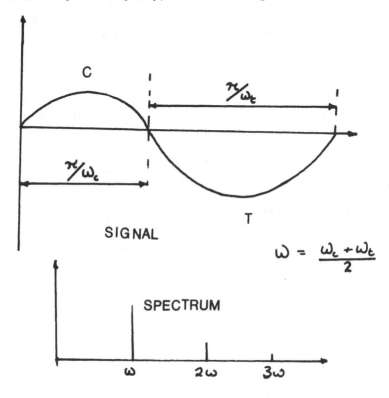

$$\omega = \frac{\omega_c + \omega_t}{2}$$

SPECTRAL PROPERTIES OF IMPULSIVE EXCITATION

It is not intended that the current paper will address the subject of impulsive excitation in any great detail. The appropriate aspects of response of multi degree of freedom systems to impulsive loading will, however, be described briefly. In particular, it is necessary to understand how the resulting spectra differ from those due to the tensile/compressive model described above.

The response of a dynamic system to a single impulse is one of the fundamental modelling tools in structural dynamics, enabling system response to be expressed in terms of the convolution integral. Such methods are well known and widely described in the literature (see for example Meirovitch [10] pp11-18). For this reason, the properties of the impulse response function will be presented without extensive justification.

The application of a single impulse to a system, in the absence of any other excitation, can be interpreted as a set of initial conditions for the differential equation describing the system. For a linear system, the solution is completely described by the complementary function of the differential equation. This solution comprises a superposition of the modes of the structure vibrating at their natural frequencies.

In describing the natural frequencies of beams it is often assumed, albeit erroneously, that the same relationships apply as for the stretched string, ie the natural frequencies are integer multiples of the fundamental. In the case of the modes of a simply supported beam, the model for the current case, this is not so. For example the natural frequencies of an encastre beam are the solutions of the transcendental equation:

$$\cosh \beta l \cos \beta l = 1$$

where l is the length of the beam.

Thus the spectral signature of the response of a linear beam to a single impulse comprises a superposition of natural modes of the beam at non-integer multiples of the fundamental frequency of the beam. This is potentially a discriminator between the impulsive assumption and the tensile/compressive model described above where the harmonics are at integer multiples of the forcing frequency.

The conditions postulated in the current context are substantially more complicated, however, in that the model postulates a train of impulses occuring at the excitation frequency and at the point of transition from tensile to compressive behaviour. Depending on the relationships between the excitation frequency and the natural frequencies of the system the impulse train may either tend to reinforce the free vibrations induced by previous impulses or may tend to suppress them. Further, depending on the degree of damping, the impulses may affect either system C only, if the damping is large, or a combination of systems C and T, if the damping is small.

Consider, alternatively, the expected system behaviour in the time domain. The impulse excites all of the modes but the high frequency modes will be damped out rapidly. Thus the amplitude trace will appear most "noisy" during the initial stages of the C portion of the cycle. (Although this is not strictly noise since it constitutes excitation of modal properties and is, consequently, more properly a (usually unwanted, but in this case significant) characteristic of the signal.

CONCLUSIONS

The paper has examined the system behaviour which might be predicted if the model of fracture behaviour proposed by Brandon and Richards [1] applies. This model had two features, namely a tensile/compressive stiffness discontinuity and an impulsive loading.

Analysing this model in the frequency domain enables discrimination between these two phenomena. In particular, the stiffness discontinuity postulated will give rise to harmonics at frequencies that are integer multiples of excitation frequency. The impulsive excitation will give rise to harmonics at the section's natural frequencies, which are not integer multiples of the fundamental.

When the model is analysed in the time domain, it is again found that the two features of the model give rise to differing, and characteristic, behaviour. The stiffness discontinuity leads to periodic behaviour with alternations between piecewise linear models. The impulsive model leads to a belief that the time domain signal will appear significantly more "noisy" in one segment of the cycle (the compressive half cycle) than in the other.

REFERENCES

1. J A Brandon and J Richards, A conjecture on the interface mechanics in fractures based on the interpretation of impulse tests, Proceedings of the Institution of Mechanical Engineers, Series H, Journal of Engineering in Medicine, 1989,203(4), pp203-205

2. I R Pocock, J A Brandon and J Richards, Vibration strategies for the monitoring of fracture healing in human long bones, American Society of Mechanical Engineers Symposium: Advances in Bioengineering, Boston, December 1987, p89-90.

3. I R Pocock, J A Brandon, J Richards and I G Mackie, Choosing a signal for vibration monitoring of fracture healing, Conference Interfaces in Medicine and Mechanics, Editor K R Williams, U C Swansea, 1988

4. J A Brandon, J Richards and I G Mackie, Vibration Methods for assessing fracture healing, Conference: Changing Role of Engineering in Orthopaedics, Institution of Mechanical Engineers/ Royal College of Surgeons, 14-15 April 1989, pp123-5

5. JA Brandon and D Macleod, An investigation into the interface mechanics in fractures, Part 2: experimental results, Conference "Interfaces 90", Bologna, September 9th-14th 1990

6. G Schmidt and A Tondl, Non-Linear Vibrations, Cambridge University Press 1986

7. A D Rosenstein, G F McCoy, C J Bulstrode, P D McLardy-Smith, J R Cunningham and A R Turner-Smith, The differentiation of loose and secure femoral implants in total hip replacement using a vibration technique: an anatomical and pilot clinical study, Proceedings of the Institution of Mechanical Engineers, Series H: Journal of Engineering in Medicine, 1989, 203, pp77-82

8. G Van der Perre, R Van der Audercke, M Martens and J C Mulier, Identification of In-Vivo Vibration Modes of Human Tibiae by Modal Analysis, Transactions of the American Society of Mechanical Engineers, Journal of Biomechanical Engineering, 1983, 105, 244-248

9. J A Brandon, Strategies for Structural Dynamic Modification, Research Studies Press, 1990

10. L Meirovitch, Analytical Methods in Vibrations, Macmillan, New York, 1967

AN INVESTIGATION INTO THE INTERFACE MECHANICS IN FRACTURES,
PART 2: EXPERIMENTAL RESULTS

J A Brandon and D Macleod
School of Engineering
University of Wales College of Cardiff
P O Box 917
Cardiff CF2 1XH, UK

ABSTRACT

The experimental work described in the paper is designed to test the observability of two characteristic non-linear vibration phenomena analysed and predicted in a companion paper. Both of the characteristic error profiles predicted are observed in the experimental tests.

INTRODUCTION

In a paper by Brandon and Richards [1] a conjecture was offered concerning the interface mechanics in fractures. This was based on the interpretation of a number of previous papers by this research group (Pocock, Brandon and Richards [2] Pocock et al [3] and Brandon, Richards and Mackie [4]) and other methods described in the literature. In the same issue of the Journal of Engineering in Medicine as the paper cited above other relevant papers presented at the 1989 Leeds Biomechanics Seminar were published.

A companion paper (Brandon and Abraham [5]) extends the analysis of Brandon and Richards [1] and predicts certain observable phenomena which the current paper is designed to test. The current paper presents experimental results which are consistent with the behaviour predicted by Brandon and Richards.

The conjecture on the interface mechanics, which is the subject of this paper [1], would, if correct, suggest distinctive features in the time domain data [5].

The suggested stiffness discontinuity would be observed with the compressive half cycle being shorter than the corresponding tensile half-cycle. The accompanying impulsive type loading effect would be observed as a "glitch" on the acceleration trace (the rate of change of acceleration is jerk). This would occur at the point of transition from the tensile to compressive half cycle.

The clear implication of the expected effects is that as the amplitude of vibration is reduced at each frequency then the stiffness discontinuity and the acceleration glitch would disappear ie there would be amplitude thresholds, not necessarily coincident, below which the stiffness discontinuity and the accompanying impulse effect would not be observed. Further, it is anticipated that the compressive-tensile behaviour would initiate at a lower threshold than that expected for the impulsive effect.

EXPERIMENTAL TEST DESIGN

Although the problem under consideration derives from biomechanics, it was regarded as unrealistic to undertake the study on organically derived biomaterials. It was decided to carry out these exploratory studies on prismatic sections of polymeric specimens whose mechanical properties were known. This approach has been used previously be Moezedi et al [6]. Two materials were used in the tests UHMWP (ultra high molecular weight polyethylene) and Perspex (poly methyl methacrylate).

Perhaps the most difficult problem encountered initially was that of obtaining a specimen that was reliably cracked. Attempts to fabricate specimens with simulated cracks, for example laminating specimens with butt joints, were found to be unsuccessful.

Success was achieved, however, by inducing actual cracks using a standard three point bending jig. It was found necessary to cool the UHWMP specimens to approximately $-60°$ Celsius to make them sufficiently brittle for crack formation. Crack initiation was induced by cutting the specimens with a razor blade. The specimen geometry is shown in figure 1.

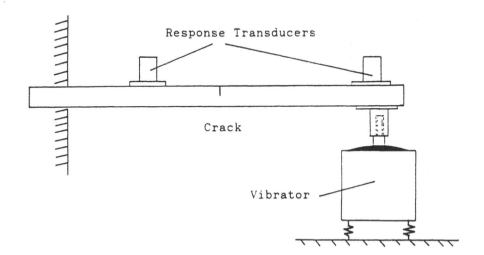

The process for the production of the test samples made of perspex was similar in that there was no material removed. The test samples were rigidly supported along their length and cracked by impacting a sharp blade perpendicularly to both the surface and the longitudinal axis of the beam. It should be noted, however, that this process was very difficult to control, especially with the smallest of samples (those of 5mm thickness or less). This relates to the critical fracture toughness of the sample.

The tests used a Ling Dynamics type V201 vibrator, with response transducers D J Birchall accelerometers type A20, with matched charge amplifiers type CA04. Signals were captured using a Thurlby Digital Storage Adaptor 511 transient data acquisition instrument and analysed using an Opus V PC microcomputer. The test configuration was as shown in figure 2.

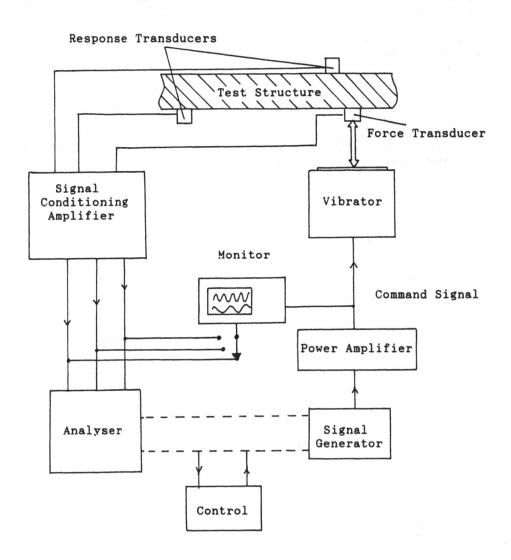

TEST RESULTS

The traces presented have not been operationally calibrated. The main objective of the presentation of the time domain data was to characterise the qualitative behaviour of the fracture, which, it was felt, obviated the need for calibration. Furthermore, in certain tests either the amplification of the signals or the scaling of the output was changed through the range of interest, for reasons of clarity. This procedure did not affect the dynamics of the operational system. Calibrations would have had to be performed each time that the amplification of the individual channels was altered and it was felt that calibration of the acceleration signals would not help in the characterisation of the dynamics of the crack interface.

For the purposes of the current paper, only the traces from a single set of tests are presented. The complete set of test data is currently only available as a departmental report [7], but is being prepared for publication in the near future. The time domain data is presented as a series of tests at the same frequency showing the behaviour of the sample as the amplitude of vibration is increased. Figure 3 shows the tests for a perspex sample at four equal amplitude steps at a frequency of 250Hz. (It should be noted that the actual test sequence was the reverse of that presented here, so as to minimise the possibility that progressive effects, due to cumulative damage to the specimen, would influence the results). In each trace the signal of larger magnitude is from the transducer on the same side of the crack as the vibrator. As can be seen, this remained sinusoidal throughout the test sequence.

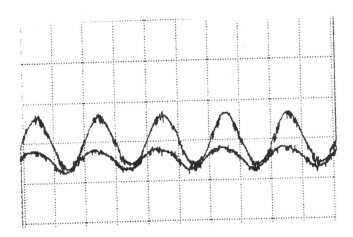

In the initial test the input amplitude was set to 25% of the stroke of the vibrator (2.5mm). The response trace comprises an alternation of a half-cycle of significantly longer period and higher amplitude than the shorter half cycle.

In the second test, where the amplitude of the vibrator is set at 50% of its maximum, the response trace of the compressive half cycle retains its integrity. The tensile half cycle exhibits a much more complex behaviour, however, consistent with the superposition of the free response of higher modes. This trace contains an abrupt change in the gradient of the acceleration, coupled with a noisy portion of the trace immediately afterwards, consistent with an impulsive effect.

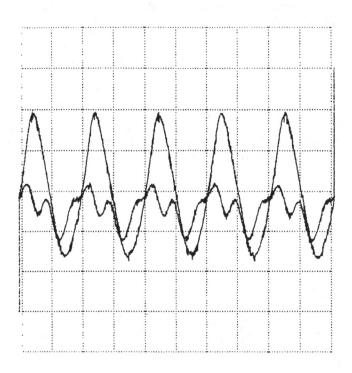

In the third test, at 75% of full stroke, the integrity of the compressive half cycle again appears to be maintained. As can be seen from the trace, the alternate half cycle can no longer be properly described as tensile, as a portion of this behaviour is compressive. It would appear that this unforced compressive phase is characterised by a higher frequency component than the forced compressive interval.

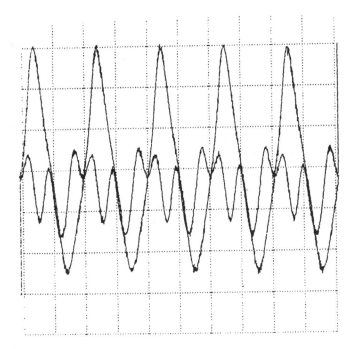

In the fourth test, with the vibrator achieving its full stroke, further
degradation of the response trace is evident. As with the third test, the
record appears to have two distinct compressive features, the first due to
the external excitation, dependent on the forcing frequency, and the second
due to the inability of the vibrator to control the response of the
specimen whilst the crack is open.

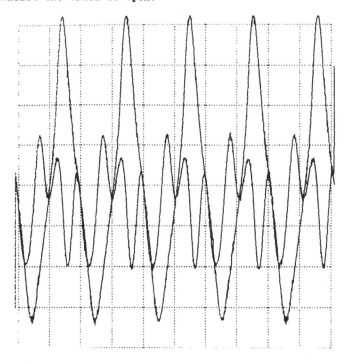

DISCUSSION

The first test provides unequivocal support for the tensile-compressive transition model and corresponds to the behaviour predicted in the companion paper [5]. The impulsive effect is less evident in the test series shown. It is perhaps most evident in the second test, at 50% of full stroke, as described above. Since the excitation is sinusoidal, however, the non-linear response of the system is established, as the response signal, although periodic at the excitation frequency, is manifestly non-sinusoidal and, consequently, contains harmonics of the excitation frequency, as would be expected from the predicted behaviour [5].

CONCLUSIONS

Analysis of the data supports the intuitive arguments put forward by Brandon and Richards on the interface dynamics of a crack. Namely, that upon closure of the crack there is an impulsive effect due to the dynamic contact of the two crack surfaces and this is accompanied by a stiffness discontinuity brought about by the increased area of contact.

REFERENCES

1. J A Brandon and J Richards, A conjecture on the interface mechanics in fractures based on the interpretation of impulse tests, Proc I Mech E, Series H, Journal of Engineering in Medicine, 1989,203(4), pp203-5

2. I R Pocock, J A Brandon and J Richards, Vibration strategies for the monitoring of fracture healing in human long bones, ASME Symposium: Advances in Bioengineering, Boston, December 1987, p89-90.

3. I R Pocock, J A Brandon, J Richards and I G Mackie, Choosing a signal for vibration monitoring of fracture healing, Interfaces in Medicine and Mechanics, Editor K R Williams, U C Swansea, 1988

4. J A Brandon, J Richards and I G Mackie, Vibration Methods for assessing fracture healing, Conference: Changing Role of Engineering in Orthopaedics, Institution of Mechanical Engineers/ Royal College of Surgeons, 14-15 April 1989

5. J A Brandon and O N L Abraham, An investigation into the interface mechanics in fractures: Part 1 theoretical models, Conference "Interfaces 90", Bologna, September 9th-14th 1990

6. M Moezedi, A Sadegh, B Liaw and A Pilla, Vibration Analysis of Bone Fractures, ASME Symposium: Advances in Bioengineering, Boston, December 1987, p91-2.

INTERFACE INTERACTIONS BETWEEN HYDROPHILIC CONTACT LENSES AND OCULAR TISSUES

A.BERTOLUZZA, P.MONTI
Dip. di Biochimica, Sez. di Chimica e Propedeutica Biochimica, Centro di
Studio Interfacolta' sulla Spettroscopia Raman, University of Bologna
Via Selmi 2, 40126 Bologna (Italy)

R.SIMONI
Dip. di Chimica "G.Ciamician", Centro di Studio Interfacolta' sulla
Spettroscopia Raman,
Via Selmi 2

R.CARAMAZZA
Istituto di Clinica Oculistica, University of Bologna
Via Massarenti 9, 40125 Bologna

ABSTRACT

Surface problems concerning hydrophilic soft contact lenses are
discussed in this note with particular regard to the different aspects
of biocompatibility. The surface structures of the lenses, deduced from
vibrational spectroscopic analysis, hydrophilicity, protein deposits and
interactions with model systems, are discussed. Since vibrational laser
Raman and ATR/FTIR spectroscopies are non-invasive and non-destructive,
they are suitable and promising techniques for this type of
investigation.

INTRODUCTION

The different fields of pure and applied physical-chemistry are greatly
interested in interface interactions, for example in heterogeneous
catalysis, in chemical adsorption and corrosion phenomena.

Every material in the solid state has surface properties different
from those of the bulk because of its finite structure. These surface
properties change from material to material (metal, alloy, oxide,
polymer), according to the type of chemical bond present in the solid,
the surface "geometry" and defects.

The quantum-mechanical studies of Tamm [1] and Sockley [2] on surface electronic states, show the existence of localized energy levels on metallic or alloy surfaces which lie in the energy gap for the electrons of the central bulk and are responsible for surface reactivity. From a qualitative view this fact is due to the unsaturation of the bonds which are on the surface of the solid.

The concept of surface electronic states can be enlarged to materials different from metals or alloys (i.e. oxides, ionic salts, inorganic and organic polymers, etc.) and, consequently, to biomaterials, depending on the types of chemical bonds. In particular, in organic polymers, which are molecular solids, the molecules interact among themselves by weak bonds, therefore they have a surface reactivity less than metals or alloys, oxides, etc. which are, on the contrary, characterized by stronger bonds of the covalent with a lack of electrons, ionic or covalent-ionic type respectively. Moreover, in the case of hydrophilic polymers which are the subject of this note, the surface is covered by a layer of water molecules which can modulate the biomaterial-tissue interactions according to the turnover between water present on the surface of the polymer and that of the surrounding biological environment.

This case represents a typical example where the surface interaction plays an important role in the correlations between the structure and properties of a biomaterial. In fact a hydrophilic polymer for ophthalmological use must have a bulk structure such as to assure suitable physical and mechanical properties (physical biocompatibility) and a surface structure such as to show the least reactivity (chemical biocompatibility). Moreover its interactions with host tissues must not lead to the formation of adducts (for example protein deposits) which can limit its use (biological biocompatibility).

On the basis of these considerations, hydrophilic soft contact lenses based on crosslinked poly-2-hydroxyethylmethacrylate (PHEMA) and polyvinylpyrrolidone (PVP) will be discussed in this note. In particular the different aspects of biocompatibility will be correlated to the structure of such biomaterials. As regards the chemical and biological biocompatibility, the surface structure of the lenses (deduced from vibrational spectroscopic analysis, hydrophilicity, protein deposits and the interactions with model systems) will be discussed.

The techniques used in this research, i.e. vibrational laser-Raman and ATR/FTIR spectroscopies, are suitable for this type of investigation since they are non-invasive and non-destructive and are complementary to each other.

MATERIALS AND METHODS

The lenses studied in this work are commercial soft contact lenses based on polyvinylpyrrolidone (PVP) and poly-2-hydroxyethylmethacrylate (PHEMA).

The Raman spectra were recorded using a JASCO R500 spectrophotometer with a Spectra-Physics Ar^+ 488 nm laser source. The hydrated lenses were examined directly in their saline solutions.

The infrared spectra were obtained with a JASCO FT/IR 7000 Fourier Transform Infrared spectrometer using, for surface spectra, a JASCO 500/M Attenuated Total Reflection (ATR) unit.

RESULTS and DISCUSSION

The hydrophilicity of a soft contact lens is one of the most important parameters for biocompatibility. In fact, water absorbed by hydrophilic polymers is the carrier which assures the permeability to metabolites, in particular to oxigen, thus allowing the normal physiological activity of the cornea. Moreover, hydrophilicity affects the elastic properties (Young's modulus) of the lens making them compatible with those of ocular tissues below.

The soft contact lenses based on PHEMA have a $\sim$ 40% w/w water amount, while those based on PVP absorb $\sim$ 70% w/w of water. Our previous differential scanning calorimetric (DSC) measurements [3,4] pointed out that different types of water (free and bound) are present both in PHEMA and PVP lenses. This swelling water completely exchanges with that of the biological environment, as Raman measurements on an explanted lens, which was previously swelled in heavy water, have shown [5]. Moreover, Raman spectroscopy allows us to study the interactions of water with both the hydrophilic and hydrophobic centres of the two polymers.

As an example, Figures 1a and 1b show the Raman spectra of a dry and hydrated PVP lens respectively. Some meaningful changes are observed going from the dry to the hydrated material. In particular a stronger

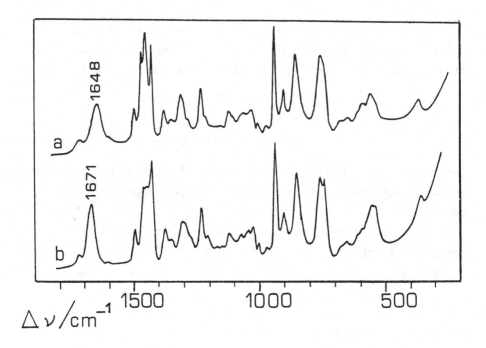

Figure 1. Raman spectra of hydrated (a) and dry (b) soft contact lens
based on PVP.

shift of the C=O stretching vibration of PVP occurs from 1671 to 1648
cm^{-1}. This shift towards the lower wavenumbers is indicative of the
formation of hydrogen bonds between the oxigens of the carbonylic groups
and the hydrogens of the water molecules. Also the bands between 1500
and 1400 cm^{-1}, which are due to the bending vibrations of CH$_2$ groups of
the polymer, modify their intensities and frequencies. These
spectroscopic trends show that water deeply modifies the polymer
structure, not only by binding itself to the hydrophilic C=O groups
through hydrogen bonds, but also by affecting the hydrophobic CH$_2$
groups, thus changing the conformation of the polymer. The same
behaviour is observed in the case of PHEMA lenses, even though, in this
case, the lower shift of the methacrylic C=O groups is indicative of a
weaker hydrogen bonding interaction.

As previously said, the bulk water affects the physical properties
of an hydrophilic lens and in particular the optical properties (for
example the refractive index). On the contrary, the surface water plays
an important role in the biomaterial-tissue interface interactions.

Measurements on bacterial adhesion of staphylococcus aureus on new
lenses have shown that bacteria adhere more to less hydrophilic lenses
(PHEMA) than to higher hydrophilic ones (PVP), whilst the adhesion on
regenerated lenses was independent from hydrophilicity [6]. This
behaviour can be explained by assuming that the different surface
reactivities of the new lenses are conditioned by the different
thicknesses of the surface water layers, which strictly depends on
hydrophilicity, whilst the almost similar reactivity of the regenerated
lenses is due to the presence of a protein layer, which cannot be
removed by the normal cleaner.

The presence of protein deposits can be shown by ATR/FTIR
technique. In Figure 2a the surface spectrum of a PVP lens which has
been worn by a patient for 15 days is reported. Figure 2b shows the
spectrum of the same lens before application on patient; the spectrum in
Figure 2c is the a-b difference spectrum which, as it can be seen,
presents two main bands, between 1700 and 1600 cm^{-1} and between 1600 and
1500 cm^{-1} respectively. Such bands, termed Amide I and Amide II, are
characteristic of proteins since they arise from peptidic group
vibrations. Nevertheless, the vibrational study of the interaction
between protein and lens surface is rather complex since it involves two
systems, i.e. polymer and protein, whose spectroscopic behaviours are
already difficult to interpret by themselves. Moreover it becomes
important to establish how the protein is adsorbed on the lens surface,
as well as according to the changes in water content. It is also
important to verify the conditions in which an irreversible adsorption
occurs with protein denaturation and consequent intolerance by patients.
A such complete study as this needs simple models, which simulate
complex biological systems and are able to give useful informations on
conformational changes in the adsorbed protein.

If on the one hand the general model of the protein-lens surface
interaction through water needs a modification in the set up of the
method, already in progress, on the other hand it is able to explain our
spectroscopic results on patients wearing highly hydrophilic soft
contact lenses in a 3 months follow-up and affected by intolerance [6].
These results point out that a correlation seems to exist between the
increased surface reactivity of the lens, which causes an increase in
protein deposits, and the loss of water occurring during the application
of the lens, experimentally measured.

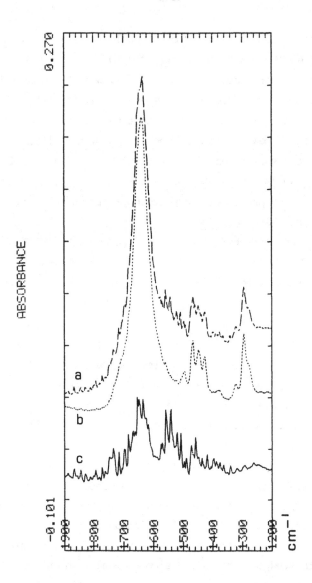

Figure 2. ATR/FTIR spectra of worn PVP lens (a), new PVP lens (b), difference spectrum a-b (c).

This investigation shows how the surface properties play an important role in the evaluation of both surface reactivity (chemical biocompatibility) and molecular interaction with host tissues (biological biocompatibility).

From this consideration a new and promising line of research on the planning of new biomaterials arises. This line of research is not so tuned to the planning of new biomaterials, which need a very long follow-up, as to a suitable surface coating of biomaterials of tested biocompatibility and suitable bulk structure.

REFERENCES

1. Tamm, I.E., A possible kind of electron binding on crystal surfaces, Phys. Z. Sowjet, 1932, 1, p.733.
2. Shockley, W., Surface states associated with a periodic potential, Phys. Rev., 1939, **56**, p.317.
3. Bertoluzza, A., Monti, P., Garcia-Ramos, J.V., Simoni, R., Caramazza, R. and Calzavara, A., Applications of Raman spectroscopy to the ophthalmological field: Raman spectra of soft contact lenses made of poly-2-hydroxyethylmethacrylate (PHEMA), J. Mol. Struct., 1986, **143**, pp. 469-472.
4. Bertoluzza, A., Monti, P., Simoni, R., Garcia-Ramos, J.V., Caramazza, R., Cellini, M., De Martino, L. and Calzavara, A., Applications of Raman spectroscopy to the ophthalmological field: Raman spectra of soft contact lenses made of polyvinylpyrrolidone (PVP), Studies in Physical and Theoretical Chemistry, vol. 45, Eds. Stepanek, J., Anzenbacher, P. and Sedlacek, B., Elsevier, Amsterdam, 1987, pp. 595-604.
5. Bertoluzza, A., Monti, P. and Simoni, R., unpublished data.
6. Arciola, C., Versura,P., Ciapetti,G., Monti,P., Pizzoferrato,A. and Caramazza,R., Evaluation of adhesion capability of Conjunctival staphylococcus Aureus on polymeric materials, in Implant Materials in Biofunction, Eds. de Putter, C., de Lange, G.L., de Groot, K. and Lee, A.J.C., Elsevier Science Publishers, Amsterdam, 1988, pp. 349-354.

STUDY OF THE BONE-BIOMATERIAL INTERFACE REACTIONS IN AN IN VITRO BONE FORMING SYSTEM: A PRELIMINARY REPORT

J.D. de Bruijn, C.P.A.T Klein, R.A. Terpstra[*]**, K. de Groot, C.A. van Blitterswijk**
Biomaterials Research Group, Ear Nose & Throat dept., University of Leiden, [*]ECN Petten, Rijnsburgerweg 10, 2333 AA Leiden, The Netherlands

ABSTRACT

An *in vitro* bone forming system is described in which interface reactions were studied with calcium phospates. A layer of hydroxyapatite or fluorapatite was applied on tissue culture plastic using the plasma-spray technique. Interactions between the bone forming cells and the extra-cellular matrix on one hand and the plasma-sprayed calcium phosphates on the other hand, were studied using light microscopy (LM), scanning electron microscopy (SEM), transmission electron microscopy (TEM) and X-ray microanalysis (XRMA).

Collagen fibrils were seen attached parallel to the plasma-sprayed calcium phosphate surface, after which needle-shaped crystals were deposited onto the materials from 4 weeks on. Frequently an electron-dense, lamina limitans-like zone was present between the newly formed bone and the calcium phosphates. The presence of this layer points to an active contribution in normal bone metabolism. Both hydroxyapatite and fluorapatite seemed to behave as nucleation sites for crystal precipitation.

The results of this paper show that by using this *in vitro* bone forming system interface reactions with biomaterials can be studied.

INTRODUCTION

When implanted into bone, certain surface-active materials can form a physico-chemical bond with this tissue. This bone bonding capacity has been described for glass-ceramics, bioglass [1,2], calcium phosphate ceramics [3,4,5,6] and recently for a copolymer [7]. To gain information about the mechanisms responsible for the formation of this bond, study of the tissue/biomaterial interface reactions is essential. Several authors published *in vivo* studies in which interfacial reactions between calcium phosphate ceramics and bone were studied. A review of these studies revealed that variation in chemical composition and crystallinity had effect on the reactions at the interface and the biodegradation rate of the implant materials [2-4,8,9]. To elicit the complexities of the *in vivo* responses after implantation, *in vitro* cell culture systems have been developed [10-12]. Several *in vivo* studies suggest that collagen bonding with the substrate and the subsequent calcification process are the fundamental stages required for the establishment of a bone bond [3,12,13]. These phenomena can be studied in an *in vitro* system by growing osteogenic cells, that are able to form a calcified extracellular matrix, onto biomaterials.

Hydroxyapatite is the most frequently used calcium phosphate because of its good bone bonding capacity and its low dissolution rate [8]. The plasma-spray technique in which calcium phosphates can be deposited onto metals is used to combine strength of the metals with the bone bonding capacity of calcium phosphates. Substitution of fluorine for the hydroxyl ions in hydroxyapatite forms fluorapatite that has a lower solubility rate than hydroxyapatite [14,15]. Hench [16] suggested that addition of fluorine to form fluorapatite might enhance bone formation and that it would be more beneficial for the surrounding bone. Another advantage of fluorapatite is stated by Lugscheider et al. [14] who found no dissociation during the plasma-spray process, while hydroxyapatite showed a slight dissociation.

The aim of the present study was twofold. First, a similar *in vitro* bone forming system as described by Maniatopoulos [10] was developed and characterized. Second, the effect of crystal structure and chemical composition was examined in a preliminary study in which the bone forming cells were cultured on either plasma-spray coated hydroxyapatite or fluorapatite.

MATERIALS & METHODS

Cell isolation and culture

Bone marrow cells were isolated by a method that has been described by Maniatopoulos [10]. Femora of 100-120 g adult Wistar rats were excised and after removal of adhering soft tissues passed through three washes of sterile phosphate buffered saline (PBS) that contained twice the concentration of antibiotics that is present in the culture medium (see below). The epifyses were removed and the marrow was collected by flushing out the diafyses with sterile PBS. The cell suspension was centrifuged for 5 minutes at 100g, the pellet resuspended with culture medium and the cells were plated in 80cm^2 plastic culture flasks. Cultures were grown in α-minimal essential medium (α-MEM DNA/RNA, Gibco) supplemented with 15% foetal calf serum (FCS, Gibco), antibiotics (100 U/ml penicillin and 100 μg/ml streptomycin; Boehringer-Mannheim, FRG) and freshly-added 10^{-7}M dexamethasone (Sigma), 10mM Na ß-glycerophosphate (Gibco) and 100μg/ml ascorbic acid (Gibco). Cultures were incubated in a humidified atmosphere of 90% air, 10% CO_2 at 37°C. The medium was changed every second day. At confluency, after trypsinisation, cells were subcultured. Cells of the third passage were used for the experiments, and plated at a density of 1×10^4 cells/cm^2 into 6-well tissue culture plates (Costar).

Plasma-sprayed materials

A coating of hydroxyapatite ($Ca_{10}(PO_4)_6(OH)_2$) or fluorapatite ($Ca_{10}(PO_4)_6F_2$) was applied using the plasma-spray technique [17] on 6-well tissue culture plates (Costar) or 12 mm round coverslips (Thermanox^{tm}). The X-ray diffraction pattern of the hydroxyapatite and fluorapatite coatings showed a crystalline structure. For culture experiments the plasma-sprayed material was sterilized by ^{60}Co gamma-irradiation (Gamma-ster, Ede, the Netherlands).

Light microscopy (LM)

Cells were fixed in 1,5% glutaraldehyde in 0.14M sodium cacodylate (pH 7.4, 4°C) for 30 minutes, before being dehydrated through a graded series of ethanol and embedded in glycolmethacrylate. Semi-thin sections were processed, routine stained with toluidin blue and embedded in depex mounting.

Alkaline phosphatase activity (ALP-activity): ALP-activity was measured by using the AZO-dye method of Gomori. The substrate was composed of 2mg/ml Na α-naphtyl and 1mg/ml Fast Blue RR salt, solved in 0.1M Na-barbiturate buffer pH 9.2. Fixed cells were incubated for 10 minutes with the substrate whereafter they were thoroughly rinsed in tap water. Specificity for ALP-activity was determined by incubating control cells with the substrate that lacked α-naphtyl.

Electronmicroscopy and X-ray microanalysis

Transmission electronmicroscopy (TEM): Cells were fixed according to the light microscopical procedures. After rinsing in 0.14M sodium cacodylate, postfixation was carried out with the modified OsO_4 fixative according to de Bruijn et al. [18] for 16 hours at 4°C. Cells were dehydrated through a graded series of alcohol and embedded in Epon. Ultra-thin sections were prepared on a LKB Ultramicrotome, stained with uranylacetate and leadcitrate and examined at 80kV in a Philips EM 201.

Scanning electronmicroscopy (SEM): Specimens were fixed according to the light microscopical procedures, after which they were dehydrated through a series of ethanol to 100% and critical point dried under carbon dioxide in a Balzers model CPD 030 Critical Point Dryer. A layer of gold or carbon was sputter coated with a Balzers sputter coater model MED 010 onto the specimens and they were examined in a Philips S 525 scanning electron microscope at an accelerating voltage of 15 kV.

X-ray microanalysis (XRMA): Unstained ultra-thin sections were used. Single spot XRMA was performed with a Tracor Northern (TN) 2000 X-ray microanalyser attached to a Philips EM 400 scanning transmission electron microscope. Spot diameter was 100 nm, accelerating voltage 80 kV and measurements were performed during 100 seconds livetime.

RESULTS

In vitro bone forming system

Primary cultures reached confluency 7 to 9 days after plating and were composed of several cell types. Spindle-shaped, osteoblast-like cells were the most prominent cells in the bone marrow culture and formed colonies. Subcultures reached confluency after 4 to 6 days and from 1 week on nodule formation was observed. Until the sixth passage the cells retained the ability to form calcified nodules. Further passages were not studied.

Lightmicroscopical observations: When stained by the AZO-dye method of Gomori, cells at the periphery of the nodules showed an intense alkaline phosphatase activity (fig.1). The nodules were composed of cells surrounded by an osteoid-like matrix. Cells present at the periphery of the nodules resembled osteoblasts, whereas cells that were embedded in the matrix showed morphological characteristics of osteocytes. The nodules gradually increased in size and became mineralized from 8 weeks on as observed with the light microscopical calcium stain.

Transmission electronmicroscopical observations: Examination of the nodules on ultrastructural level revealed cells surrounded by an extracellular matrix (ECM). This matrix consisted of well banded fibrillar collagen with a 64-67 nm periodicity (fig.2). Cells in the matrix often contained deposits of intracellular glycogen and fat droplets (fig.3). Whereas cells at the periphery were rich in endoplasmic reticulum, golgi complexes and microfilaments. In contrast

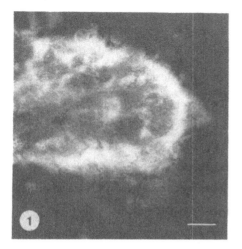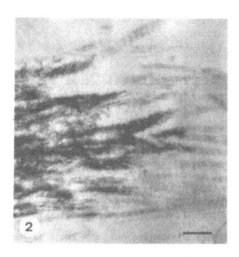

figure 1. Negative photograph of part of a nodule 10 weeks after culture, stained for ALP-activity. Note high activity at periphery of nodule. bar= 50 μm

figure 2. TEM showing partially calcified, well-banded collagen with a 64-67nm periodicity. Note parallel arrangement of needle-like crystals with collagen fiber axis. 8 weeks, bar= 0.26 μm to the light microscopical observations, TEM revealed that extracellular matrix calcification started from 4 weeks on with the formation of small mineralization spots. These spots gradually increased in size and were composed of needle-shaped crystals. Crystals were observed to be arranged parallel to the axis of the collagen fibrils (fig.2). They were composed of calcium and phosphorus as was shown with X-ray microanalysis. After 8 weeks of culture osteocyte-like cells were observed that were surrounded by a heavily calcified extracellular matrix (fig.4). An uncalcified osteoid zone was frequently seen between the osteocyte-like cells and the calcified extracellular matrix (fig.3). Decalcified sections showed that the periphery of the calcified extracellular matrix was composed of electron dense material, similar to the lamina limitans that is present in bone (fig.4).

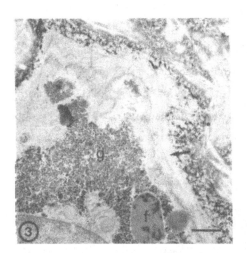 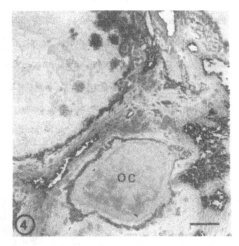

figure 3. Decalcified TEM section, 8 weeks of culture. Note osteoid zone (arrow), intracellular glycogen (g) and fat droplets (f) in osteocyte. bar= 0.8 μm
figure 4. TEM section, 8 weeks of culture. Osteocyt (OC) in a highly calcified ECM bar= 2.6 μm

Interactions between bone forming cells and calcium phosphates
Figure 5 shows the surface of hydroxyapatite after it has been plasma-sprayed onto coverslips. Individual particles are melted togeter to form a relatively smooth surface.
No sign of increased cell death could be detected when cultures were grown onto the plasma-sprayed materials. Cells adhered and spread out over the ceramic materials, and initially no distinct difference could be seen between the two plasma-sprayed materials and the control. Cell multi-layers were formed within two weeks of culture and a birefringent extracellular matrix was distinguished using polarized light microscopy.

figure 5. SEM of plasma-sprayed hydroxyapatite layer. bar= 19 μm

Scanning electronmicroscopical examination: SEM revealed that cells and fibrous tissue were adhered onto the ceramic surfaces (figs.6,7). No striking difference was observed between hydroxyapatite and fluorapatite. When cells were grown onto the ceramic materials, the normally smooth surface of the ceramics sometimes had a granular appearance (fig.6). On uncoated control coverslips the multi-layer was easily stripped off using a fine pair of tweezers. On hydroxyapatite and fluorapatite coated coverslips however, the cell multi-layer was firmly bound and could only be scraped off with force. This resulted in the presence of well bound particles into the cell multi-layer, that had loosened from the coating (fig.7).

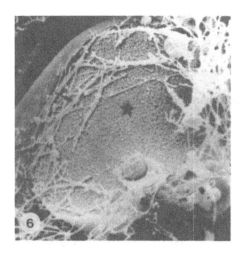 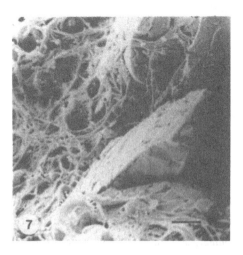

figure 6. SEM of fluorapatite after 4 weeks of culture. Note collagen fibrils (arrow) and granular surface (★) of fluorapatite. bar= 3 μm

figure 7. SEM of hydroxyapatite (HAP) after 8 weeks of culture. Cell layer has been scraped off the HAP-coating. Note well bound collagen fibrils (arrow) and loosened HAP particles. bar=2 μm

426

Transmission electronmicroscopical observations: Examination of the cultures on ultrastructural level showed that from 1 week on well banded collagen fibrils were seen running parallel to the surface of the ceramic materials. These fibrils were frequently seen in contact with the ceramic surface (fig.8). From 4 weeks on a calcified matrix could be seen deposited onto the ceramic materials (figs.9,10). Collagen fibrils were distinguished in this matrix and at the periphery of this calcified matrix an electron dense mineralization zone was observed. This zone was also frequently seen in conjunction with the ceramic surface (fig.10) and resembled a lamina limitans. X-ray microanalysis showed that the needle shaped crystals that were deposited either on collagen fibrils or on the ceramic surface were composed of calcium and phosphorous (fig.11). Small detached particles were frequently seen phagocytosed by surrounding cells (fig.12).

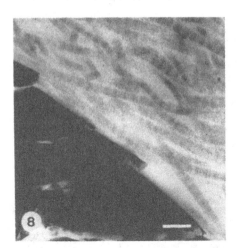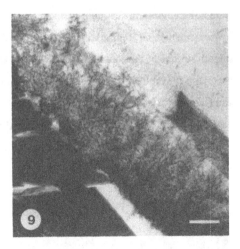

figure 8. TEM interface of HAP after 2 weeks of culture. Collagen fibrils are parallel to, and in contact with the HAP surface. bar= 0.2 µm
figure 9. TEM interface of HAP after 4 weeks of culture. Needle-shaped are deposited onto HAP. bar= 0.14 µm

DISCUSSION

Several *in vitro* systems have been described in literature in which interactions between biomaterials and cultured bone cells are studied [12,19-21]. However, to study the physicochemical bonding phenomena of biomaterials, an *in vitro* system is required that resembles the *in vivo* situation. Therefore we used the *in vitro* system described by Maniatopoulos et al. [10] in which rat stromal bone marrow cells form a calcified extracellular matrix. The osteogenic

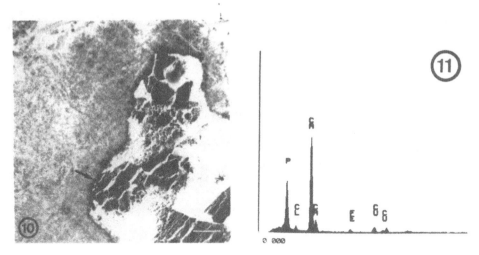

figure 10. TEM section showing a highly calcified ECM deposited onto HAP after
4 weeks of culture. Note lamina limitans at interface (arrow). bar= 0.6 μm
figure 11. XRMA on calcified ECM in fig.10 showing the presence of calcium (Ca)
and phosphorous (P).

potential of this system is displayed by Maniatopoulos by showing the presence of bone Gla-
protein (osteocalcin), osteonectin and collagen type I. Except ALP-activity, our observations
mainly refer to morphological characteristics by showing the presence of intracellular deposits
of glycogen and fat droplets in osteoblasts [22], a 64-67 nm collagen band periodicity, the
presence of a lamina limitans and the presence of an osteoid zone.

    This *in vitro* system shows that collagen is deposited parallel to the surface of the calcium
phosphates. This is in accordance to *in vivo* observations done by Hench et al. [1], who found
collagen fibrils juxtaposed to glass-ceramic and parallel to its surface. The parallel arrangement
of the collagen fibrils can be explained as an interaction with the PO_4^{3-} groups of the calcium
phosphate ceramics [1]. In fact this arrangement is the same as we study the plane of needle-like
crystals and the collagen fibrils (see fig.2). We did not observe any difference in interface
reactions between hydroxyapatite and fluorapatite. With the observation that mineralization starts
onto the ceramic materials only when cells are present, crystal deposition seems to be caused by
the osteogenicity of these cells. However, the presence of the needle-shaped crystals could also
be due to partial dissolution of the ceramics, followed by reprecipitation and secondary
nucleation [23]. Confirmation for bone specificity however, can only be performed by displaying
bone specific proteins in this zone. All observed interface reactions must somehow be cell

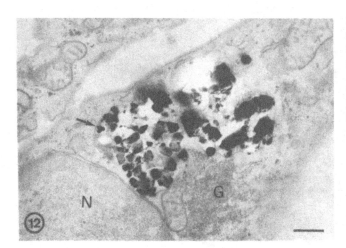

figure 12. TEM section showing phagocytosed HAP (arrow) in a residual body of an osteoblast-like cell. 8 weeks of culture.
bar= 0.4 μm
N=nucleus
G=glycogen

mediated, because incubating the plasma-sprayed materials with culture medium in a cell-free environment results in an unaltered ceramic surface. The presence of a lamina limitans-like structure at the periphery of the calcification spots and at the interface between calcium phosphates and the bone-like tissue, is similar to *in vivo* observations done by van Blitterswijk et al. [6,24], who stated that the presence of this layer points to an active contribution in normal bone metabolism.

All these data suggest that the available *in vitro* bone forming system can be used to study interface reactions between biomaterials and bone-like tissue. Currently interactions between this system and a series of calcium phosphates are under investigation.

Acknowledgements:
The authors gratefully acknowledge CAM-Implants B.V. for plasma-spraying the materials, Mr. J.G.C. Wolke for technical assistence and Mr. L.D.C. Verschragen and Mr. B. van der Lans for preparing the photographic material.

REFERENCES

[1] Hench, L.L., Splinter, R.J., Allen, W.C. and Greenlee, T.K., Bonding mechanisms at the interface of ceramic prosthetic materials, J. Biomed. Mater. Res. Symp., 1971, 2(partI), 117-41.

[2] Fujiu, T. and Ogino, M., Difference of bone bonding behaviour amoung surface active glasses and sintered apatite, J. Biomed. Mater. Res., 1984, 18, 845-59

[3] Jarcho, M., Calcium phosphate ceramics as hard tissue prosthetics, Clin. Orthop. Rel. Res., 1981, 157, 259-78.

[4] Osborn, J.F. and Newesely, H., Dynamic aspects of the implant-bone-interface, In Dental Implants, ed. G. Heimke, Carl Hansen Verlag, 1980, pp. 111.

[5]   Tracy, B.M. and Doremus, R.H., Direct electron microscopy studies of the bone-hydroxylapatite interface, J. Biomed. Mater. Res., 1984, 18, 719-26.

[6]   Blitterswijk, C.A. van, Hesseling, S.C., Grote, J.J., Koerten, H.K. and Groot, K. de, The biocompatibility of hydroxyapatite ceramic: A study of retrieved human middle ear implants, J. Biomed. Mater. Res., 1990, 24, 433-53.

[7]   Bakker, D., Blitterswijk, C.A. van, Hesseling, S.C., Daems, W.Th. and Grote, J.J., Tissue/biomaterial interface characteristics of four elastomers. A transmission electron microscopical study, J. Biomed. Mater. Res., 1990, 24, 277-93.

[8]   Klein, C.P.A.T., Calcium phosphate implant materials and biodegradation, Ph.D. Thesis, University of Amsterdam, The Netherlands, 1983.

[9]   Blitterswijk, C.A., Calcium phosphate middle-ear implants, Ph.D. Thesis, University of Leiden, The Netherlands, 1985.

[10]   Maniatopoulos, C., Sodek, J. and Melcher, A.H., Bone formation in vitro by stromal cells obtained from bone marrow of young adult rats, Cell Tiss. Res., 1988, 254, 317-30.

[11]   Davies, J.E., Tarrant, S.F. and Matsuda, T., Interaction between primary bone cell cultures and biomaterials part I: Method; The in vitro and in vivo stages, In Biomaterials and clinical applications, ed. A. Pizzoferrato, Elsevier Science Publishers BV, Amsterdam, 1987.

[12]   Harmand, M.F., Bordenave, L., Duphil, R., Jeandot, R. and Ducassou, D., Human differentiated cell cultures: in vitro models for characterization of cell/biomaterial interface, In Biological and biomechanical performance of biomaterials, ed. P. Christel, Elsevier Science Publishers BV, Amsterdam, 1986.

[13]   Davies, J.E. and Matsuda, T., Extracellular matrix production by osteoblasts on bioactive substrata in vitro, Scan. Microsc., 1988, 2(3), 1445-52.

[14]   Lugscheider, E., Weber, Th. and Knepper, M., Production of biocompatible coatings of hydroxyapatite and fluorapatite, Presented at National thermal spray conference, Cincinatti, 1988.

[15]   Heling, I., Heindel, R. and Merin, B., Calcium-fluorapatite. A new material for bone implants, J. Oral. Implantol., 1981, 9, 548-55.

[16]   Hench, L.L. and Ethridge, E.C., Biomaterials, an interfacial approach, Acad. Press, Biophys. and Bioeng. series, Vol. 4, 1982.

[17]   Groot, K. de, Geesink, R., Klein, C.P.A.T. and Serekian, P., Plasma sprayed coatings of hydroxyapatite, J. Biomed. Mater. Res., 1987, 21, 1375-81.

[18]   Bruijn, W.C. de, Glycogen, its chemistry and morphological appearance in the electron microscope I: A modified OsO_4 fixative which selectively contrasts glycogen, J. Ultra Struct. Res., 1973, 42, 29-50.

[19]   Gregoire, M., Orly, I. and Menanteau, J., The influence of calcium phosphate biomaterials on human bone cell activities. An in vitro approach, J. Biomed. Mater. Res., 1990, 24, 165-77.

[20]   Matsuda, T. and Davies, J.E., The in vitro response of osteoblasts to bioactive glass, Biomaterials, 1987, 8, 275-84.

[21]   Uchida, A., Nade, S., McCartney, E. and Ching, W., Growth of bone marrow cells on porous ceramics in vitro, J. Biomed. Mater. Res., 1987, 21, 1-10.

[22]   Scott, B.L. and Glimcher, M.J., Distribution of glycogen in osteoblasts of the fatel rat, J. Ultrastruct. Res., 1971, 36, 565-86.

[23]   Daculsi, G., Legeros, R.Z., Heughebaert, M. and Barbieux, I., Formation of carbonate-apatite crystals after implantation of calcium phosphate ceramics, Calc. Tiss. Int., 1990, 46, 20-27.

[24]   Blitterswijk, C.A. van, Grote, J.J., Kuijpers, W., Blok-van Hoek, C.G.J. and Daems, W.Th., Bioreactions at the tissue/hydroxyapatite interface, Biomaterials, 1985, 6, 243-51.

GROWTH HORMONE STIMULATES OSTEOID FORMATION AT A BONE-CERAMIC INTERFACE.

Rachel Hann, Mike Kayser, *Christel Klein and Sandra Downes.

Institute of Orthopaedics (U.C.L.), Stanmore, Middx., England, HA7 4LP.

*Biomaterials Research Group, Dept. of Biomaterials, Leiden University,

School of Medicine, Leiden, Holland.

ABSTRACT

Tri-calcium phosphate (TCP) ceramic pins have been loaded with human growth hormone (GH) and inserted into rabbit femora. A new method for processing bone with biomaterial implants has enabled histomorphometric analysis of the tissue response to insertion of the TCP pins at an intact interface.

This in vivo study demonstrates that loading TCP ceramic with human growth hormone significantly increases the percentage of osteoid tissue that forms on the bone surface adjacent to the ceramic, when compared to a normal bone-ceramic interface. We propose that growth hormone stimulation of early osteoid formation at the interface could lead to increased remodelling of bone and improve the success rate of implantation of ceramic coated prostheses.

INTRODUCTION

Ceramic coated implants have been widely used in surgery since Smith [1] reported favourable results with these materials.

Calcium phosphate ceramics are considered to be 'bioactive' materials, ceramic dissoloution being a causative factor of bone tissue growth enhancement [2].

Downes et al., [4] have demonstrated that growth hormone can be released from bone cement and that it stimulates osteoid formation at a bone-cement interface, one month after implantation. In this work we use ceramic to deliver human growth hormone and investigate the in vivo response in a rabbit model.

Until recently difficulties encountered in preparing material for morphological and ultrastructural examination have hindered observation at a cellular level of the bone tissue reaction to implantation with these materials. Using the technique of Kayser et al. [3], observation of morphological and ultrastructural integration of bone with biomaterials at an intact interface is possible.

In this study we have examined the early remodelling that occurs when TCP ceramic is implanted into bone. Histomorphometric analysis has been carried out to quantitate the tissue response to plain TCP ceramic and compare this with growth hormone loaded ceramic. Although the full effect of growth hormone on osteoblasts is not yet known, we are able to report, through histomorphometric analysis of intact bone-ceramic interfaces, that the release of growth hormone stimulates osteoid formation in vivo. This implies that not only are ceramics "bioactive' but, when loaded with growth hormone, they become truly osteogenic.

MATERIALS AND METHODS

Preparation of TCP pins: Tricalcium phosphate ceramic, $Ca_3(PO_4)_2$, obtained from BDH was used to make calcium phosphate ceramic discs. The powders were compressed in a stainless steel die up to an applied pressure of 80 MN/m^2. After the powder compact was pushed out, it was first sintered at 800°C, then pins with diameter 2.5mm and length 5mm were manufactured and sintered at 1150°C. Tricalcium phosphate ceramic pins were loaded by placing five pins in a solution of growth hormone (12U/ml) for 24 hours. The pins were allowed to dry and stored at 4°C.

Rabbit model: A mature Sandy Lop rabbit of at least 3.5 Kg weight was used in this study. Three holes were drilled, using 3mm diameter drill bits, into the lateral cortex of each femur and growth hormone loaded pins were implanted on one side. Three plain pins were inserted in to the contralateral femur as a control. The rabbit was kept unrestrained for one month. After sacrifice the femora were dissected out and each was divided into three sections, each containing a pin surrounded by intact bone.

Processing of blocks: One pair of pins was processed for histomorphometric analysis by fixation in 70% alcohol. Two pairs of pins, processed for electron microscopical studies, were fixed in 2.0% glutaraldehyde in 0.1M sodium cacodylate buffer at pH 7.2 for 48 hours at and then washed in buffer. All were dehydrated then through a graded series of alcohols (70%, 90%, 100%) and impregnated with a 1:1 absolute alcohol/Spurrs' resin mixture for 12 hours. Neat resin was changed every 5 days for a total of 21 days with intermittent vacuum impregnation at 150mbar for several hours every day. Whole sections were blocked and polymerised at 70°C for 18 hours. Excess resin was trimmed away and sections 0.5mm thick were sliced using a diamond cutter, washed in absolute alcohol and soaked in Spurrs' resin for a further 24 hours. The slices were placed in embedding moulds with fresh resin and polymerised at 70°C for 18 hours. Selected areas were removed and the ceramic trimmed leaving just sufficient ceramic at the interface for these studies. The blocks were then re-embedded using flat silicone rubber moulds and polymerised at 70°C for 18 hours. 1 micron sections were cut on a Reichert ultracut E. For further details of the methods employed see Kayser [3].

Treatment of sections: One micron sections were stained with methylene blue-azure II-basic fuchsin [5]. These were viewed using an Olympus BH-2 photomicroscope with Merz graticule [6] for histomorphometric analysis. Ten sections (five from each side) of the plain and GH loaded pins were analysed in a blind test completed by four readers to give an average result using a x40 objective lens. Scores were counted along the bone surface parallel to the ceramic pin. At the point that the graticule line crossed the bone surface a count was made of the tissue proximal to the bone. This was either classed as osteoid, cells, if the line crossed a cell immediately adjacent to the bone surface, or mineral if neither osteoid or cells were in immediate contact with the bone surface. Values were calculated as a percentage of the total points scored along the bone interface.

RESULTS

Histology
Neither the GH loaded or plain ceramic pins were in direct contact with the bone. The gap contained various cell types including marrow cells and fibroblasts. Cells were also seen lining the ceramic surface. At the bone surface there were areas of plain mineral where the bone abruptly apposed the space. Occasionally cells were adjacent to the mineral surface but, more usually, the bone was covered by a layer of osteoid tissue that frequently had more cuboidal cells, assumed to be active osteoblasts, incorporated into the layer. Comparison of the interface between ceramic and bone clearly showed that there was increased osteoid formation along the bone surface that was close to the growth hormone loaded TCP-pin. (Fig: a and b). The GH-loaded pin also had a densely pink staining tide-mark of new mineral parallel to much of the osteoid tissue. This was rarely seen to be associated with the plain pin.

Histomorphometric analysis.
Histomorphometric analysis of the matrix adjacent to the mineral surface demonstrated that two thirds of the bone surface was osteoid in the GH-loaded pin, whereas close to the plain pin from the contralateral femur, less than one third of the bone surface appeared to be covered by osteoid tissue.

TABLE 1
Percentage of the bone surface covered with osteoid, cells or exposed as mineral.

	Osteoid	Mineral	Cells
LOADED	66.5%	29.5%	3.9%
UNLOADED	29.7%	67.8%	2.5%

DISCUSSION

Calcium phosphate ceramics have been described as bioactive materials thought to stimulate new bone formation at implant interfaces. The osteogenic potential of such materials may be improved by the incorporation of osteoinductive factor(s). GH released from the ceramic into the tissues around the implant may stimulate new bone formation and in this work we have shown that one month after implantation of ceramic pins there is significantly greater osteogenesis occurring due to growth hormone stimulation than purely from the bioactive stimulus of the TCP-ceramic. We therefore suggest the growth hormone released stimulates osteoid formation by increased osteoblast activity. The gap that was observed between both loaded and non-loaded TCP pins and the bone was due to

433

Figures a and b.

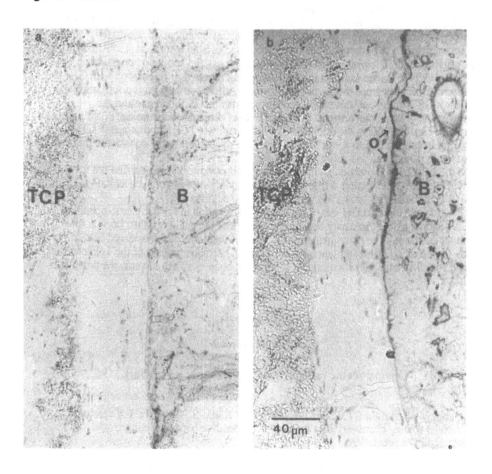

Fig: a.   Plain tricalcium phosphate (TCP) ceramic plug inserted into
bone (B).

Fig: b.   Growth hormone loaded tricalcium phosphate (TCP) ceramic
plug inserted into bone (B). Note the presence of osteoid (o)
lining the bone surface.

Micrographs are magnified by the same scale.

the 0.5mm difference in diameter between pin and hole drilled. With time bone remodelling could be expected to fill this gap.

The in vitro and in vivo release of growth hormone from polymethylmethacrylate cement has been studied by Downes et al. [4]. GH impregnated cement demonstrated a similar effect as proximal to both media there was increased osteoid formation stimulated by the growth hormone. The percentage of GH stimulated osteoid formation on the bone interface in TCP-loaded samples is greater than that reported in association with GH loaded cement after a one month period; however, direct comparison is not feasible due to different GH loading concentrations and methods for the TCP-ceramic and the methylmethacrylate bone cement. A higher level of osteogenesis from the TCP ceramic might be expected due to the duel effect of the 'bioactive' TCP and osteogenic GH working in conjunction to stimulate bone growth. It should also be considered that the the release parameters will differ from the two delivery agents.

The mechanism of this action is unknown and further work is being carried out to elucidate whether the action is direct or by a secondary messenger such as IGF-1. Local IGF-1 stimulation has been observed in osteoblast cultures [7].

This work suggests that GH may induce osteoid formation in the early post-operative period and with time there may be an improvement of the quality of then bone at the ceramic interface. The early formation of bone at the interface is considered favourable for long-term fixation of the implant to bone. Growth hormone loading of the TCP ceramic could stimulate osteoid formation and subsequent bone ingrowth towards the prosthetic implant.

REFERENCES

1. Smith, L., Ceramic-plastic material as a bone substitute. Arch Surg., 1963, **87**: 653-661.

2. Ducheyne P, Beight J, Cuckler J, Evans B and Radin S., The effect of calcium phosphate coating characterisitics on early postoperative bone tissue ingrowth. Biomaterials in press 1990.

3. Kayser MV, Downes S and Ali SY., A new technique for intact Interface Studies of Bone and Biomaterials using Light and Electron Microscopy. Interfaces 1990, Bologna. 1990

4. Downes S, Wood DJ, Malcolm AJ, and Ali SY., Growth hormone in Polymethylmethacrylate Cement. Clin. Orthop. 1990, **252** 294-298.

5. Humphrey CD and Pittman FE. A simple methylene blue-azure II-basic fuchsin stain for epoxy embedded tissue sections. Stain Technol. 1974., **49**: 9-14.

6. Merz, WA., Die Streckenmessung an gerichteten Strukturen im Mikroskop und ihre Anwendung zur Bestimmung von Oberflachen-Volumen-Relationen im Knochengewebe. Mikroscopie. 1967 **22** 132.

7. Stracke, H., Schultz., Moeller, D., Rossol, S., and Schatz, H. Effect of growth hormone on osteoblasts and demonstration of Somatomedin C/IGF-1 in bone organ culture. Acta Endocrinol. (Copenhagen) 1984 **107** 16.

AN INTERFACE COMPARISON BETWEEN A PRESS-FIT AND HA-COATED COMPOSITE HIP PROSTHESIS

Joseph A. Longo, MD, Frank P. Magee, DVM
Steven E. Mather, MD, Janson E. Emmanual,
James B. Koeneman, Ph.D., and Allan M. Weinstein, Ph.D.
Harrington Arthritis Research Center
1800 E. Van Buren, Phoenix, Arizona, 85006, USA

ABSTRACT

In an attempt to circumvent the problems of loosening and adverse bone remodeling associated with metal femoral stems, carbon fiber composite femoral implants have been developed. A canine femoral implant was constructed with a carbon fiber core and braid and encased in a thermoplastic polysulfone. A proximal particulate hydroxylapatite coating was used in half of the implants, and the remainder were implanted as a press-fit composite stem. There were no clinical differences between the two groups. Positive bone remodeling responses were seen in both groups at fourteen months. Histological examination showed stable bone/implant interfaces in both groups with a propensity for more bone contact proximally in the HA-coated group. The uncoated implants produced a stable fibrous tissue interface with cortical bone maintenance suggesting constructive bone remodeling.

INTRODUCTION

Porous-coated metal femoral implants currently in use for total joint replacement have demonstrated adverse bone remodeling in various canine [1,2,3] and clinical studies [4]. Concerns over the long-term potential proximal bone loss with these metal implants has led to the development of a carbon fiber/thermoplastic femoral prosthesis [5]. The use of a composite material offers the ability to alter the elastic characteristics of a femoral hip implant via changes in materials or designs in order to provide a better mechanical match with the host bone [6]. Although it is not known whether direct attachment to bone is necessary for the long-term success of weight-bearing implants, recent developments in hydroxylapatite (HA) coatings appear promising as a mechanism of attachment and fixation [7]. It is the purpose of this study to compare the host to implant interface between an HA-coated carbon composite femoral stem with a press-fit uncoated composite stem in a canine model.

MATERIALS AND METHODS

The canine femoral implant was constructed using a uni-directional carbon fiber (CF)/polysulfone (PSF) core, surrounded with a 45° bidirectional braid and encased in PSF [5]. The overall geometry of the gently tapered, curved, antetorted stem allowed for excellent press-fit in the proximal femur. The HA coating consisted of particulate HA (200-400 micrometers in diameter) airbrushed onto the proximal half of the implant after the PSF surface was softened with methylene chloride. The implant was then oven-dried at 130° F for 24 hours to remove any remaining solvent.

Eight purebred greyhounds with an average age of 27 months and an average weight of 30 kilograms were utilized. The animals were selected by pre-operative radiographic templating to ensure proper femoral implant fit. Four canines received a unilateral femoral hip implant with the HA-coated composite stem, and four received an uncoated press-fit composite stem. Standard AP and lateral radiographs taken monthly and CT scans performed quarterly were evaluated and all data tabulated chronologically by animal for each radiographic follow-up and reviewed for any changes over time. At sacrifice at fourteen months post-implantation, the femora were harvested, fixed in 10% buffered formalin, serially dehydrated in graded alcohols, and embedded whole in Spurr's embedding medium. Two millimeter transverse sections were cut on a modified water-cooled diamond saw for ground transmitted light histology and contact radiography. Contact radiographs of the sections were made using Kodak SO343 high resolution film exposed at 45 kilovolts and 10 milliamperes for 30 minutes.

RESULTS

All animals recovered from surgery uneventfully and were fully ambulatory in the first post-operative week. There were no dislocations, fractures or infections.

Radiographic analysis demonstrated all implants to be in a neutral to slight varus position, and all had a good proximal femoral fit. A radiodense line was seen around both types of implants by 6 weeks, but appeared more dense and pronounced particularly on the proximal portion of the HA-coated implant interface. On clinical radiographs there were no observed changes in bone density when compared to the unoperated contralateral control femora. The CT scans confirmed a good fit in the proximal femur for both groups. Within the resolution of the CT scan, there appeared to be more bone immediately surrounding the HA-coated implants.

Histological analysis showed the implants to be intact, with no signs of wear, particulate debris, delamination or separation of the HA or polysulfone coating. There were no inflammatory cells or areas of inflammatory resorption evident at the interface in either group. In the proximal sections, a thickened trabecular ring of bone was in direct contact with the HA-coated implant throughout the circumference of the implant (Figure 1).

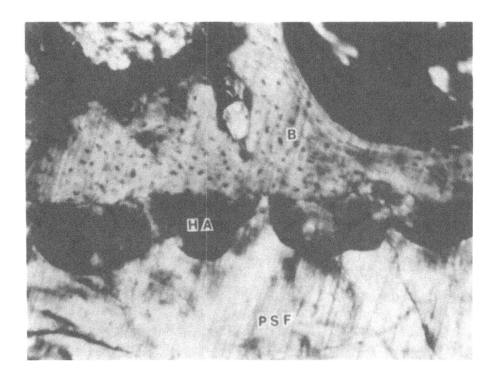

Figure 1. Photomicrograph of ground section of HA-coated implant and bone interface. Note the integrity of the HA and Polysulfone (PSF) coating and the intimate contact with the trabecular bone ring (B). (Original magnification x 175).

A slight increase in cortical porosity in the endosteal surface was evident in the areas of HA-coated implant proximity. In contrast, there was no change in cortical porosity with direct endosteal bone contact in the proximal sections of the press-fit component. Active bone remodeling of the trabecular bone was seen in both groups by fluorescent microscopy of the double tetracycline labelled specimens. In the proximal sections of the press-fit implants, there was occasional direct bone (both endosteal and trabecular) contact; however, the majority of the perimeter demonstrated a 50-100 micrometer fibrous tissue interposition between the polysulfone and the dense trabecular ring surrounding the implant (Figure 2).

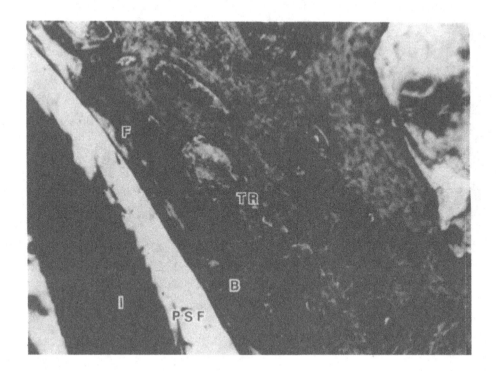

Figure 2. Photomicrograph of a ground section of uncoated implant and bone interface. Note the bone (B) as well as fibrous tissue (F) interface, and the thickened trabecular ring of cancellous bone (TR). (Original magnification x 62.5).

The fibrous tissue was lamellar, stable in appearance and had a circumferential orientation with no inflammatory cells. Outside of the trabecular ring in both groups were normal marrow elements and trabecular bone patterns. In the distal sections, all implants of both groups were surrounded by a shell of bone with a fibrous tissue interposition that was generally thicker in the uncoated press-fit implant sections. No differences were seen in cortical porosity between the groups in the distal sections.

The contact radiograph findings paralleled the histological differences between the two groups (Figures 3,4). The thickness and total area of cortical bone in all sections were similar between the groups and were nearly identical to the contralateral control femora. Occasionally a radial strut formation of trabecular bone was noted jutting from the dense trabecular ring surrounding the HA-coating. This was more prevalent in the areas in which the HA-coated implants were in close proximity to the endosteal surface (Figure 3).

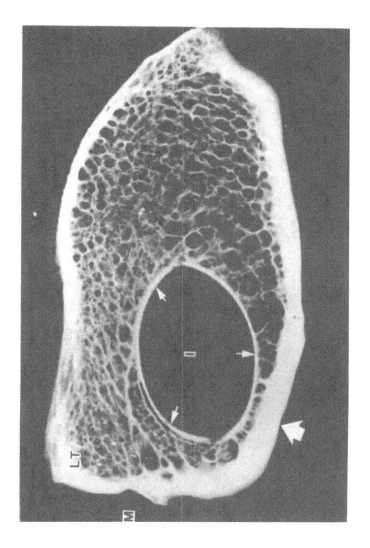

Figure 3. Contact radiograph of the section at the level of the lesser trochanter (LT) with HA-coated implant. Note the intimate trabecular bone contact with the HA-coating (small arrows). Radial struts of trabecular bone jutt from the implant coating to the endosteal surface (large arrows).

440

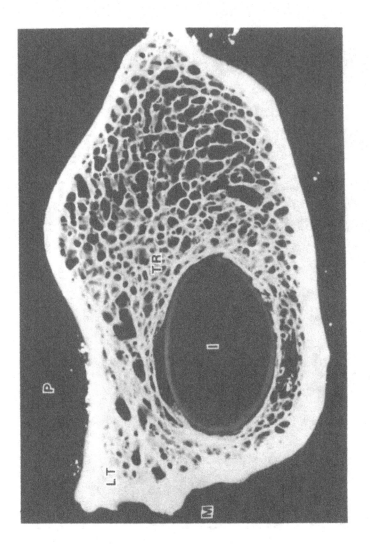

Figure 4. Contact radiograph of a section at the level of lesser trochanter (LT) with an uncoated press-fit implant. Note occasional contact of the implants (I) with trabecular bone and the increased density of the trabecular bone ring (TR).

DISCUSSION

This study has demonstrated the feasibility of using carbon fiber/polysulfone composite implants for weight-bearing orthopaedic applications. Both the uncoated press-fit implant and the proximally HA-coated implant of identical geometry and structure functioned well and were clinically indistinguishable at 14 months in this canine hip model. Cortical thickness and area were maintained in both groups which contrasts sharply with the adverse cortical remodelling seen even by six months in canine studies utilizing metal stems with porous ingrowth fixation techniques [1,2,3]. The cortical bone maintenance as well as the densification of the trabecular ring surrounding the composite implants suggests constructive bone remodeling. Although the HA-coated implants produced an interface with more direct bone contact, there was also an increase in cortical porosity in high contact areas. In contrast, the uncoated press-fit implants produced a stable fibrous tissue interface with cortical bone maintenance and no change in porosity. This is similar to the stable fibrous interface seem with some metal implants [8]. It is interesting that these two very different histological responses obtained the same clinical result at over one year. Although 14 months is a relatively short time, more substantial differences might be observed between these two type of implants at longer term follow up. The results of this study suggest that direct implant bonding to bone might not be a prerequisite to implant success and, in fact, could lead to adverse bone remodeling, as is seen in porous coated metal hip implants.

CONCLUSIONS

This study demonstrates that the functional interface of a load-bearing femoral hip implant can be modulated by the application of an hydroxylapatite coating. The functional interface between the implant and bone is quite different in terms of extent and amount of direct bone contact between the HA-coated implant and the uncoated press-fit implant. The HA-coated implant demonstrated more direct bone contact and little, if any, fibrous tissue contact with the coating. Clinically, both types of implants behaved similarly and in fact were radiographically indistinguishable at fourteen months following implantation. The complex design of composite materials and the specific geometry of this tailored implant to the inner supporting structures of the proximal femur and neck have undoubtedly enhanced the stability of this implant with or without coating. The addition of a coating such as hydroxylapatite can modulate the interface in a predictable manner. However, it is unclear at this time how this will effect the local and regional remodelling of cortical and cancellous bone and ultimately long-term implant stability in the proximal femur.

ACKNOWLEDGMENT

The authors would like to thank Lifecore Biomedical Inc., Minneapolis, Minnesota, USA for their partial support of this study, Ed Dueul and Norm Brown for their preparation of the photographs, and Caren Davis and Debbie Fink for their help in preparation of the manuscript.

REFERENCES

1. Bobyn, J.D., Pilliar, R.M., Binnington, A.G. and Szivek, J.A., The effect of proximally and fully porous-coated canine hip stem design on bone modeling. J. Orthopedic Research, 1987, 5, 393.

2. Sumner, D.R., Turner, T.M., Urban, R.M. and Galante, J.O., Long-term femoral remodelling as a function of the presence, type and location of the porous coating in cementless THA. Trans. Orthopedic Research Society, 1988, 13, 310.

3. Turner, T.M., Sumner, D.R., Urban, R.M. and Galante, J.O., Cortical remodelling and bone ingrowth in proximal and full-length porous-coated canine femoral stems. Trans. Orthopedic Research Society, 1988, 13, 309.

4. Engh, C.A., Bobyn, J.D. and Glassman, A.H., Porous-coated hip replacement: The factors governing bone ingrowth, stress shielding and clinical results. J. Bone Joint Surgery, 1987, 69B, 45.

5. Magee, F.P., Weinstein, A.M., Longo, J.A., Koeneman, J.B. and Yapp, R.A., A canine composite femoral stem: An in vivo study. Clinical Orthopedic Rel. Research, 1988, 235, 237-252.

6. Longo, J.A., Magee, F.P., Koeneman, J.B. and Weinstein, A.M., Long-term performance of a canine carbon composite femoral hemiarthroplasty. Trans. Society Biomaterials, 1990, 16, 126.

7. Cook, S.D., Thomas, K.A., Kay, J.F. and Jarcho, M., Hydroxyapatite - coated titanium for orthopedic implant applications. Clinical Orthopedic Rel. Research, 1988, 232, 225.

8. Kozinn, S.C., Johanson, N.A. and Bullough, P.G., The biologic interface between bone and cementless femoral endoprosthesis. J. Arthroplasty, 1986, 1, 249.

NAPROXEN AND SKELETAL IMPLANT INTERFACES

Longo, J.A., Magee, F.P.,
Van De Wyngaerde, D.P., Mather, S.E., Lane, N.E.*
Harrington Arthritis Research Center
1800 East Van Buren
Phoenix, AZ 85006, USA

Syntex Laboratory *
Palo Alto, CA 94303 USA

ABSTRACT

Nonsteroidal anti-inflammatory drugs, including naproxen are commonly prescribed for the symptoms of arthritis. Many patients with arthritis will undergo total joint replacement with porous-coated implants during their treatment with these medications. We have studied the effects of naproxen on the early fixation of porous-coated implants in bone in a transcortical canine model. There was a significant decrease in interface strength and bone ingrowth at six weeks postoperatively when naproxen therapy was started immediately postoperatively. By delaying the start of naproxen therapy for two weeks after surgery, the bone interface quantity and strengths were comparable to those of a control group at six weeks. This data suggests that nonsteroidal anti-inflammatory drugs such as naproxen should not be used in the immediate postoperative period (7-14 days), but suggests that they can be used after this period in patients receiving implants destined for bone ingrowth fixation.

INTRODUCTION

Nonsteroidal anti-inflammatory medications such as naproxen are commonly prescribed for patients with arthritis conditions. As the arthritis becomes more severe, many of these patients will undergo total joint replacement. In the past, the fixation of these total joint prostheses to bone has been with the use of cement. However, because of numerous problems with the use of cement in active people, such as loosening, bone lysis, cement particle debris and wear: alternative non-cemented techniques of implant fixation have been developed and employed [1]. These alternative techniques of fixation rely on bone regeneration and tissue ingrowth to secure the implant to bone. The quantity and quality of bone regeneration and security of the tissue ingrowth for fixation depends on the initial press-fit, the design of the prosthesis andvarious biologic factors of the host [1,2]. It is the purpose of this study to determine the effects of perioperative naproxen therapy on the biologic response at the interface of implants destined for fixation by tissue ingrowth in a canine model.

MATERIALS AND METHODS

Forty adult greyhounds with an average age of 28 months and an average weight of 30 kgs were divided into 5 groups including one untreated control group. The remaining 4 groups all received naproxen, 4mg/kg/day orally beginning either the 1st, 4th, 8th or 15th postoperative day and continued until sacrifice at 6 weeks. Each animal received 3 fully porous transcortical cylinders and 1 fully porous transmetaphyseal implant in each femur at the time of surgery. The transcortical cylinders were fully porous cobalt-chromium beaded implants, 6.4mm in diameter and with an average pore size of 325 microns and 30% porosity, so as to simulate current porous-coating in clinical use. The transmetaphyseal implants were fully porous polyethylene cylinders, 6.4mm in diameter and with an outer pore size of 150 microns and 30% porosity. These facilitated histological evaluation of the rate and type of ingrowth into the cancellous bone of the distal femoral metaphysis [3,4].

At 6 weeks, the femora were harvested and sectioned to allow for mechanical testing of the cortical bone/implant interface in a transcortical manner, within 12 hours. The endosteal surface was mated to a supporting receptacle within 500 microns of the interface junction. The implants were pushed out with an hydraulic test system at a displacement rate of 1.2 mm/minute, and load versus displacement was recorded. Cortical thickness was measured in order to determine the average cortical contact area for each implant. Maximum interface strength was determined by dividing the maximum load to failure by the cortical contact area. Several representative samples from each group were embedded in methacrylate for evaluation by back-scattered electron microscopy imaging of the implant-bone interface. The sections were surface polished, gold sputter-coated and imaged on a JEOL scanning electron microscope in back-scattered electron imaging mode.

Contact radiographs were taken of 1mm cross-sections of the porous polyethylene implants in the distal femoral metaphyses. Image analysis was performed to determine the percent area of mineralized ingrowth. One way analysis of variance and subsequent Student-Newman-Keuls test were used to compare within and between groups.

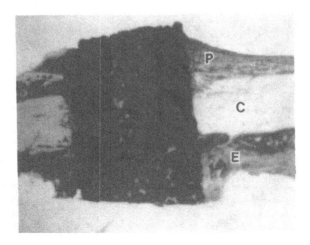

Figure 1. Low magnification photomicrograph of transcortical porous CoCr cylinder demonstrating the alignment of the implant in cortical bone (C). Note the new bone formation at the periosteal (P) and endosteal (E) tissue/implant interfaces.

The depth of penetration of the bone ingrowth in the immediate naproxen-treated group and even in the 4th day naproxen-treated group was typically 1-2 bead layers deep (Figure 2). The majority of the bone ingrowth in the control and 2 week naproxen withheld groups extended into the pores 3-5 bead layers deep (Figure 3). The quality of the bone and tissue ingrowth was similar in all of the groups as demonstrated by the back-scattered electron imaging of the transcortical implants, and thin-section histology of the porous polyethylene implants. In all the polyethylene sections, trabecular bone, woven bone, and osteoid extended from the perimeter interface into the pores in a centripedal fashion. Normal marrow elements and connective tissue constituents including a well organized vascular stromal tissue were evident in close association with the bone ingrowth. The percent area of mineralized bone ingrowth in the immediate naproxen-treated group was significantly less than the 4 other groups (TABLE 1).

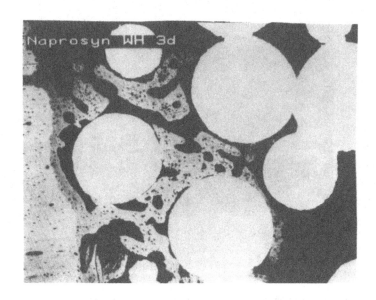

Figure 2. Back-scattered elctron photomicrograph at the cortical bone/implant interface of a 4th day naproxen-treated group demonstrating bone ingrowth to a depth of only 1-2 bead layers at 6 weeks. (Magnification x 65)

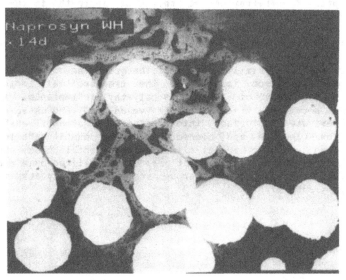

Figure 3. Back-scattered electron photomicrograph at the cortical bone/implant interface of a 15th day naproxen-treated group with bone ingrowth to a depth of 3-4 bead layers at 6 weeks. (Magnification x 100)

RESULTS

The groups were similar in age, weight at time of surgery, and average weight gain after surgery of 0.5 kgs. The interface strength of the transcortical implants at 6 weeks in the control group was 14.2 ± 3.00 megapascals, and the strength of the immediate naproxen treated group was 10.2 ± 2.6 megapascals. The interface of the control and 15th day naproxen-treated groups were significantly stronger than the 4th and 8th day naproxen groups ($P<0.05$), which were in turn significantly stronger than the immediate naproxen group ($P<0.05$) (TABLE 1).

TABLE 1

Transcortical Interface Strengths and Percent
Area of Mineralized Ingrowth

Group	Interface Strength (MPa)	% Area Ingrowth
Control	14.2 ± 3.0	76 ± 14
Immediate Naproxen	10.2 ± 2.6	56 ± 19
4th Day Naproxen	12.1 ± 2.4	86 ± 9
8th Day Naproxen	12.7 ± 3.3	84 ± 11
15th Day Naproxen	15.1 ± 2.0	79 ± 8

All failures of the push-out specimen occurred at the bone-implant interface, that is the porous cobalt-chromium beaded cylinders remained structurally intact. The trabecular pattern of bone ingrowth was consistent in all groups and in close association with the cortical, endosteal, and periosteal perimeters (Figure 1). The fit between the original drill-hole and the transcortical implant was consistent in size with an average gap of 50 microns and no difference between the groups. This gap was consistently filled with new bone in a trabecular woven pattern with growth usually evident perpendicular to the orientation of the cortical bone striations.

Marked subperiosteal and endosteal new bone formation was also evident at the borders of the transcortical porous cylinders (Figure 1).

DISCUSSION

The transcortical model employed in this study allows for a comparative mechanical and histological examination of a large number of bone-implant interfaces while minimizing surgical variables such as fit and design. The transcortical implants are not directly loaded, and therefore, micromotion at the interface is essentially nonexistent, thus providing optimum conditions for bone ingrowth [1,2]. By minimizing technique, design, and population variables, the biological factors responsible for local bone regeneration and tissue stabilization of porous implants can be comparatively studied between similar groups. In this canine transcortical study we have shown that immediate naproxen therapy causes a significant reduction in the strength and amount of bone regeneration into a porous metal component at 6 weeks. However, if naproxen therapy is not initiated until 2 weeks post-skeletal-implantation, there is no reduction in strength or quantity of bone regeneration in this animal model. The first 2 weeks following surgical implantation is the most sensitive period for establishing bone and tissue ingrowth fixation. Inflammatory responses, cellular migration and vascular invasion are primary events that occur during this critical period [3,4]. This study suggests that withholding naproxen therapy during this critical period even for 3 -7 days postsurgical implantation minimizes any retardation of bone regeneration and ingrowth strength. If the mechanism of modulation of the early interface by naproxen is via its anti-inflammatory mechanisms, then judicious use of potentially all nonsteroidal anti-inflammatory drugs in the immediate post-operative skeletal reconstruction period is suggested.

ACKNOWLEDGMENT

The authors wish to thank Syntex laboratories, Palo Alto, California, USA, for their support of this study, Jeannie and Janson Emmanual for their technical assistance with the histological preparations, and Caren Davis for preparation of the manuscript.

REFERENCES

1. Engh, C.A., Bobyn, J.D. and Glassman, A.H., Porous-coated hip replacement: The factors governing bone ingrowth, stress shielding and clinical results. J. Bone Joint Surg., 1987, 69B, 45.

2. Bobyn, J.D., Pilliar, R.M., Cameron, H.U. and Weutherly, G.C., The optimum pore size for the fixation of porous-surfaced metal implants by the ingrowth of bone. Clin. Orthop. Rel. Res., 1980, 150, 263-270.

3. Longo, J.A., Weinstein, A.M. and Hedley, A.K., The effects of collagen on tissue growth into a porous polyethylene ingrowth model. In Biological and Biomechanical Performance of Biomaterials, ed. P. Christel, A. Meurier and A.J.C. Lee, Elsevier Science Publishers, Amsterdam, 1986, p 483.

4. Spector, M., Flemming, W.R., Sauer, B.W. and Kreutner, A., Bone ingrowth into porous high-density polyethylene. J. Biomed. Mater. Res. Symp., 1976, 10, 595.

CHRONIC INDOMETHACIN THERAPY AND THE INTERFACES OF SCREWS AND POROUS IMPLANTS IN BONE

Joseph A. Longo, MD, Frank P. Magee, DVM
Robert D. Poser, DVM
Harrington Arthritis Research Center
1800 East Van Buren
Phoenix, Arizona, USA 85006

ABSTRACT

Anti-inflammatory medications, including indomethacin, are commonly used perioperatively in patients undergoing total joint reconstruction as well as skeletal repair following trauma. In this study we compared the effects of several pre-operative indomethacin withdrawal regimes upon the early biomechanical performance of porous CoCr beaded cylinders and AO/ASIF screws in bone utilizing a canine model. Chronic indomethacin given at a dose of 0.5mg/kg/day for 10 weeks prior to surgery and 6 weeks following surgery caused a 20% reduction in interface strength at 6 weeks. By withholding (interrupting) the indomethacin therapy for 3-5 weeks prior to surgery, the resultant interface strengths of the transcortical porous implants were the same as an untreated control group at 6 weeks post-operatively. Cortical and cancellous screw pullout strengths at 6 weeks post-implantation were not affected by chronic indomethacin therapy in this experimental model.

INTRODUCTION

The stability of skeletal implants used in reconstruction or trauma depends upon early mechanical interlock of the implant to the bone. This interlock can be achieved by screw fixation or ultimate tissue ingrowth into a porous structure to achieve long-term implant stability [1]. It is felt that the early post-implantation time is most critical to establish stable fixation and to minimize micromotion that may ultimately lead to interface failure. It is well known that early biologic growth into a porous structure involves vascular and mesenchymal cellular invasion, proliferation, maturation and remodeling [2,3]. The placement of a screw in bone also elicits a local response and bone remodeling [4]. Many patients undergoing reconstructive surgery requiring either porous or screw fixation of skeletal implants are on chronic anti-inflammatory drug therapy. It is not known how anti-inflammatory drugs such as indomethacin affect the early biologic response and biomechanical interface performance of these types of skeletal fixation techniques in the early critical post-operative period. It is the purpose of this study to determine the biomechanical effects of various perioperative treatment regimes of indomethacin on the 6 week performance of tissue ingrowth and screw fixation in a canine model.

MATERIALS AND METHODS

Twenty-three male greyhounds with an average age of 27.5 months (range 18-48) and an average weight of 30.5 kgs (range 26-34) were divided into 1 control and 3 indomethacin (IND) treated groups. The treated groups were given IND in the morning with food at a dose of 0.5mg/kg/day for 10 weeks in order to establish a chronic level. One group had IND given continually prior to surgery, one group had IND withheld 3 weeks prior to surgery, and one group had IND withheld for 5 weeks prior to surgery in order to simulate clinical situations. All 3 IND treated groups had treatment started again on the first post- operative day, and all 4 groups were sacrificed at 47 days following surgery. Serum samples were drawn 1.5 hours post treatment once during the pre and post-op period to verify serum drug levels.

Surgery was performed under general anesthesia utilizing a standard transcortical technique for each femur. Two fully porous CoCr cylinders, 6.4mm x 22mm, with a pore size of 250-400 micrometers and 35% porosity were implanted transversely in each femoral shaft. Additionally, using standard AO technique, a 4.5mm x 22mm cortical screw was placed in each femoral midshaft, and a 6.5mm x 25mm fully-threaded cancellous screw placed in the metaphyseal region of each distal femur. At sacrifice, the femora were harvested and kept moist in normal saline until mechanical testing within 10 hours. The portion of the femurs containing the porous CoCr cylinders were cut sagitally to yield 2 push-out specimens per cylinder. A closed-loop servo-controlled hydraulic test system with a displacement rate of 1.2mm per minute was utilized, and load versus displacement was recorded during the push-out tests. Maximum load to failure was divided by the cortical interface contact area to determine the interface strengths. The screws were pulled out in a uniaxial direction at a rate of 2.4mm per minute utilizing a specially designed fixture and a supporting washer with an internal diameter of 20.5mm. One way ANOVA and the Student-Newman-Keuls test were used to compare among and between groups.

RESULTS

All animals recovered uneventfully from surgery and were fully ambulatory within 1 week. The animals given IND tolerated the drug well, and serum IND levels averaged 0.42 mcg/ml (range 0.1-1.1 mcgs/ml). Mechanical test results at 47 days post-implantation are listed in Tables 1 and 2. Interface strengths of the transcortical push-outs of the continuous IND treated group were significantly lower than those of the control untreated group (p<0.001). The interface strengths of the control group, however, were not significantly different from the 3 week and 5 week pre-op IND withheld groups (Table 1). Consistent corticalbone regeneration and ingrowth were seen in all of the groups; and the failure occurred at the regenerated bone/beaded implant interface, as determined by macroscopic evaluation post push-out.

TABLE 1
Transcortical Interface

Group	Strengths (MPa)
Control	14.6 ± 1.3
Indomethacin (continuous)	11.8 ± 1.9
Indomethacin 3 week (Pre-op withheld)	13.5 ± 4.5
Indomethacin 5 week (Pre-op withheld)	15.5 ± 3.5

Although there was a trend towards lower cortical screw uniaxial fixation strengths with the IND treated groups compared to those of the control group (Table 2), the difference between the strengths of the cortical or cancellous screw fixation between the untreated, continuous IND treated, or the 3 and 5 week IND withheld groups at 47 days were not significant.

TABLE 2
Screw Pull-out (Newtons)

Group	Cancellous	Cortical
Control	1154 ± 226 n-11	1511 ± 219 n-11
Indomethacin (continuous)	1139 ± 188 n-11	1171 ± 393 n-10
Indomethacin 3 week (Pre-op withheld)	1044 ± 291 n-12	1360 ± 310 n-12
Indomethacin 5 week (Pre-op withheld)	995 ± 255 n-7	1327 ± 379 n-8

DISCUSSION

Skeletal implant stability relies on early mechanical interlock and regenerative biologic tissue attachment for early fixation following the trauma of surgical implantation [1]. Early regeneration of tissue following an injury (such as surgical implantation) involves a sequence of vascular and mesenchymal cellular invasion, proliferation, maturation and eventual remodeling at the site [2,3]. Cortical and cancellous bone regeneration to fill gaps around surgical implants and to provide growth into porous structures involves similar biologic repair sequences. Indomethacin, which is a commonly prescribed anti-inflammatory medication, has been shown to inhibit cortical bone remodeling [5,6]. We have shown that chronic indomethacin therapy can inhibit early cortical bone regeneration and attachment strength in a canine transcortical model. Indomethacin given continuously in the perioperative period, including 10 weeks pre-operatively, reduced cortical bone ingrowth attachment strength by 20%. Thus, indomethacin can deleteriously modify the early process of tissue infiltration and regeneration at the interface of an ingrowth prosthesis destined for ingrowth fixation. Careful modifications of indomethacin therapy should be considered in the perioperative period of reconstructive surgery utilizing ingrowth fixation techniques.

ACKNOWLEDGMENT

This work was supported by the Flinn Foundation of Arizona, USA, Grant #051-267-010-87. The authors would also like to thank James B. Koeneman, Ph.D. for technical assistance in mechanical testing, Barbara Capwell for assistance with animal care, and Caren Davis for assistance with preparation of this manuscript.

REFERENCES

1. Hulbert, S.F., Cooke, F.W., Klawitter, J.J., Leonard, R.B., Sauer, B.W., Moyle, D.D. and Skinner, H.B., Attachment of prosthesis to the muskuloskeletal system by tissue ingrowth and mechanical interlocking. J. Biomed. Mater. Res. Symp., 1973, 4, 1-23.

2. Sauer, B.W., Weinstein, A.M., Klawitter, J.J., Hulbert, S.F., Leonard, R.B. and Bagwell, J.G., The role of porous polymeric materials in prosthesis attachment. J. Biomed. Mater. Res. Symp., 1974, 5, 145-153.

3. Spector, M., Flemming, W.R. and Sauer, B.W., Early tissue infiltrate in porous polyethylene implants into bone: A scanning electron microscopy study. J. Biomed. Mater. Res., 1975, 9, 537-542.

4. Martin, R.B., Osteonal remodeling in response to screw implantation in canine femora. J. Orthop. Res., 1987, 5, 445-452.

5. Keller,   J., Bunger, C., Andreassen, T.T., Bak, B. and Lucht, U., Bone
    repair inhibited by indomethacin, Effects of bone metabolism and
    strength of rabbit osteotomies.   Acta Orthop. Scand., 1987, 58,
    379-383.

6. Sudman,   E. and Bang, G., Indomethacin induced inhibition of haversion
    remodelling in rabbits. Acta Orthop. Scand., 1979, 50, 621-627.

CHARACTERIZATION OF LOCAL ANISOTROPIC ELASTIC PROPERTIES OF FEMORAL AND TIBIAL DIAPHYSIS USING ACOUSTIC TRANSMISSION MEASUREMENTS AND ACOUSTIC MICROSCOPY.

A. MEUNIER[°], O. RIOT[°], P. CHRISTEL[°], J.L. KATZ[°°].
[°] Laboratoire de Recherches Orthopédiques, URA CNRS
Faculté Lariboisière-St Louis, Université PARIS VII,
10 Avenue de Verdun, 75010, Paris, France.
[°°]School of Engineering, Case Western University, Cleveland,
OHIO, 44106, USA.

ABSTRACT

Using two different acoustic techniques, the present work had three major purposes: (i) to investigate inhomogeneity patterns in anisotropic elastic constants of both human tibia and femur using a standard transmission method, (ii) to investigate the entire cross section of the same cortical bones using a low-frequency acoustic microscope and (iii) to correlate the acoustic impedance values obtained with this microscope with the c33 (longitudinal direction) found in transmission. 80 femoral and 120 tibial specimens prepared from 20 pairs of femur and tibia obtained at autopsy were investigated. The longitudinal elastic stiffness was significantly higher in the tibia when compared to the femur while all other elastic constants were identical. The elastic stiffness patterns were completely different for the two bones and in good agreement with the average acoustic impedance found with the acoustic microscope in the same anatomical locations.

INTRODUCTION

Human cortical bone is an anisotropic composite material which mechanical properties are related to a complex hierarchical structure (1). Age, sex, physical activity, bone type as well as spatial location play a major role in the local anisotropic elastic properties of this material. For example, Ashman (2) has demonstrated, using a continuous acoustic wave technique, that elastic properties of human and canine femur vary along the entire diaphysis and depend upon the anatomical locations. Cortical bone is a living tissue which continuously adapts its geometry and microstructure to the physiological mechanical environment. For this reason, any variation in the local elastic properties of cortical bone is directly related to its in-vivo mechanical load history. To study local mapping of anisotropic elastic properties of mineralized tissues is an indirect way to investigate the relationship between mechanical stresses and bone remodelling. However, the acoustic transmission methods previously used to determine anisotropic elastic properties of cortical bone require the machining of relatively large specimens (about 3 to 5 mm on a side) and only gross gradients in elastic properties can be described. For this reason, we have developed a simple low-frequency scanning reflection acoustic microscope (frequency range: 3-30 MHz) specially dedicated to the study of the local changes in bone elastic properties (3). This experiment had three main purposes: (i) to compare, using standard acoustic transmission technique, the anisotropic elastic constants distribution of the mid-diaphysis of human tibias and femurs, (ii) to obtain acoustic impedance mapping of the same bone areas using the acoustic microscope and (iii) to calibrate this microscope and compare the bulk transmission values (c33) and the scanned reflection data.

MATERIALS AND METHODS

The mid-diaphysis of 20 pairs of right tibias and femurs (six females and 14 males) were obtained at autopsy and kept frozen (-18°c) until further preparations. The specimens' age ranged from 48 to 92 years and the weight from 40 to 90 Kg.

Using a low-speed diamond saw, two thin slices (thickness =1 mm) and a 20 mm block were obtained from each mid-diaphysis. One of the thin slices was polished with 0.5 μm alumina slurry and used for acoustic microscopy. The second one was embedded in PMMA, microradiographed and then stained with PARAGON for later histomorphometry. Using the same diamond saw, each anatomical aspect (anterior, medial, lateral and posterior for the femur and anterior, medial, lateral, postero-medial, postero-lateral and posterior for the tibia) of the thick blocks was cut in parallelepipedic samples (≈4x4x10 mm) along the principal directions of the haversian microstructure (dir.3= longitudinal, 2= transverse, 1=radial). A total of 80 femoral and 120 tibial samples were machined. The density of each sample was measured using an immersion method and the longitudinal and transverse acoustic waves velocities were measured in transmission. The acoustic microscope used a spherically focused transducer (frequency= 30 MHz) and the maximal scanned area was 35 mm on a side. This system has already been presented (3) and will not be described in this paper.

RESULTS

Acoustic transmission: assuming an orthogonal symmetry of cortical bone, the anisotropic elastic constants (GPa) are presented in table I:

BONE	C11	C22	C33	C44	C55	C66	C12	C13	C23
FEMUR Ant.	17.5 ±2.9	19.0 ±2.3	27.8 ±2.8	5.90 ±.74	5.50 ±.63	4.15 ±.61	9.8 ±1.8	9.4 ±2.5	10.2 ±1.3
FEMUR Med.	19.6 ±2.3	20.7 ±2.0	28.7 ±2.1	6.31 ±.58	5.92 ±.53	4.72 ±.69	10.7 ±1.3	10.7 ±1.8	10.9 ±1.2
FEMUR Post.	16.7 ±3.4	17.3 ±2.6	26.1 ±3.1	5.26 ±.83	5.16 ±.89	3.80 ±.85	9.3 ±1.6	9.1 ±2.3	9.6 ±1.8
FEMUR Lat.	18.6 ±2.6	19.4 ±2.4	27.8 ±2.8	5.79 ±.77	5.49 ±.65	4.23 ±.74	10.5 ±1.3	10.6 ±1.8	10.7 ±1.5
TIBIA Ant.	15.8 ±2.3	15.8 ±1.9	28.0 ±2.9	4.97 ±.80	5.12 ±.92	3.28 ±.77	9.2 ±0.9	8.5 ±1.5	8.9 ±1.2
TIBIA Med.	17.7 ±2.8	19.9 ±2.1	29.9 ±3.0	5.98 ±.67	5.31 ±.57	4.13 ±.46	10.4 ±2.1	10.4 ±2.4	11.2 ±1.6
TIBIA P-Med	18.9 ±2.6	20.0 ±1.9	30.0 ±2.7	5.79 ±.67	5.62 ±.62	4.11 ±.62	11.2 ±1.3	11.0 ±2.0	11.7 ±1.2
TIBIA Post.	19.2 ±2.5	20.4 ±1.9	30.1 ±1.8	5.78 ±.55	5.59 ±.56	4.26 ±.51	11.2 ±1.5	11.3 ±2.0	12.0 ±1.4
TIBIA P-Lat	18.7 ±2.1	20.1 ±1.7	30.4 ±1.9	5.81 ±.55	5.40 ±.56	4.19 ±.57	11.0 ±1.3	11.3 ±1.7	11.8 ±1.3
TIBIA Lat.	18.1 ±2.3	20.7 ±1.9	30.3 ±2.7	5.96 ±.73	5.26 ±.68	4.07 ±.61	11.1 ±1.3	11.0 ±1.7	12.0 ±1.1
FEMUR avg.	18.1 ±3.0	19.1 ±2.6	27.6 ±2.8	5.80 ±.81	5.52 ±.73	4.22 ±.78	10.1 ±1.6	10.0 ±2.2	10.4 ±1.6
TIBIA avg.	18.1 ±2.7	19.5 ±2.5	29.8 ±2.6	5.71 ±.74	5.38 ±.67	4.01 ±.67	10.7 ±1.6	10.6 ±2.1	11.3 ±1.7

TABLE 1 (Cij in GPa- means ± standard-deviations)

All elastic constants are strongly correlated (r>.85) with the samples' density. For example, Figure 1 presents the evolutions of some of the C_{ijkl} of the femoral specimens versus density.

C33, averaged over all the anatomical aspects, is significantly higher for the tibia when compared to the femur, while all other elastic constants are identical. Variations of C33 v.s. density are presented in Figure 2.

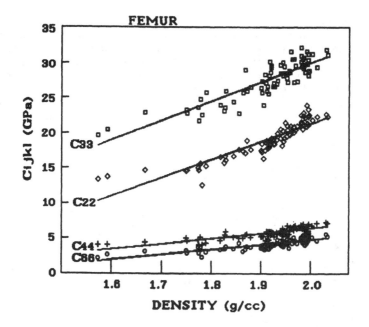

Figure 1

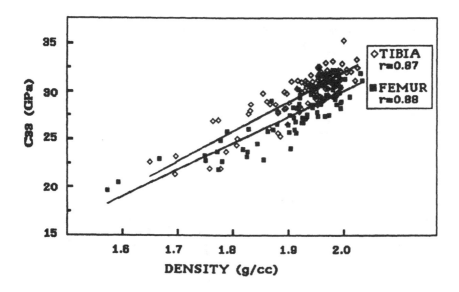

Figure 2

Anatomical aspect locations significantly affect all the elastic constants. For example, Figures 3 and 4 show the C11 and C33 patterns for the femur and tibia. It is clear that cortical bone behavior is not unique, each bone (femur or tibia) exhibiting a specific pattern of elastic constants with a marked decrease in the posterior aspect of the femur and the anterior aspect of the tibia. Moreover, all the elastic constants do not present the same level of statistical significance when comparing the various aspects of each bone.

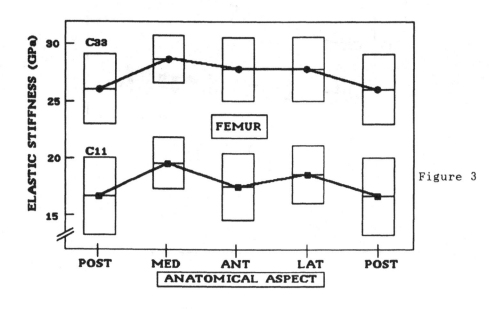

Figure 3

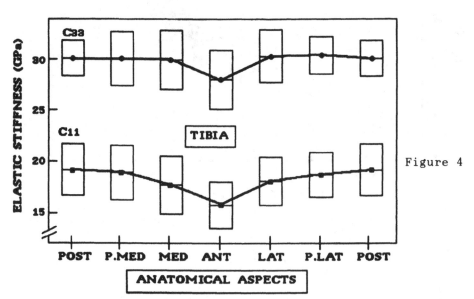

Figure 4

Acoustic imaging: Due to space limitation it is not possible to present all images obtained with the acoustic microscope. Two black and white prints of reflection coefficient variations found in femoral sections are shown in Figure 5 (male, 48 y.o) and 6 (male ,79 y.o). It is clear from these figures that inhomogeneities distribution in bone acoustic properties is complex and not limited to the artificially chosen anatomical aspects. Gradients of acoustic properties are present in areas much smaller than the average area of samples used in transmission and cannot be detected with this technique.

Moreover, these images show that the exact location of the transmission specimens is a critical parameter which may affect the elastic constants measurement. An other observation, consistent for all images, is that a "low quality bone" (i.e a bone exhibiting very low average acoustic characteristics) still presents numerous spots of high acoustic properties. This finding indicates that decrease in elastic properties due to bone remodelling is not a uniform process through all the cortex but occurs in preferential areas while some parts remain intact for a long period of time.

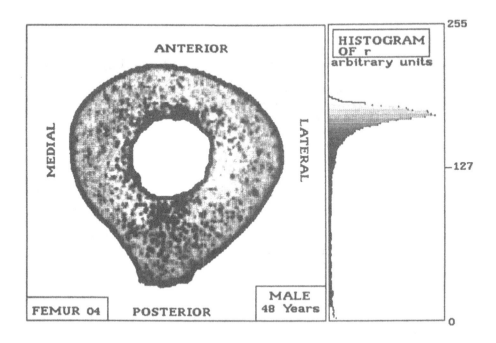

Figure 5

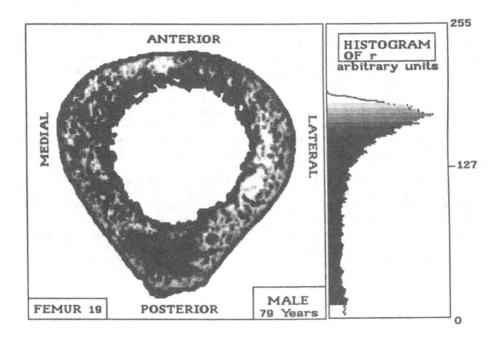

Figure 6

Comparison of acoustic transmission and imaging: Acoustic
microscopy provides mapping of bone coefficient of reflection
(from which the acoustic impedance can be computed). However,
this technique is an interesting tool only if the data
contained in these images are directly related to more
commonly used elastic properties such as C33. For this
reason, we tried to calibrate the apparatus and correlate the
reflection coefficient averaged over areas corresponding to
the location of transmission samples with the C33 measured in
transmission. We first defined the theoretical correlation
between reflection coefficient and C33 for cortical bone.
      When the liquid used for sound transmission is always the
same, the reflection coefficient is directly and uniquely
related to the acoustic impedance of the specimen through an
hyperbolic function. The acoustic impedance is the product of
the specimen's density and the longitudinal acoustic velocity
while the elastic stiffness is the product of the density and
the squared velocity. In general, if the specimen is made of
various materials, no correlation between acoustic impedance
and elastic stiffness can be found. For example, lead
exhibits a large density and a low velocity while glass has
inversed properties. Thess two materials have a similar
acoustic impedance but very different elastic stiffnesses.
However, cortical bone properties vary in a small range and
the elastic properties are correlated with the material's
density. In this case, the acoustic impedance (as well as the
reflection coefficient) exhibits a strong correlation with

the elastic stiffness. Figure 7 presents the 2nd degree regression (r=.99) found between the reflection coefficinet and the corresponding C33 for the femur. These values are computed from the transmission measurements. Figure 8 shows the correlation (r=0.85) found between the coefficients of reflection computed from the transmission data and those measured with the acoustic microscope in areas closed to the transmission specimens locations.

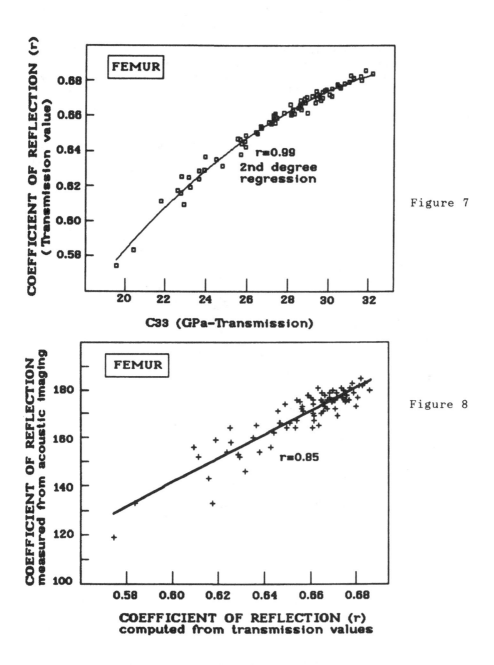

Figure 7

Figure 8

From these two correlations, on may compute the elastic stiffness C33 corresponding to a measured reflection coefficient. Figure 9 shows the correspondance between C33 measured in transmission and those computed from the acoustic images data. The scattering has been found to be ± 2.5 GPa.

This relatively large errors find their origin in (i) locating the exact areas of the transmission samples on the acoustic images and (ii) the exact zero setting of the acoustic microscope electronics. The last parameter is critical when investigating materials exhibiting very similar acoustic properties.

From the calibration curves found for the femur, we computed the C33 averaged in all directions perpendicular to the cortex. In this case the measurement's origin was the center of gravity of the image and the 0 direction corresponded to the posterior aspect. Figures 10 shows the computed distribution of C33 as well as the measurements performed in transmission in the four anatomical quadrants.

The good agreement between both curves indicates that variations found in acoustic images describe the true local longitudinal elastic stiffness of both bones.

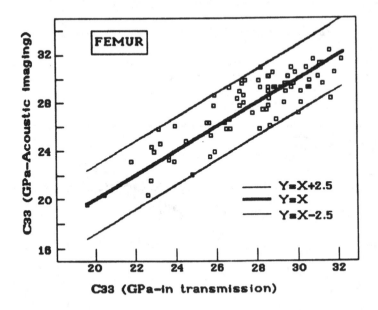

Figure 9

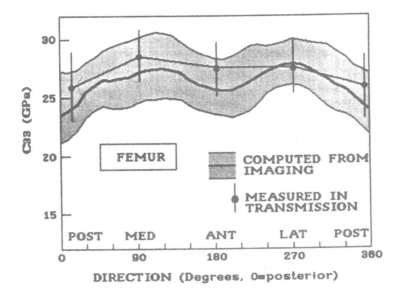

Figure 10

DISCUSSION AND CONCLUSION

Both acoustic techniques provide interesting and complementary data concerning evolution of cortical bone elastic properties. Transmission method allows an accurate computation of all the anisotropic constants on large specimens while acoustic microscopy only deals with longitudinal properties,however with a much higher resolution. Calibration of the acoustic microscope through the transmission data increases the field of application of this method and brings a valuable insight into bone physical properties. However, in order to understand the bone form-function relationship, this investigation must be completed using other techniques such as histomorphometry and mineral content analysis.

REFERENCES

1. Katz J.L., Anisotropy of Young's modulus of bone. <u>Nature</u>, 1980, **283**, 106-107.
2. Ashman R.B. et al., Acontinuous wave technique for the measurment of the elastic properties of cortical bone. <u>J. Biomech.</u>, 1984, Vol. 17, 5, 349-361.
3. Meunier A. et al., A reflection scanning acoustic microscope for bone and bone-biomaterials interfaces studies. <u>J. Orthop. Res.</u>, 1988, 5, 770-775.

AKNOWLEDGEMENTS
This research is supported by INSERM grant #879012.

AN IN-VITRO COMPARATIVE STUDY OF THE WALL SHEAR RATE
PRODUCED BY THE CENTRAL AXIS PROSTHETIC HEART VALVE
USING A LASER DOPPLER ANEMOMETER

Y.HAGGAG, A.NASSEF, A.SALLAM, A.MANSOUR

King Saud University - College of Applied Medical Sciences
P.O.Box 10219 - Riyadh 11433
Kingdom of Saudi Arabia

ABSTRACT

A new mechanical prosthetic heart valve is tested in-vitro,
and compared with other four common prosthetic cardiac valves.
All valves studied have the same orifice diameter, and were
installed inside a mitral test chamber which enables wall
shear rates measurements using a laser Doppler anemometer,
under the steady state flow conditions. The steady state flow
tests are essential to predict certain flow characteristics
before conducting more complicated, expensive, and difficult
to interpret pulsatile flow tests. All experiments were in
vitro, at steady volumetric flow rates of 10 lit/min to 30
lit/min, and using a blood analog fluid.

Results showed that at high flow rates the wall shear rate may
attain over 100×10^3 sec-1 in the very near vicinity of the
mitral prostheses studied. However, at a downstream distance
about 125 mm from the prostheses, the wall shear rate did not
exceed 30×10^3 sec-1. The Starr-Edwards SE 6120 showed the
highest values for wall shear rate. The St. Jude Medical valve
offers the minimum wall shear rate. The central axis valve
comes second to the Starr-Edwards valve for such type of
measurements, and followed by the Bjork-Shiley valve and the
Medtronic-Hall valve.

INTRODUCTION

Three decades have elapsed since the introduction of the first
serially produced prosthetic heart valves. In spite of periodic improvements in design and fabrication technology, the
present prostheses are neither ideal systems nor do they perform as well as the natural healthy biological valves. The
perfect, or universal, prosthetic valve remains elusive.

Historically, many different designs and materials have been used in an effort to improve the performance of prosthetic cardiac valves. Some designs have evolved and are still in use today; others have been abandoned. This selection process has not always been the result of systematic studies. The haphazard nature of valve recovery programs and the lack of a standard for presentation of data and tests for the various commercially available valve models, have considerably delayed effective dissemination of data regarding long-term problems in such implants[1].

Even after thirty years of experience the postoperative problems associated with the prosthetic heart valves have not been totally eliminated, and the analysis of laboratory and clinical data of available heart valve prostheses today show that the optimum valve does not yet exist. The use of anticoagulants (to avoid thrombus formation) is still requested for the mechanical prostheses, whereas the tissue valves have other problems in terms of questionable longevity, durability and calcification[1].

In this paper a new mechanical valve is introduced (the Central Axis Prosthetic Heart valve) and compared with four commercial mechanical type valves.

EXPERIMENTAL INVESTIGATIONS

In the present study, it has been concentrated on one specific aspect of heart valve in vitro testing: The measurement of the wall shear rate in the very near vicinity of the prostheses under the steady state flow conditions using a laser Doppler anemometer. The wall shear rate is an important factor among many other parameters affecting the overall performance of a prosthetic heart vavle. All the tested prostheses are installed in the mitral position. The study is performed considering the maximum open position, this condition being the worst case in terms of flow separation and shear stresses in the fluid. The occluder of the prosthesis remains in the maximum open position during the best part of the diastolic cycle, and the higher Reynolds number occurs under this condition. As the maximum open position is achieved during a very short period of time, compared with the whole diastolic cycle, the flow can be considered steady. One can assume that steady state occurs when the mitral valve is fully open and velocity is maximal.

The valves investigated are listed in table (1) together with their year of introduction, sewing ring diameter, primary orifice diameter, and maximum opening angle. All the tested valves have almost the same orifice diameter of 22 mm.

TABLE 1

Prosthetic heart valves investigated(Mitral prostheses)

Name of Valve and year of introduction	Sewing Ring diameter(mm)	Primary Orifice diameter(mm)	Opening angle
Starr-Edwards (6120-34 M) 1966 to present	34	22	-
Central-Axis T-T (CA-M27) Under study	27	22	-
Bjork-Shiley (S27 MBUMS) 1982 to present	27	22	70°
Medtronic-Hall (27 MHK) 1977 to present	27	22	70°
St. Jude Medical (27M-101) 1977 to present	27	22.3	85°

CENTRAL AXIS VALVE DESCRIPTION

The new valve consists of an occluder, an orifice and a T-shape solid bar. The occluder is a low profile hydrodynamically shaped poppet formed geometrically like a top-toy. The occluder rests on an annular ring when closed, and is guided through a central axis with adequate tolerance during the valve operation. One end of this axis has a suitable bulb-shaped ball to limit the valve poppet displacement. The other end of the axis is supplied with a perpendicular bar, located diametrically inside the orifice for fixation purposes. The all T-shape bar as well as the orifice should be fabricated from a single material piece. Welding is prohibited to avoid sudden cracks and mechanical dysfunction. The preferred material for the poppet is a graphite substrate coated with pyrolytic carbon. This material is known for its combination of blood compatibility and high resistance to degradation, wear and fatigue[2].

This new type of valve is characterized by the absence of a cage, a place very prone to the formation of a thrombus. Furthermore, the occluder is allowed to rotate with angular velocity around the axis during operation, thus providing a desirable constant washing action in the opening and closing valve phases. This rotation movement also assures uniform wear between the axis and inner surface of the poppet. This new mechanical central axis prosthetic heart valve is a promising concept. Previous studies [3] have demonstrated that the pressure drop across the new prosthesis compares well with other most commonly used mechanical prostheses today.

MATERIALS AND METHODS

Test Fluid

In this experimental work, a blood analog fluid (solution of 36.7% glycerin by volume mixed with distilled water) is used to mimic the viscosity of the blood. At 20°C room temperature, this solution has a density of 1.10 gm/cm^3, and a viscosity of about 3.5 centipoise. Human blood viscosity is approximately 3.5 to 4 centipoise at 37°C and specific gravity about 1.058. This aqueous glycerol solution has the advantage to be transparent, non-toxic and inexpensive. The blood analog solution contained a sprinkling of cornstarch particles which were 10-12 μm in diameter, during the wall shear rate measurements. These particles were suspended in the solution so as to scatter the laser light. The concentration of the cornstarch in the solution was very small. These particles helped to obtained a stronger signal to noise ratio particularly when the laser measuring system was used in the backward scatter mode.

Test Chamber

The prosthetic cardiac valves were tested under steady state flow conditions in a mitral test chamber which was first introduced by Wieting[4] and accepted in the biomedical engineering literature for pressure drop assessment and observation of other flow characteristics. Different valve sizes could be accommodated in the chamber by using special adaptors. The test chamber is made of a transparent material (plexiglass) in order to visualize the fluid flow in the vicinity of each studied valve. The test chamber was mounted on a special movable table which has three degree of freedom in x,y and z directions. This permitted the measurement of the axial velocity at any location inside the flow channel without moving the measuring control volume of the laser anemometer. The design of the test chamber was slightly modified to allow better penetration of the laser beams particularly in the annular area between the valve poppet and the test chamber inner wall.

Steady State Flow Circuit

Figure (1) illustrates the steady flow circuit and its components.

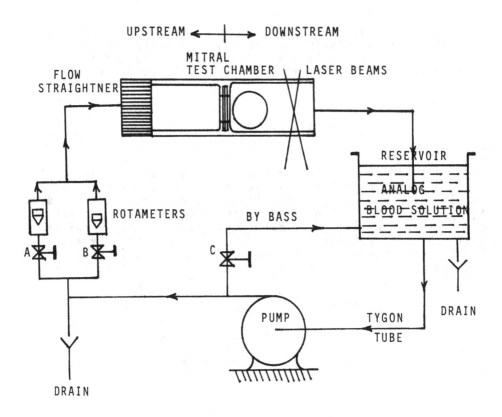

Figure 1. The steady state flow circuit

The flow loop consists of a) a centrifugal pump of 1/3 hp with a maximum flow rate of 35 liters per minute; b) two rotameters connected in parallel, one for low flow rates and one for higher flow rates. The final accurate flow adjustment is done by manipulating the three valves A,B and bypass valve C as shown in figure (1); c) a flow rectifier for control of vortices and other flow perturbing effects at the test chamber inlet; d) an inlet tube of 50 diameter length ahead of the test section to assure fully developed flow; e) a mitral test chamber for the prosthesis; and f) a reservoir containing the blood analog solution. All circuit components are connected with tygon tubes.

Wall Shear Rate Measurements

Wall shear rate data were obtained with a laser-Doppler anemometer (LDA). The laser Doppler anemometer is an opto-electronic measuring system without the need of calibration and without any influence on the flow field caused by probes, holders, etc. Because of these features, it is a unique tool for measuring the flow field characteristics in the vicinity of heart valve prostheses.

The basic components of the LDA used in this study consists of a He-Ne laser source (15 mW), a transmitting optical arrangement with beam expander and frequency shift to provide two beams of focused light, a light receiving system which can define the measuring control volume, a photomultiplier to convert the optical signal to an electronic signal, and a tracker-type signal processor system (TSI Model 1090) to convert frequency to a voltage proportional to the measuring velocity. The forward and backward scattered modes were used. Backward scattered technique was only used when the fully open valve occluder prevents the passage of the laser beams to the other side of the installation.

The LDA system was operated in the dual beam mode with a focal length of 120 mm and beam angle $10.750°$. The frequency shifter assembly was added to the actual laser system in order to measure accurately small velocities near the test chamber wall. Finer adjustments for optimizing the quality of the signal were made by monitoring the signal on the oscilloscope, sampling rate per second of the tracker, and filtering the signal by suitable high pass and low pass filters.

It was necessary to measure velocities very close of the flow channel in order to obtain a realistic assessment of the wall shear rates. The dimensions of the ellipsoidal measuring volume, resulting from the intersection of the two laser beams under the above system condition, were 0.30 mm (length) by 0.04 mm (width). A velocity measurement taken at a point x,y, and z implies that the center of the measuring volume was at these coordinates. Therefore, the closest possible measuring point in the liquid from the inner wall was at a distance of 0.15 mm (i.e. half the length of the measuring volume). Any closer approach to the wall led to interference of the Doppler signal from the liquid with that from the inner surface wall. Therefore, in order to allow measurements closer to the wall than a distance of half the length of the measuring volume, the unwanted signal produced by the wall was attenuated to a large extent by a powerful high pass filter. This permitted measurements of velocities to a distance of only 0.040 mm from the inner surface of the wall.

Therefore with the present experimental apparatus, it was possible to measure velocities as close as 0.040 mm from the flow channel wall. From these measurements the wall shear rates were calculated. The velocity measurements made adjacent to the walls were within the laminar sub-layer and therefore the wall shear rate is given by

$$W.S.R./_{max} = \frac{du}{dr}\Big|_{wall} \quad \frac{cm}{sec}/cm = sec^{-1}$$

It is to be noted that all measurements were obtained in the axial direction in the horizontal plane through the center of the channel. Wall shear rate was measured by assuming that the velocity increased linearly from the wall to the point of measurement closest to the wall (i.e. 0.040 mm from the wall).

RESULTS AND DISCUSSION

Table (2) gives typical values of the wall shear rates in sec^{-1} measured in the downstream section of the flow channel for the different types of prostheses studied under the steady state flow rates of 10 lit/min and 30 lit/min. X is the distance downstream from the front end of the valve sewing ring. Wall shear rates were measured at distance of X = 15, 25, 40 and 125 mm.

The reliability and accuracy of the laser Doppler anemometer to measure wall shear rate and velocity fields in the very near vicinity of prosthetic heart valves were previously demonstrated by many investigators[5],[6]. Blackshear[7],[8] contends that the shear rate required in the bulk of the flow to damage red blood cells is about 1142×10^3 sec^{-1}. Neveril and his co-workers [9] contend, however, that this value could be as low as 42.8×10^3 sec^{-1}. In vitro experiments[10],[11] have also shown that platelets could be damaged by shear rates of the order of 2800 to 14280 sec^{-1}. A formed element such as a red blood cell which, however, adheres to a vessel wall or to a foreign surface may be damaged by shear rates as small as 285 to 2850 sec^{-1} [7],[8],[12]. A red blood cell will not stick to the intact endothelial lining of a vessel wall. If, however, the vessel wall is damaged and the endothelial lining is not intact, red blood cells could adhere onto the vessel wall. If the adhered red blood cell is exposed to shear rates of the order 285 to 2850 sec^{-1} it may be morphologically damaged or hemolyzed in the extreme case[7],[8],[12]. Red blood cells contain ADP and a clot-promoting factor known as erythrocytin. Both of these substances are released from a red blood cell into the plasma when it is hemolysed[13],[14],[15]. When the ADP and the erythrocytin are released into the plasma, both platelet-adhesion aggregation and coagulation may be initiated, resulting in thrombus formation.

Due in large part to turbulent flow, cavitation, shear stress, reflux and mechanical crushing effect, some destruction of red blood cells (hemolysis) occurs with prosthetic heart valves. Chemical interaction does not take place because the materials developed for use in these valves are sufficiently inert. It is clear that modern mechanical valves with their improved flow characteristics cause less hemolysis than previous models. All prostheses cause a minor degree of hemolysis and some damage can still be demonstrated. Such hemolysis that does occur is well compensated by bone marrow hyperfunction and anemia is not a problem. It should be noted that the major problems usually associated with mechanical prosthetic heart valve replacement are thrombus formation, thromboembolism, and hemorrhagic complications.

TABLE 2

Wall shear rates for different prostheses at 10 lit/min and 30 lit/min flow rates (Mitral prostheses).

Valve Type	X mm	Wall Shear Rate $Sec^{-1} \times 10^3$ $Q = 10$ lit/min	Wall Shear Rate $Sec^{-1} \times 10^3$ $Q = 30$ lit/min
Starr-Edwards (SE 6120)	15	36.85	105.17
	25	30.14	87.75
	40	26.25	80.37
	125	7.12	30.22
Central-Axis (T-T) (Under study)	15	30.85	100.66
	25	24.42	77.35
	40	10.20	36.96
	125	5.35	16.11
Bjork-Shiley (B-SH)	15	16.78	46.10
	25	16.20	44.80
	40	8.19	21.95
	125	4.72	14.03
Medtronic-Hall (HK)	15	14.77	40.47
	25	14.52	38.97
	40	7.20	20.03
	125	3.45	13.18
St. Jude Medical (SJM)	15	13.24	38.42
	25	12.90	36.01
	40	6.06	17.96
	125	3.10	10.10

CONCLUSIONS

The new mechanical prosthetic heart valve is a promising concept. In the present study, it has been concentrated on one specific aspect of heart valve in vitro testing in the mitral position: measurement of wall shear rate in the very near vicinity of the prosthesis under steady state flow condition by means of a laser Doppler anemometer. The wall shear rate is an important factor among many other parameters affecting the overall performance of a prosthetic heart valve.

The present conducted steady state flow test is essential in order to predict certain flow characteristics, before conducting more expensive, complicated and difficult-to-interpret pulsatile flow tests. The Starr-Edwards SE 6120 occupies the top of the list, while the Saint Jude Medical offers the best flow characteristics from the shear rate point of view. The new valve has a better assessment for such measurements over the Starr-Edwards valve, but the last three prostheses have shown a better performance than the Central Axis prosthesis. A minor degree of hemolysis may be demonstrated for all valves tested particularly in the regions very close to the valve sewing ring. It should be noted that such hemolysis may be well compensated by bone marrow hyperfunction and anemia is not a problem.

The material suggested for fabrication the new valve is the pyrolytic carbon. This material appears to be a promising material for prosthetic heart valves. In fact, pyrolytic carbon is ideal for a variety of prostheses requiring blood and tissue compatibility, chemical inertness, surface smoothness, low wear rates and high fracture strength. The highly polished, electro-negative carbon surface contributes to the low incidence of valve related thromboembolism.

Some modification and design alterations to the new valve may appear to be necessary in the light of other hydrodynamic results obtained following a proposed complete heart valve developing program which may take as long as 5-10 years.

REFERENCES

1. Haggag, Y., Selection of a prosthesis for heart valve replacement: Analysis and future development. Journal of Clinical Engineering, 1988, 13(3), 217-223, U.S.A.

2. Haggag, Y., An overview of materials considerations for prosthetic cardiac valves. Journal of Clinical Engineering, 1989, 14(3), 245-253, U.S.A.

3. Haggag, Y., The central axis prosthetic cardiac valve: An in vitro study of pressure drop assessment under steady state flow conditions. Journal of Biomedical Engineering, 1990, 12(1), 63-68, England.

4. Wieting, D.W., Dynamic flow characteristics of heart valves Ph.D. thesis, University of Texas at Austin, Texas, USA,1969.

5. Woo, Y.R. and Yoganathan, In vitro pulsatile flow velocity and turbulent shear stress measurements in the vicinity of mechanical aortic heart valve prostheses. Life Support Systems, 1985, 3, 283-312.

6. Chandran, K.B., Cabell, G.N., Khalighi, B., and Chen, C.J., Laser anemometry measurements of pulsatile flow past aortic valve prostheses. Journal of Biomechanics, 1983, 16, 865-873.

7. Blackshear, P.L., Hemolysis at prosthetic surfaces. In Chemistry of Biosurfaces, Vol. 2, ed. M.L. Hair, Marcel Dekker Publishers, New York, N.Y., 1972, pp. 523-561.

8. Blackshear, P.L., Mechanical hemolysis in flowing blood. In Biomechanics its Foundations and Objectives, ed. Y.C. Fung, N. Perrone, and M. Anliker, Prentice-Hall, Englewood Cliffs, New Jersey, 1972, pp. 501-528.

9. Nevaril, C.G., Hellums, J.D., Alfrey, C.P., and Lynch,E.C., Physical effects in red blood cell trauma. Am. Inst. Chem. Eng. J., 1969, 15, 707-711.

10. Hellums, J.D., and Brown III, C.H., Blood cell damage by mechanical forces. In Cardiovascular Flow Dynamics, ed., N.H.C. Hwang and N.A. Normann, University Park Press, Baltimore, Maryland, U.S.A., 1977, pp. 799-823.

11. Hung, T.C., Hochmuth, R.M., Joist, J.H., and Sutera, S.P., Shear-induced aggregation and lysis of platelets. Trans. Am. Soc. Int. Organs, 1976, 12, 285-290.

12. Mohandas, N., Hochmuth, R.M., and Spaeth, E.E., Adhesion of red cells to foreign surfaces in the presence of flow J. Biomech. Mat. Res., 1974, 8, 119-136.

13. Hellem, A.S. The adhesiveness of human platelets in vitro. Scand. J. Clin. Lab. Invest., 1960, 12 (Suppl. 51), 1-117.

14. Harrison, M.J.G. and Mitchell, J.R.A., The influence of red blood cells on platelet adhesiveness. Lancet, 1966, 2, 1163-1164.

15. Crexells, C., Aerichide, N. et al., Factors influencing hemolysis in valve prostheses, American Heart Journal,1970, 84, 161.

MICROSCOPE IMAGE ANALYSIS IN BIOMATERIAL TESTING

Holger Schreiber, Hans-Peter Kinzl

Institute of Forensic Medicine,
District Hospital Gera,
Straße des Friedens 122, Gera, GDR

ABSTRACT

The use of the computer-based image analysis belongs to the
new methods of biocompatibility testing. When materials were
implanted subcutaneously in animals the cells of the connective tissue capsule can give a good standard for the evaluation of biocompatibility.The use of image processing systems
allows the automation of a great number of measurings and
test techniques. It is also possible to get quantitative
information about cell- and hämocompatibility testing. With
the help of the automatic microscopic image analysis accuracy
of morphometric methods increased and scientists time was
saved.

INTRODUCTION

The use of the microscope image analysis belongs to the new
methods of biocompatibility testing. Biomaterials are variously used in modern substitution medicine, for example as
hip cups, osteosynthesis, pacemakers, vascular protheses and
stomatological implants. It is not only necessary to grasp
the material features of the implants. The behaviour of the
biomaterials in the organism is also very important. Usual
test methods as for example a differentiated histological
analysis of the subcutaneous tissue partly depend on subjective factors. Histological investigations were described by
Knöfler et al. (7), (8) and Keller et al. (5), (6) in 1984.
Different test systems for the evaluation of biocompatibility
have been developed, Anderson (1), Antian (2), Fischer (4),
Meachim (9) and Daniels (3).

Subjective factors are among other things founded in the
histological knowledge of the observer and in the visual man

system. Image analysis was often used in medicine to exclude the subjective factors (12), (10), (11).

With the help of the automated microscope analysis it became possible to get quantitative information about .the subcutaneous implantation test. When materials were subcutaneously implanted in animals, the cells of the connective tissue capsule can give a good standard for the evaluation of biocompatibility. The histological analysis consists of a visual counting and classifying of cells by using a haemocytometer.

MATERIAL AND METHODS

Subcutaneous implantation was carried out in rats, exactly into the dorsal region and parallel to the vertebra. Titan- and silicone samples were prepared into implants of 5 x 5 x 2 mm. 50 animals were sequentially sacrificed after 8, 14, 30, 90 and 200 days in groups of 10 animals. The connective tissue capsules for histological investigation were sectioned at 5 um, stained with haematoxylineosin and examined under a "Jenaval" microscope. The "BVS 6471" and the "Robotron-Mini-BVS-System" were used to analyse different cells of the connective tissue capsule as one criterion of biocompatibility.

HARDWARE REQUIREMENTS

Computer: AT or XT compatible
        PCs with 640 KB,
        1 floppy disk drive
        1 hard disk with 20MB min.

Frame grabber/graphic sub system:
        PIP 1024 of Matrox
        IL 7067.15 of Robotron
        incl. interface

Peripheral: Microscops. Video Camera. Printer. Mouse/
        Trackball/Digitizer

SOFTWARE REQUIREMENTS

- MS DOS 3.2. or 3.3.
- Microsoft CS.1, if own programs shall be written
- Mouse or digitizer driver regarding the model used

The software packages will be delivered on disks with full documentation.

A special software package called "GSCAN IMAGE-C-BIO" was created by scientists of our group to verify small differences of biocompatibility. In the module "Optical Conditions" we used no optical filter, a magnification of 1000 and immersion oil. In the module search parameters we gave the computer the order which objects must be found. In this case the computer was trained to find the nuclei of the cells of the connective tissue capsule. In the next module "Classifier learning" we gave the computer the information about 3 different groups of nuclei.

Figure 1. Cells of the connective tissue capsule titan implanted in rats, 7 days after implantation. Cell cores after automated microscopic image analysis. Long cells like fibrocytes and fibroplasts. Round cells like lymphocytes, macrophages, plasma cells, granulocytes and conglomerates of cells.

Figure 2. Cells of the connective tissue capsule titan implanted in rats, 200 days after implantation. Cell cores
after automated microscopic image analysis. Long cells like
fibrocytes and fibroplasts. Round cells like lymphocytes,
macropha- ges, plasma cells, granulocytes and conglomerates
of cells.

After the self learning process the computer is able to find nuclei automatically. The 3 groups of cells are long cells, round cells and conglomerates. Long cells are cells like fibrocytes and fibroblasts. Round cells are lymphocytes, plasma cells, macrophages and granulocytes. Combined cell cores were classified as conglomerates. Poor quality images may require preprocessing prior to automatic analysis, for instance if there are touching particles to be counted or areas of poor contrast. This can be effected either by image editing or by grey level image processing.

For some samples which do not lend themselves to automatic grey level image analysis, the alternative approach is to use a semiautomatic interactive system utilising a graphic tablet or the mouse. The alternative is the automatic system of cell classificating and counting. You can storage the statistical data analysis of your own labor standard.

Histograms show the distribution of special parameters. Colours were used for denoting areas of interest.

"GSCAN-IMAGE-C-Bio" provides the measurement of area, size, shape, optical density and many more complex parameters. The great differences between the factors of shape of the round cells and the long cells illustrate the kindness of the classificator.

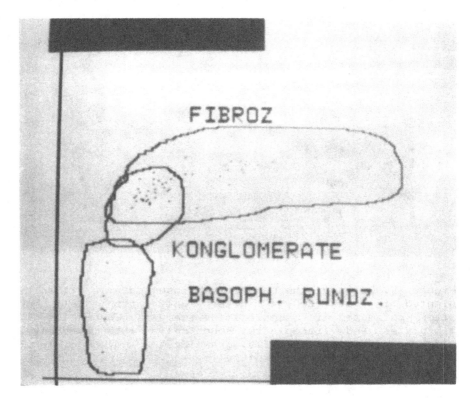

Figure 3. Differences of the factor of shape after cluster analysis: long cells, round cells, conglomerates

RESULTS
Usual histological investigations showed that the number of
cells of the connective tissue capsule is the best criterion
to verify the biocompatibility of materials. By using a haemocytometer a relatively smaller number of "inflammation
cells" was measured.

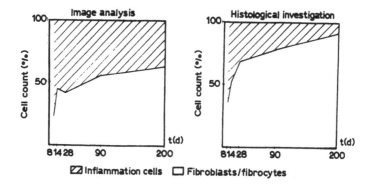

Figure 4. Relation of "Inflammation cells" to fibroblasts/
fibrcytes after image analysis (left side) and after histological investigation (right side). Titan implanted in rats,
     time of implantation 7, 14, 30, 90 and 200 days.

"Inflammation cells" are in this study round cells and conglomerates, (after image analysis), or lymphocytes plasma
cells, monocytes and granulocytes, (after histological investigation). With the help of the microcopic image analysis
all cells were counted. This leads to a relatively greater
number of inflammation cells which are important for the
hiocompatibility. The results of cell counting, (Fig.: 5)
demonstrate that the GSCAN-IMAGE-C-BIO is able to verify
small differences of biocompatibility which usual histological methods can't find.

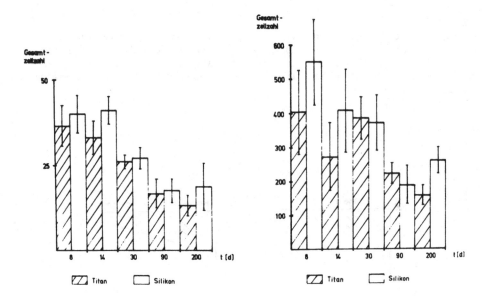

Figure 5. Number of cells of the connective tissue capsule after visual cell counting (left side) and after image analysing (right side). Titan and silicone implanted in rats, time of implantation 7, 14, 30, 90 and 200 days.

USING POSSIBILITIES OF THE PROGNOSTIC METHOD

The method can be variously used in a clinical orientated material research.

Because biomaterials were nearly equalized with pharmaceutics institutes for pharmaceutic testing are potential users.

The biologic testing is also done in experimental departments of the clinics as well as in anatomic and pathologic institutes. In this way an additional part of users needs the image analysis.

Following clinical fields need the test for the biocompatibility of implant materials:

Surgery, cardiac surgery (osteosynthesis materials, hip cups, suture materials, arterial prosthesis, valve, artifical heart)

dental surgery, facial surgery, stomatology (stomatologic implants, filler)

opthalmology (intra-ocular lens)

internal medicine (insulin pumps, membrana, implantable catheter systems, hormone- and drug applicators)

otolaryngology (ear ossiculum prosthesis, plastic operations)

DISCUSSION

In regard to long time routine cytometry a considerable saving of time was achieved. In contrast to usual histology

analysis the results are exactly reproducible. In this way a biologic test of implant materials can be better correlated with also reproducable exactly material-featured results. The method contributes to the prognostic recognition of material-based damages and to their prevention.
Small differences of the biocompatibility became verifiable, which other methods (2), (5),(6), (7), (8) can't find. By using the Microscopic Image Analysis cell conglomerates were devided. Combined objects will be automatically separated. The representation of a microscopic image leads to shading effects. Preparationally determined shading is not always suppressable. Shading correction can be carried out to be still able to determine the level important for contour searching. Characteristics like form factors and areas of the objects allow the distinction of the different nuclei as well as the recognition of artefacts (11), (12). The results of the Image Analysis became reproducable and comparable. With the help of the Expertsystem "Biomaterial testing" it was possible to find significant characteristics for a potential biocompatibility of implantable materials.

CONCUSIONS

"GSCAN-IMAGE-C-Bio" contributes to the prognostic recognition of material-based damages and to their prevention. Special software solutions can be used in the field of haemocompatibility- or cell testing.

REFERENCES

1. Anderson, J.M., The biocompatibility of human implants
   In Fundamental Aspects of Biocompatibility (Ed.D.F. Williams) Vol II, CRC Press, Boca Raton, Florida, 1981, pp 107 - 144

2. Autian, J., Toxicological aspects of implantable plastics
   F. Williams) Vol II, CRC Press, Boca Raton, Florida, 1981, pp 205 - 218

3. Daniels, A.U.; Applied Biomaterials, J. Biomed. Mater. Res, New York, 21, 1987, pp 143 - 146

4. Fischer, J.P., Burg, K., Fuhge, P., Heimburger, N.; Physical and biochemical characterization of biomaterials with respect to blood compatibility, RSAO. Proc. V, VIII, 1981, pp 81 - 85

5. Keller, F., Knöfler, W., Schreiber, H., Zur Biokompatibilität von Implantaten mit und ohne Fluorkohlenwasserstoff-Glimmpolymerbeschichtung, IV Entwicklung und Ableitung eines Histokompatibilitätsindexes für die Bindegewebsreaktion auf implantierte Materialien, Z. exp. Chir. Transplant. Künstl. Organe, Leipzig, 18, 1984, pp 9 - 18

6. Keller, F., Knöfler, W., Schreiber, H., Heß, J., Wohlgemuth, B., Zur Biokompatibilität von Implantaten mit und
   ohne Fluorkohlenwasserstoff – Glimmpolymerbeschichtung
   III Mathematische Modellierung des Abklingprozesses der
   Bindegewebsreaktion auf implantierte Materialien, Z. exp.
   Chir. Transplant. Künstl. Organe, Leipzig, 17. 1984, pp
   325 – 329

7. Knöfler, W., Wohlgemuth, B., Schreiber, H., Keller, F.,
   Zur Biokompatibilität von Implantaten mit und ohne Fluorkohlenwasserstoff – Glimmpolymerbeschichtung, II Automa-
   tisierte metrische Erfassung der Gewebsreaktion, Z. exp.
   Chir. Transplant. Künstl. Organe, Leipzig, 17, 1984, pp
   325 – 329

8. Knöfler, W., Wohlgemuth, B., Schreiber, H., Keller, F.,
   Heß, J., Zur Biokompatibilität von Implantaten mit und
   ohne Fluorkohlenwasserstoff – Glimmpolymerbeschichtung,
   I Histologische und semiquantitative Beurteilung der
   Reaktion des Subkutangewebes von Meerschweinchen, Z. exp.
   Chir. Transplant. Künstl. Organe, Leipzig, 17, 1984, pp
   316 – 324

9. Meachim, G., Pedley, B.R., The tissue response at implant
   sites. In Fundamental Aspects of Biocompatibility (Ed.D.F.
   Williams), Vol I, CRC Press, Boca Raton, Florida, 1981, pp
   107 – 144

10. Sakai, K., Katori, H., Chromosome analysis in a case of
    Klinefelders syndrome, J. Ipn. Chromosome 2, Tokyo, No 1,
    1984, pp 5 – 8

11. Scharz, H., Neue Anwendungsmöglichkeiten der Bildanalyse
    in Forschung und Routine, CIT Fachz. Lab. 28, 1984, pp
    1128 – 1136

12. Simon, H., Kunze, K.D., Voss, K., Hermann, W.R., Automati-
    sche Bildverarbeitung in Medizin und Biologie, Th. Stein-
    kopf Verlag Dresden, 1975, pp 45 – 48

INDIRECT MEASUREMENT OF ARTERIAL PRESSURE, BLOOD FLOW AND
VOLUME ELASTIC MODULUS IN HUMAN FINGERS
USING ELECTRIC IMPEDANCE-CUFF

HIROSHI ITO*, HIDEAKI SHIMAZU*, ATSUSHI KAWARADA*,
HIROKO KOBAYASHI*, JUN MASUDA, and TERUHITO AMANO**

Department of Physiology(*) and Critical Care Center &
Primatology(**), Kyorin University School of Medicine,
20-2, Shinkawa 6 chome, Mitaka-shi, Tokyo, Japan

ABSTRACT

A new plethysmograph, the electric impedance-cuff, was designed
for the indirect measurement of blood pressure and flow, volume
elastic modulus (Ev) and compliance (C) in human finger
arteries. The device comprises a compression chamber filled
with electrolyte solution and a tetrapolar electric impedance
plethysmograph whose electrodes are embedded inside the
chamber; the compression chamber for controlling transmural
arterial pressure (Pt) and the impedance plethysmograph for
detecting total finger volume (Vo), mean arterial volume ($\overline{Va}$)
and its variations (ΔV). Systolic (Pas) and mean (Pam) arterial
pressures in fingers were determined by detecting pulsatile
impedance variations during the gradual increase (or decrease)
in Pc using volume oscillometric technique. Diastolic pressure
(Pad) and pulse pressure (ΔP) were calculated using the
equations: Pad = (3Pam - Pas)/2 and ΔP = Pas - Pad. Ev and C
defined as Ev = $\Delta P/(\Delta V/\overline{Va})$ and C = $\Delta Va/\Delta P$ were measured at
various Pt levels controlled by Pc. Arterial pressures, blood
flow and Ev were measured in 171 healthy subjects. In some
subjects, the change in pressure, blood flow and Ev of finger
arteries were detected during acupuncture.

INTRODUCTION

Plethysmograph is a commonly used technique for the noninvasive
measurement of volume change and blood flow in human body
limbs. It was originally developed by Glisson in 1622 and
Swammerdam in 1737. Since then, many types of plethysmographs
have been reported (1).
    This technique has been used for the indirect measurement

of arterial pressure in human limbs since the time of Marey (1876) and Huethle (1896) (2). Janssen (3) employed electric impedance plethysmography combined with an occlusive cuff to determine systolic and diastolic pressure in human upper arms. Yamakoshi et al. (4, 5) developed the "volume oscillometric technique" to measure systolic and mean pressure in human finger arteries by photoelectric plethysmography. Penaz (6), Yamakoshi et al. (7, 8) and Wesseling et al. (9) succeeded in recording beat-to-beat arterial pressure waveforms in human finger and rat tail arteries by the "volume compensation technique" using photoelectric plethysmography. Plethysmographs are commonly used for the measurement of blood flow by venous occlusive technique.

Elastic properties of human arterial wall were also indirectly determined by plethysmography. Nakayama and Azuma (10) recorded arterial compliance at fingertip using a pneumoplethysmograph with a cuff. Shimazu et al. (11) recorded beat-to-beat compliance and volume elastic modulus in human finger and forearm arteries using electric admittance plethysmograph with the volume oscillation technique. Further more, Shimazu (12) developed a simpler but more precise technique to measure volume elastic modulus in the basal phalanx of human fingers by transmitted infra-red photoelectric plethysmography.

Recently, Shimazu et al. (13) developed a new plethysmography called the "electric impedance-cuff". This consists of a compression cuff filled with electrolyte solution and a tetrapolar impedance plethysmograph whose electrodes are embedded inside the cuff. Systolic and mean arterial pressure, blood flow and elastic properties of arterial wall were recorded indirectly and continuously. Through the study, the data by during the monitoring of these cardiovascular indices in resting and stress tests such as acupuncture are shown.

PRINCIPLE & METHODS

Electric Impedance Cuff for Finger
A schematic diagram of the electric impedance-cuff for finger is shown in Fig. 1. It consists of a compression chamber filled with 0.25% NaCl solution and a tetrapolar impedance plethysmograph. The chamber was made of an acrylic cylinder, 4.0cm long, 3.0 cm diameter and 0.3 cm thick. Two pairs of silver-silver chloride wire electrodes, 0.5 mm diameter, were placed in the recesses(1.0 mm thick and 1.0 mm wide) cut into the inside wall 0.5 cm and 1.5 cm from each end of the cylinder. Thus, the distances between the outside and inside pair of the electrodes were 3.0 cm and 1.0 cm, respectively. The outside electrodes were for passing electric current(I1 and I2), while the inside ones were for picking up the voltage drop (E1 and E2).

A rubber tube, 5.5 cm long, 2.7 cm diameter and 0.3 mm thick, was inserted into the cylinder, and its flanges were fixed at both ends to form the compression chamber using pieces of elastic tape. The chamber was then filled with NaCl solution. It has three side connections: one at the ceiling for an air trap, the others at the side walls for a pressure controller via a depulsation tank and a strain-gauge pressure transducer (P23Db, Statham). The length of the compression chamber and the distance between the voltage pick-up electrodes were determined according to Alexander et al.(14).

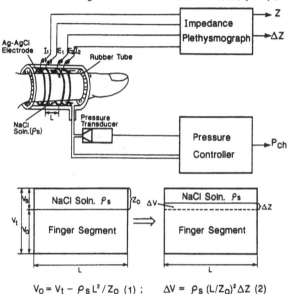

$$V_o = V_t - \rho_s L^2 / Z_o \quad (1) ; \qquad \Delta V = \rho_s (L/Z_o)^2 \Delta Z \quad (2)$$

Figure 1. Schematic diagram of the electric impedance cuff. In the insets at the bottom of the figure, the principle and equations for measuring total limb volume (V_o) and its variation (ΔV) are illustrated. For symbols and details see text.

Principle to Determine Total Finger Volume and Its Variation
    Electrolyte volume and its variation in the chamber were measured by electric impedance technique. A 50 kHz 1 mA (p-p) current was passed through the chamber via the outside electrodes. The voltage drop across the inside electrodes was fed into an AC amplifier (50 $\pm$ 10 kHz), a demodulator and then a DC amplifier to obtain total impedance (Z_o). Impedance variation (ΔZ) was detected by leading the output signals into another AC amplifier (0.3 - 30 Hz).
    The principle of measuring total limb volume (V_o) and its variation (ΔV) is illustrated in the insets of the bottom of Fig. 1. Assuming the finger inserted between the inside electrodes is regarded as a cylinder and completely covered with the rubber chamber, V_o is expressed as

$$V_o = V_t - V_s. \tag{1}$$

where V_t is the total volume of the acrylic cylinder and V_s the volume of the electrolyte column between the inside electrodes. This equation is arranged as

$$Vo = Vt - \rho (L^2/Zo) \tag{2}$$

where ρ is the resistivity of the NaCl solution (225 ohm-cm at 25 °C), L the distance between the inside electrodes (1.0 cm). Vt was 7.07 ml. The volume variation of the finger between the inside electrodes is

$$\Delta V = \rho (L/Zo)^2 \Delta Z \tag{3}$$

where ΔZ is the impedance variation.

Measurement of Blood Pressure by Volume Oscillometric Technique

Figure 2 shows an example of simultaneous recordings of (1) compression pressure (Pc; upper trace), (2) DC record of impedance variation (Z; middle trace) and its AC component (ΔZ; bottom trace) obtained at the basal phalanx of a right index finger of a normal male (36 years old).

With the gradual increase of Pc from about 50 to 180 mmHg (3 - 5 mm Hg/heart beat), the finger volume represented by Z curve decreased; its AC component represented by ΔZ signals gradually increased in amplitude, reached maximum, then decreased until at least disappeared. According to the principle of the volume oscillometric technique (4, 5), systolic (Pas) and mean (Pam) arterial pressure were determined at Pc corresponding respectively to the disappearance and maximum point of the impedance variation. Diastolic pressure (Pad) was calculated by a conventional equation:

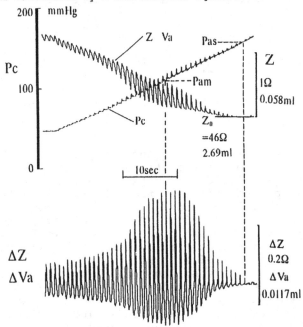

Figure 2. Example of simultaneous recordings of chamber pressure (Pc; upper trace), DC component of impedance variation (Z; middle trace) and its AC component (ΔZ; bottom trace).

Pad = (3Pam - Pas)/2   (4)

Systolic, mean and diastolic pressure were Pas = 155, Pam = 111, and Pad = 89 mmHg, respectively, in this record.

Measurement of Elastic Properties of the Arterial Wall

Elastic properties of the arterial wall in fingers were evaluated using indices, compliance (C) and volume elastic modulus (Ev) which are defined as

$$\Delta C = \Delta V / \Delta P \tag{5}$$

$$Ev = \Delta P / (\Delta V / \overline{Va}) \tag{6}$$

where ΔV is the pulsatile volume change in the artery, and $\overline{Va}$ the mean arterial volume at a controlled transmural pressure level. $\overline{Va}$ was determined as a difference in arterial volumes detected at two Pc levels, desired and systolic (see the middle trace in Fig. 2). Transmural pressure Pt was defined as

$$Pt = Pam - Pc \tag{7}$$

Measurement of Finger Blood Flow

Blood flow in fingers was measured by a conventional venous occlusive method. the impedance cuff was placed on the basal phalanx of an index finger and a cuff (8 cm wide; for children use) for a conventional sphygmomanometry on the wrist. Volume change of the finger following the cuff inflation was recorded. The blood flow was calculated from the initial slope of the volume change. The occlusion pressure and time were 40 - 50 mmHg and about 10 sec, respectively.

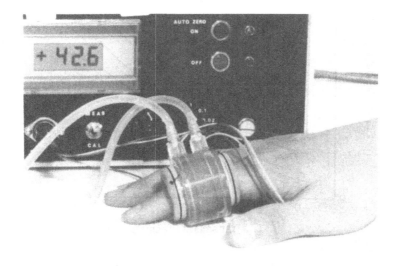

Figure 3. Photograph taken during the measurement.

Subjects and Measurements
     Elastic properties of the finger arteries were measured in
171 healthy subjects (111 men, 19 - 63 years old, and 60 women,
18 - 65 years old) of the resting state. In some of these
subjects, the effects of acupuncture on Pas, Pam, Pad, Ev and
blood flow were evaluated by comparing the stimulus point LI 10
and LI 4, respectively. The point LI 10 locates at the venter
of the brachioradial muscle and is noted for the point to
decrease the systemic blood pressure. The point LI 4 locates at
the middle of the first dorsal interosseus muscle (between
thumb and index finger) and is the point for the acupuncture
anesthesia. Either the same or the contralateral side of the
measuring finger was selected for the stimulus point. All the
experiments were performed at a room temperature of 20 - 25 °C.
A sheet of picture taken during the measurement is shown in
Fig. 3.

                            RESULTS

In Fig. 4, Ev values obtained from three different age groups
are plotted against Pt within the range 0 - 40 mmHg. Open and
closed circles and open triangles denote the averages obtained
in 18 - 39 (n = 39), 40 - 64 (n = 50) and over 65 (n = 39)
years old men and women. Vertical bars represent ± 1.0 SE. This
shows that the Ev values nonlinearly increases with Pt and that
these values increases with age.

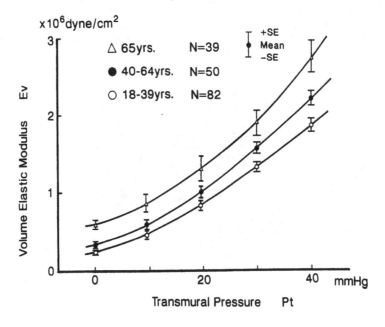

Figure 4. Volume elastic modulus (Ev) plotted against
transmural pressure (Pt) at three different age groups.

In Fig. 5, relations between Ev values and age are shown at three different Pt levels, 0, 30, and 40 mmHg in men (n = 111; upper panel) and women (n = 60; lower panel). Linear regression relations are indicated by solid lines and 95% confidence limits by dotted lines. The equations and correlation coefficients are shown in the figure. This shows that the Ev values tend to increase with age in both men and women. The change in Ev may be more markedly detectable at higher Pt level. Men showed a higher Ev values than women when compared at the same age and Pt level.

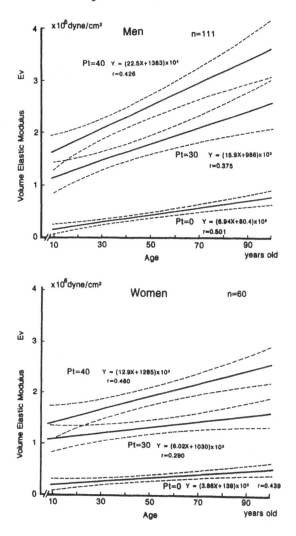

Figure 5. Relation of volume elastic modulus (Ev) <u>versus</u> Age. Data are displayed at three different transmural pressure levels, 0, 30, and 40 mm Hg. Upper panel for men, and lower one for women.

In Fig. 6 is shown the change of Ev values during acupuncture at LI 10. The differences of Ev values between the control (ordinate; in x 10^6 dyne/cm^2) are plotted against time (abscissa; in min). These Ev values were determined at Pt = 40 mmHg. Closed circles denote the average of 16 subjects and vertical bars $\pm$ 1.0 SD. Symbols, * and **, indicates p < 0.05 and p < 0.01, respectively. The control was recorded 3 min before the insertion of acupuncture needle and insertion period was 2 min. This record shows the decrease in the Ev values during and after the acupuncture. Pas, Pam and Pad decreased following the insertion. Acupuncture at another point such as LI 4 showed no change or in some cases a little increase in the Ev values and blood pressures.

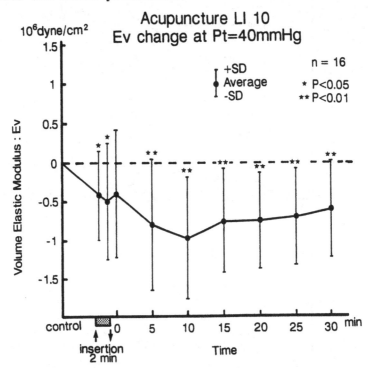

Figure 6. Changes in volume elastic modulus in finger caused by acupuncture. The differences of Ev values from the control are plotted against time. Vertical bars denote $\pm$ 1.0 SD. * p < 0.05 and ** p < 0.01. Needle insertion is indicated by the shaded column at the bottom.

In Fig. 7 is an example of the blood flow measurement by this method. The upper trace shows the change in the finger volume due to venous occlusion, and the lower trace the cuff pressure. The value of the blood flow determined from the slope of tangent line in the upper record was 1.13 ml/min which corresponds to 23.5 ml/min-100ml finger volume.

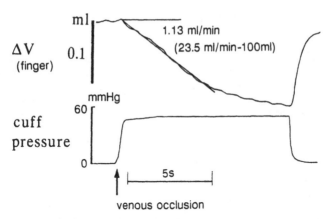

Figure 7. Measurement of blood flow. Upper record shows the
finger volume change, and lower record occlusive cuff pressure.
Blood flow calculated from the slope of the tangent line on
the volume change were 1.13 ml/min which correspond to 23.5
ml/min-100 ml finger volume.

In Fig. 8 is shown the changes of the finger blood flow
following acupuncture. Closed and open rectangles are the
averages (n = 22 and n = 16) of the relative changes (in %)
obtained during the acupuncture of LI 10 and LI 4,
respectively. Vertical bars denote ± 1.0 SE and symbols, * and
**, p < 0.05 and p < 0.01, respectively. This shows the finger
blood flow significantly increased following the acupuncture at
LI 10, while it did not show significant change by the
insertion at LI 4.

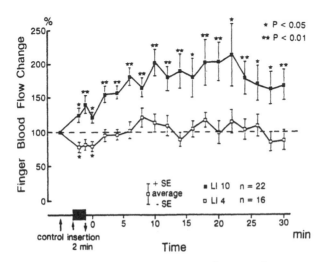

Figure 8. Change in blood flow in finger by acupuncture.
Relative blood flow (in %) are plotted against time. Vertical
bars denote ± 1.0 SE. * p < 0.05 and ** p < 0.01.
Acupuncture insertion is indicated by the shaded column at
the bottom.

DISCUSSION & CONCLUSIONS

In a previous study (13), we showed that the following cardiovascular indices can be noninvasively measured by the impedance-cuff: (1) arterial pressure, Pas and Pam, (2) arterial and venous C and Ev, (3) blood flow, (4) venous pressure. Through this study, continuous monitoring of arterial pressure, flow, and Ev were carried out by this method during acupuncture .

Mechanical compliance of this plethysmograph---the total of the compliance of the acrylic cylinder and that of the finger and rubber tube---was 1×10^{-5} (Pc > 25 mmHg) and 1×10^{-4} (10 mm Hg < Pc < 25 mmHg). Thus, the measurement error of finger volume caused by the mechanical compliance should be minute. For this reason, its frequency characteristics should presumably high enough to detect more than 8th harmonics of 120 beat/min.

Volume change and blood flow can be directly expressed in ml/min-100 ml finger volume, because L and Vo are known. Transmural pressure Pt can be controlled by Pc. Thus, the elastic properties of an artery with a known length L are recorded at a desired Pt level. This should be an advantage of this technique.

This method does not require electrodes to be placed directly on the skin, as happens with conventional impedance plethysmography. Therefore volume change can be measured without passing an electric current through a human limb segment. Thus, the volume value should not be affected by the individual difference of blood resistivity of the subjects.

In conclusion, the impedance-cuff is a precise technique for the noninvasive long-term monitoring of cardiovascular information.

Acknowledgment: The authors wish to thank Professor K. Tsuchiya, Department of Mechanical Engineering, Waseda University, and Dr. Ken-ichi Yamakoshi, Associate Professor, The Research Institute of Applied Electricity, Hokkaido University. A part of this study was reported at the 2nd International Symposium on Biofluid Mechanics and Biorheology in Large Blood Vessels, June 25 - 28, 1989, Munich, West Germany, XXXI International Congress of Physiological Sciences, July 9 - 14, 1989, Helsinki, Finland,.

REFERENCES

1. Woodcock, J.P., Theory and Practice of Blood Flow Measurement, 1st edn., Butterworth & Co. Ltd., London, UK, 1975, pp. 1 - 274.
2. Geddes, L.A., The Direct and Indirect Measurement of Blood Pressure, Year Book Publ., INC., Chicago, 1970, pp. 1 - 196.

3. Janssen, F.T., The rheographic determination of systolic and diastolic blood pressure. Digest 7th Intn'l. Conf. Med. & Biol. Engng., Stockholm, Sweden, pp. 221, August 1967.

4. Yamakoshi, K., Shimazu, H., Shibata, M. and Kamiya, A., New Oscillometric method for indirect measurement of systolic and mean arterial pressure in the human finger. Part 1: model experiment. Med. Biol. Eng. & Comput., 1982, 20, 307 - 313.

5. Yamakoshi, K., Shimazu, H., Shibata, M. and Kamiya, A., New Oscillometric method for indirect measurement of systolic and mean arterial pressure in the human finger. Part 2: correlation study. ibid, 1982, 20, 314 - 318.

6. Penaz, J., Photoelectric measurement of blood pressure, volume and flow in the finger. Digest 10th Intn'l. Conf. Med. & Biol. Engng., Dresden, German Democratic Republic, pp. 104, August 1973.

7. Yamakoshi, K., Shimazu, H. and Togawa, T., Indirect measurement of instantaneous arterial blood pressure in the rat. Am. J. Physiol., 1979, 237, H632-H637.

8. Yamakoshi, K., Shimazu, H. and Togawa, T., Indirect measurement of arterial blood pressure in the human finger by the vascular unloading technique. IEEE Trans., 1980, BME-27, 150 - 155.

9. Wesseling, K.H., De Wit, B., Settels, J.J. and Klawer, W.H., On the indirect registration of finger blood pressure after Penaz. Funkt. Biol. Med., 1982, 1, 245 - 250.

10. Nakayama, R. and Azuma, T., Noninvasive measurements of digital arterial pressure and compliance in man. Amer. J. Physiol., 223, H168 - H179, 1977.

11. Shimazu, H., Fukuoka, M, Ito, H. and Yamakoshi, K., Noninvasive measurement of beat-to-beat vascular viscoelastic properties in human fingers and forearms. Med. Biol. Eng. & Comput., 1985, 23, 43 - 47.

12. Shimazu, H., Yamakoshi, K. and Kamiya, A., Noninvasive measurement of the volume elastic modulus in finger arteries using photoelectric plethysmography. IEEE Trans., 1986, BME-33, 795 - 798.

13. Shimazu, H., Kawarada, A., Ito, H. and Yamakoshi, K., Electric impedance cuff for the indirect measurement of blood pressure and volume elastic modulus in human limb and finger arteries. Med. Biol. Eng. & Comput., 1989, 27, 477 - 483.

14. Alexander, H., Cohen, M.L. and Steinfield, L.S., Criteria in the choice of an occluding cuff for the indirect measurement of blood pressure. Med. Biol. Eng. & Comput., 1977, 15, 2 - 10.

ELECTRICAL IMPEDANCE PLETHYSMOGRAM IN PERIPHERAL BLOOD FLOW

KYOKO SASAOKA

Department of Physiology, Tokyo Dental College,
1-2-2 Masago, Chiba City 260, Japan

KEIKITSU OGAWA

Nihon Kohden Corporation, 1-31-4, Nishiochiai,
Shinjuku-ku, Tokyo 161, Japan

ABSTRACT

We have previously reported the results of experiments on electrical
impedance plethysmography and electromagnetic flowmetry in the central
caudal arteries of anesthetized rats.   Although the theoretical formula
we proposed was successfully applied, we did not clarify correlations
between CR and mean blood flow obtained in other experiments.   In this
study, we used the method of least squares to calculate α and β and
obtained results more satisfactory than our previous ones.

INTRODUCTION

We have already demonstrated that the following equation may be applied
to the relationship between flow F(t) and either the electrical impedance or the photoelectric plethysmogram Z(t) in the rat peripheral
artery [1],[2]:

$$F(t) = \alpha \, dZ(t)/dt + \beta Z(t) \qquad (1)$$

Where α and β are coefficients, and the following relation exists:

$$CR= \alpha \,/\, \beta \tag{2}$$

C and R stand for the blood vessel compliance and the peripheral resistance respectively.
In the equation (1), $F(t)$ and $Z(t)$ are supposed to be periodic functions
of t with a period T and have the following relationships:

$$F(t)=f_0 + f(t) \tag{3}$$

$$Z(t)=z_0 + z(t) \tag{4}$$

where f_0 and z_0 are mean values of $F(t)$ and $Z(t)$, i.e.

$$f_0 = \frac{1}{T} \int_0^T F(t)\,dt \tag{5}$$

$$z_0 = \frac{1}{T} \int_0^T Z(t)\,dt \tag{6}$$

Then, rewriting equation (1) with $f(t)$ and $z(t)$ gives

$$f(t)= \alpha \, \dot{z}(t) + \beta\, z(t) \tag{7}$$

Although understanding peripheral hemodynamics is very important, at
present it is impossible to separate R and C in CR.
　　Previously we calculated α and β according to an older method as
follows [1]:
writing $\dot{z}(t)$ for $dz(t)/dt$ gives

$$f(t_1)= \alpha \, \dot{z}(t_1)+\beta\, z(t_1) \tag{8}$$

$$f(t_2)= \alpha \, \dot{z}(t_2)+\beta\, z(t_2) \tag{9}$$

at $t=t_1$ and $t=t_2$.　α and β may then be obtained by solving the simultaneously equations:

$$\alpha = \frac{f(t_1)\, z(t_2) - f(t_2)\, z(t_1)}{\dot{z}(t_1)\, z(t_2) - \dot{z}(t_2)\, z(t_1)} \tag{10}$$

$$\beta = \frac{f(t_2)\,\dot{z}(t_1) - f(t_1)\,\dot{z}(t_2)}{\dot{z}(t_1)\,z(t_2) - \dot{z}(t_2)\,z(t_1)} \tag{11}$$

Therefore, the value of CR=$\alpha\,/\,\beta$ may be obtained as follows:

$$CR = \frac{f(t_1)\,z(t_2) - f(t_2)\,z(t_1)}{f(t_2)\,\dot{z}(t_1) - f(t_1)\,\dot{z}(t_2)} \tag{12}$$

Our new system, however, employs the method of least squares to calculate α and β.

If sampling time (ts) is determined by actual measurement, there will be n combinations of sample values in one heart beat (provided, $n \cong T\,/\,ts$):

$$(t_i, \ f(t_i), \ \dot{z}(t_i), \ z(t_i)) \quad i = 1, 2, \cdots, n$$

for these n combinations, α and β in equation (15) may be obtained by means of the method of least squares:

$$\alpha = (RU - QV)\,/\,D$$

$$\beta = (PV - QU)\,/\,D$$

Where

$$P = \Sigma\ \{\dot{z}(t_i)\}^2, \qquad Q = \Sigma\ \dot{z}(t_i)\cdot z(t_i)$$

$$R = \Sigma\ \{z(t_i)\}^2, \qquad U = \Sigma\ f(t_i)\cdot \dot{z}(t_i)$$

$$V = \Sigma\ f(t_i)\cdot z(t_i), \qquad D = PR - Q^2$$

In theese processes, we assumed that $\bar{f}$, $\bar{\dot{z}}$ and $\bar{z}$ in the next expressions are equal to zero, because the means of f(t) and z(t) in a period are zero.

$$\bar{f} = \frac{1}{n}\Sigma\ f(t_i), \qquad \bar{\dot{z}} = \frac{1}{n}\Sigma\ \dot{z}(t_i), \qquad \bar{z} = \frac{1}{n}\Sigma\ z(t_i)$$

In this report, either electrical impedance plethysmography or photoelectric plethysmography was used to obtain values for CR and mean

blood flow with electromagnetic flowmetry in the central caudal artery
of rats under anesthesia.

MATERIALS AND METHODS

Seven male Wistar rats weighing between 400 and 750 gm each were anesthetized with sodium pentobarbital (30-40 mg/Kg, intraperitoneal). The
maneuvor used to measure the electrical impedance plethysmogram is the
same one that we reported previously. To prepare for photoplethysmogra
phy, the tail of the supine rat was extended horizontally on a plate,
and the skin was incised along the median line about 10-25 mm from the
anus (Figure 1).

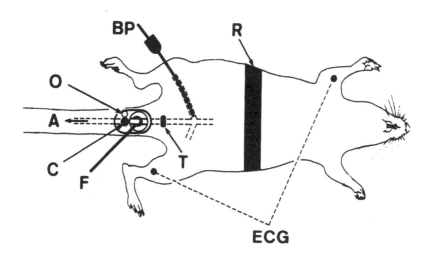

Figure 1. Placement of electrodes and transducers. A: central caudal
artery; BP: blood pressure transducer; F: electromagnetic flowmeter;
ECG: electrocardiogram electrodes; R: respiration transducer; T:
temperature transducer; C: CdS photoconductive cell; O: optic-bundle
tip.

About 10 mm of the central caudal artery was exposed in the operation area. The probe (inner diameter 0.3 mm) of the electromagnetic
flowmeter (Nihon Kohden) was fitted to the artery. A CdS photoconductive cell (Moririka, MPB2-4H48), a lamp (Hamai, LNS-OMB62S) and a
fiber-optic bundle (Nakanishi) were used in measuring the photoelectric

plethysmogram. The cell and the tip of the optic bundle were placed on the subcutaneous tissue surface about 5 mm distal from the probe of the electromagnetic flowmeter.

A catheter for measuring blood pressure was inserted into the femoral artery.

Electrocardiogram, respiration, and rectal temperature too were recorded simultaneously. All data were taken in through an A/D converter and stored in a computer memory (PC-9801, NEC) for calculations of each trial. A single trial-time was 99.84 sec. Values were continuously measured at 5 msec intervals; 19968 samples were taken for each item in a single trial. Each trial contained between 300 and 600 heart beats. There were from 18 to 45 sampling points in a single heart beat. Each value for α, β, and CR in each heart beat was calcurated according to both the old and the new methods.

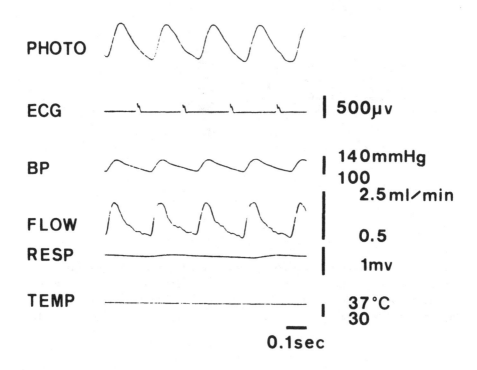

Figure 2. Sample polygram. PHOTO: photoplethysmogram; ECG: electrocardiogram; BP: blood pressure; FLOW: blood flow; RESP: respiration; TEMP: rectal temperature

Room temperature was maintained at 25°C ± 2°C.

499

As polygraphic estimations show, the anesthetized animals' physiological conditions were kept as stable as possible.

RESULTS

Values for α, β, and CR in each heart beat were calcurated according to both the old and the new methods.   In the trial shown in Figure 2, a sample polygram, with the old method, in CR (sec), mean value was 0.0095 and standered deviation was 0.0057.   With the new method, however, the corresponding values were 0.0087 and 0.0024.

DISCUSSION

Results of calcurations showed that the new method for arriving at α and β is better than the old one.   With the new method, standard deviation was approximately halved.   In the old method, 2 times (t_1 and t_2) are chosen to caluculate α and β; but in the new method, all times in a given single heart beat are used in the calculation.   When physiological conditions as assumed from polygrams remain comparatively stable, a smaller standard of deviation is thought to represent greater accuracy of estimation.

Generally, our experimental method combined electrical impedance plethysmography and electromagnetic frowmetry, although occasionally, to avoid hemorrhage, we had to make do with photoelectric plethysmography. We have already demonstrated the possibility of substituting photoelectric plethysmography for electrical impedance plethysmography with experimental [2] application of equation (1).

When mean blood flow (f) increased, average values and scattering in CR tended to decrease.   Adequate correlation coefficients between f and CR did not necessarily result from the measurements.   Furthermore, the source of CR scattering remains unknown.   In the future, we intend to apply our new method in studying this point.

REFERECES

1. Sasaoka, K. and Ogawa, K., Analytical study of electrical impedance plethysmography in peripheral blood flow. Med & Biol. Eng. & Comput., 1987, 25, pp.386-90.

2. Sasaoka, K. and Ogawa, K., Study of the blood flow in the lingual artery of rat. Bull. Tokyo dent. Coll., 1986, 27,   pp. 115-26.

Printed in the United States
By Bookmasters